T0257976

Biomedical Science and Technology

Biomedical Science and Technology

Edited by **Mark Walters**

LANRYE
INTERNATIONAL

New Jersey

Published by Clanrye International,
55 Van Reypen Street,
Jersey City, NJ 07306, USA
www.clanryeinternational.com

Biomedical Science and Technology
Edited by Mark Walters

International Standard Book Number: 978-1-63240-086-4 (Hardback)

Printed in the United States of America.

Contents

Preface

This book is an attempt to provide a comprehensive account of biomedical science and biotechnology. Biomedical science is a vast field dealing with disease progression, paradigms and therapeutic measures. For instance, physiological systems engineering in medical assessment is analyzed under biomedical sciences. Biotechnology covers study of biological sciences, efficacy of pharmaceuticals, application of genetic engineering and much more. Mathematical drafting of physiological systems and their assessment comes under physiological engineering. Even in hospital management, both biomedical science and engineering are required in order to run hospitals efficiently. This book will prove to be a valuable source for clinicians, scientists and students.

This book has been the outcome of endless efforts put in by authors and researchers on various issues and topics within the field. The book is a comprehensive collection of significant researches that are addressed in a variety of chapters. It will surely enhance the knowledge of the field among readers across the globe.

It is indeed an immense pleasure to thank our researchers and authors for their efforts to submit their piece of writing before the deadlines. Finally in the end, I would like to thank my family and colleagues who have been a great source of inspiration and support.

Editor

Part 1

Biotechnology

Hydrophobic Interaction Chromatography: Fundamentals and Applications in Biomedical Engineering

Andrea Mahn
Universidad de Santiago de Chile,
Chile

1. Introduction

Hydrophobic interaction chromatography (HIC) a powerful technique used for separation and purification of biomolecules. It was described for the first time by Shepard & Tiselius (1949), using the term "salting-out chromatography". Later, Shaltiel & Er-el (1973) introduced the term "hydrophobic chromatography". Finally, Hjerten (1973) described this technique as "hydrophobic interaction chromatography", based on the retention of proteins on weakly hydrophobic matrices in presence of salt. Owing of its high versatility and efficiency, HIC is widely used for the separation and purification of proteins in their native state (Porath et al., 1973), as well as for isolating protein complexes (Chaturvedi et al., 2000) and studying protein folding and unfolding (Bai et al., 1997).

HIC has been applied in separating homologous proteins (Fausnaugh & Regnier, 1986), receptors (Zhang et al., 2008), antibodies (Kostareva et al., 2008), recombinant proteins (Lienqueo et al., 2003) and nucleic acids (Savard & Schneider, 2007). HIC shows similar capacity to ion exchange chromatography (IEC) and a high level of resolution. Since it exploits a different principle than IEC and other separation techniques it can be used as an orthogonal method to achieve the purification of complex protein mixtures (Haimer et al., 2007). In this chapter, the theoretical principles underlying macromolecule retention in HIC are reviewed and discussed in sight of their application for predicting macromolecule behavior in HIC. Besides, novel applications of HIC are discussed regarding their suitability on Biomedical Engineering.

2. Theoretical principles underlying macromolecule retention in Hydrophobic Interaction Chromatography

2.1 Thermodynamics fundamentals

Hydrophobicity can be defined, in general terms, as the repulsion between a non-polar molecule and a polar environment, such as that conferred by water, methanol, and other polar solvents. Two hydrophobic molecules (non-polar) located in a polar environment will show a trend to minimize the contact with the polar solvent. This is accomplished by coming in contact with each other thus minimizing the molecular surface exposed to the solvent. This phenomenon is known as "hydrophobic interaction". Hydrophobic interaction is the most common macromolecular interaction in biological systems. It is also the driving force of several biological and physicochemical processes, such as protein folding, antigen-antibody recognition, stabilization of enzyme-substrate complexes, among others.

From a thermodynamic point of view, the interaction between hydrophobic molecules is an entropy-driven process, based on the second law of Thermodynamics and considering that temperature (T) and pressure (P) remain constant during the process, in this case, the hydrophobic interaction between two biological molecules. Considering equation (1), when a non-polar molecule enters in contact with a polar solvent (usually water), an increase in the degree of order of the solvent molecules that surround the hydrophobic molecule is observed, producing a decrease in entropy ($\Delta S < 0$). Given that enthalpy (ΔH) does not suffer a significant increase in this kind of processes (constant temperature) in comparison with $T\Delta S$, an overall positive change in the Gibbs energy ($\Delta G > 0$) is produced. Hence, the dissolution of a non-polar molecule in a polar solvent does not occur spontaneously, since it is thermodynamically unfavorable.

$$\Delta G = \Delta H - T\Delta S \tag{1}$$

The thermodynamics situation changes when two or more non-polar molecules are located in a polar environment. In this case, the hydrophobic molecules spontaneously aggregate because of hydrophobic interaction, and in this way the hydrophobic surfaces of the macromolecules become hidden from the polar surrounding. Entropy increases ($\Delta S > 0$) owing to a displacement of the highly structured solvent molecules surrounding the exposed surface of the hydrophobic molecules towards the solvent bulk consisting of less structured molecules. As a consequence, the Gibbs energy decreases ($\Delta G < 0$), and therefore, hydrophobic interaction becomes a thermodynamically favorable process. In conclusion, the hydrophobic interaction between two or more non-polar molecules in a polar solvent solution is a spontaneous process governed by a change in entropy. Accordingly, hydrophobic interactions can be weakened by raising temperature or by modifying the solvent polarity through the addition of another solute.

2.2 Retention mechanisms in Hydrophobic Interaction Chromatography

Macromolecule retention in HIC occurs due to hydrophobic interactions between the hydrophobic ligands immobilized on a stationary phase and the hydrophobic moieties on the macromolecule surface (Queiroz et al., 2001). There is a variety of stationary phases used in HIC, corresponding to organic polymers or silica. Their main characteristics are being chemically modifiable, highly porous, and of high moisturizing power. Among them, the most commonly used are polyacrylamide (BiogelP™), cellulose (Cellulafine™), dextran (Sephadex™), agarose (Sepharose™), and others. These supports are further modified by linking hydrophobic ligands that become a sort of "active group" that allows hydrophobic interaction with the macromolecule to be separated from a solution. The ligand is linked to the support through a spacer arm (usually glycidyl ether), so that there is no steric impediment for macromolecule-ligand interaction, and avoiding hydrophobic interaction between the ligands. Figure 1 depicts the retention of a protein to a HIC stationary phase.

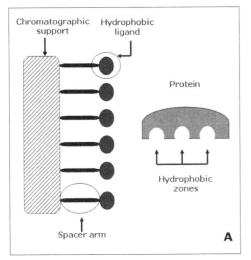
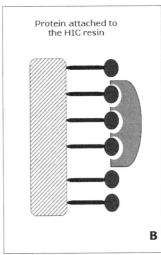

Fig. 1. Protein retention mechanism in HIC. (A) The basic structure of a HIC resin is depicted, and a protein is schematized highlighting the hydrophobic zones on the protein surface. (B) The protein gets in contact with the hydrophobic ligands of the resin, suffering a spatial reorientation. The hydrophobic ligands of the matrix interact with the exposed hydrophobic zones of the protein, and thus the protein is reversibly attached to the resin.

The most common hydrophobic ligands are alkyl or aryl groups of 4 to 10 carbons (Jennissen, 2000). The length of the carbon chain usually does not exceed 10 units in order to avoid self-folding. The nature of the hydrophobic ligand determines the performance of a HIC process. Figure 2 shows a scheme of stationary phases used in HIC and the chemical structure of the most commonly used alkyl and aryl groups, such as butyl (four carbons), octyl (eigth carbons) and phenyl (aromatic ring that promotes π-π interactions with the aromatic residues on a proteins surface). The hydrophobic interaction is directly proportional to the length of the alkyl chain. The most commonly used ligands in HIC resins are butyl, octyl and phenyl, in the following order in terms of relative interaction strength:

Phenyl > Octyl > Butyl

In the HIC process, retention is reinforced by the presence of a neutral salt. When a neutral salt is added to a solution consisting of a polar solvent, i.e. water, and a non-polar macromolecule, such as a protein, a competition for the water molecules that hydrate the macromolecule is observed, being more favorable to the salt. As a consequence, high salt concentration will reduce the number of solvent molecules that surround the macromolecules, thus favoring the hydrophobic interaction between them. Furthermore, if such solution comes in contact with a HIC resin, the interaction between the macromolecule and the hydrophobic ligand on the resin surface will be promoted, resulting in the adsorption of the macromolecule to the HIC stationary phase. From a process point of view, it is essential to choose the right type of salt and a concentration that minimizes macromolecule precipitation due to solubility decrease in the presence of high salt concentration ("salting-out").

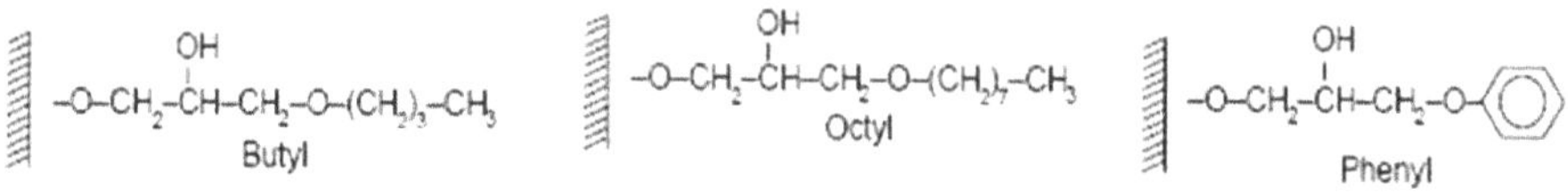

Fig. 2. Schematic representation of stationary phases used in HIC. Butyl is the shortest carbon chain used as HIC ligand and therefore the less hydrophobic one; octyl exhibits an intermediate hydrophobicity, and phenyl offers the strongest hydrophobic interaction.

The effect of different types of salt on macromolecule retention in HIC follows the Hofmeister (or lyotropic) series according to their positive influence in increasing the molal surface tension of water (Melander & Horvath, 1977). Besides, anions and cations exhibit cosmotropic or chaotropic properties. The salts at the beginning of the series are known as "cosmotropic" or "antichaotropic", since they promote hydrophobic interactions (as well as protein precipitation due to the "salting-out" effect) because of their water structuring ability. On the other hand, the salts at the end of the series, called "chaotropic", tend to randomize the structure of water and therefore they disfavor hydrophobic interactions. The salts ammonium sulfate and sodium chloride are most preferred in HIC.

Hofmeister series

$$Na_2SO_4 > K_2SO_4 > (NH_4)_2SO_4 > Na_2HPO_4 > NaCl > LiC > KSCN$$

Increasing cosmotropic effect

Anions: PO_4^{3-}, SO_4^{2-}, $CH_3 \cdot COO^-$, Cl^-, Br^-, NO_3^-, CLO_4^-, I^-, SCN^-

Cations: NH_4^+, Rb^+, K^+, Na^+, Cs^+, Li^+, Mg^{2+}, Ca^{2+}, Ba^{2+}

Increasing chaotropic effect

Once the macromolecule of interest is attached to the stationary phase, it is necessary to detach it in order to recover it as a bio-product. Desorption is most commonly accomplished by reducing the ionic strength in the mobile phase, by building a decreasing gradient of salt concentration (Fausnaugh et al., 1984). In this stage, the hydrophobic interaction between the macromolecule and the ligand is weakened as salt concentration diminishes in the mobile phase. As a consequence, the macromolecule is desorbed when a specific salt concentration is reached. This salt concentration, or ionic strength, depends on the physicochemical properties of the macromolecule. In this way, HIC can be used to selectively detach different macromolecules in a solution, thus becoming a powerful separation process.

Protein retention in HIC has been interpreted in the light of the underlying thermodynamic phenomena, by considering the effect of salt. Melander et al. (1989) proposed a thermodynamic model that describes protein retention in terms of electrostatic and hydrophobic interactions. This model describes protein retention due to only electrostatic interactions (case of ion Exchange Chromatography), only hydrophobic interactions (case of HIC), and both types of interactions (case of a weakly hydrophobic support or a chromatographic support bearing both hydrophobic and charged ligands). Simplifications of this model have been used to develop methodologies to predict protein retention in HIC. This model is described below.

2.3 Thermodynamic model for protein retention in HIC

The thermodynamic model proposed by Melander et al. (1989) to describe the effect of salt concentration on macromolecule retention in chromatography (IEC and HIC) can be applied to any stationary phase consisting of a highly hydrated surface modified with charged ligands (in the case of IEC), weakly hydrophobic moieties (in the case of HIC), or both. Electrostatic and hydrophobic interactions between the macromolecule and the stationary phase are treated separately. Electrostatic interaction is modeled based on the Manning's counter ion condensation theory (Manning, 1978), whereas hydrophobic interaction is treated by considering an adaptation of the Sinanoglu's solvophobic (Sinanoglu, 1982) theory that relates the salting out of proteins with their retention in HIC (Melander & Horvath, 1977). Figure 3 depicts protein retention due to hydrophobic interactions, electrostatic interactions, and both hydrophobic and electrostatic interactions.

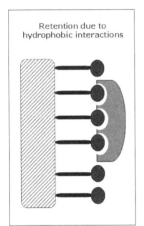 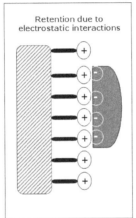 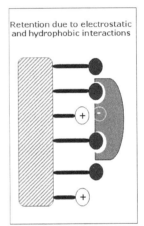

Fig. 3. Protein retention due to hydrophobic, electrostatic, and both interactions. Electrostatic interactions are long-range interactions, and then moieties with opposite charges do not need to be in physical contact. Hydrophobic interactions are short-range, and then interacting hydrophobic moieties must be in contact. As a consequence, when hydrophobic and charged moieties are present, both types of interactions may occur.

The main assumptions considered in the model are listed below:
i.   The dimensions of the pores of the support are large with respect to the macromolecule size, and their shape is approximately a cylinder of infinite radius and the size-exclusion effects are negligible.
ii.   The immobilized charges or hydrophobic moieties are uniformly spaced and equally accessible at the pore wall.
iii. The macromolecule is spherical and presents uniformly distributed and equally accessible fixed charges and hydrophobic patches on its surface.
iv.   The macromolecule does not suffer conformational changes during the adsorption-desorption process.
v.   Only a small fraction of the binding moieties on the stationary phase are occupied by the macromolecule.
vi.   There are no specific interactions between the salt and the macromolecule.

2.3.1 Electrostatic interaction

The Gibbs energy of binding ($\Delta G^0{}_{es}$) of the macromolecule to a stationary phase in presence of salt (that acts as a counter ion) is given by equation (2). Here m_s is the molal salt concentration, N_{Av} the Avogadro's number, "e" the base of the natural logarithm, "b" the average spacing of fixed charges on the surface, δ_p the thickness of the condensation layer over the surface of the stationary phase where each fixed charge occupies an area of b^2, δ_s the layer thickness of salt counter ion, Z_p the characteristic charge of the protein, Z_s the valence of the salt counter ion, and ξ a dimensionless structural parameter that characterizes the charged surface. R is the universal constant of gases and T the absolute temperature.

$$\frac{-\Delta G_{es}^0}{2.3 \cdot R \cdot T} = \log\left(\frac{N_{Av} \cdot b^2 \cdot \delta_p}{1000 \cdot e}\right) + \frac{Z_p}{Z_s}\log\left(\frac{1000 \cdot e}{\left(N_{Av} \cdot b^2 \cdot \delta_s \cdot m_s\right) \cdot \left(1 - Z_s \cdot \xi\right)}\right) \tag{2}$$

2.3.2 Hydrophobic interaction

The contact between the hydrophobic patches on the macromolecule surface that are exposed to the solvent and the hydrophobic ligands on the stationary phase, trigger the retention due to hydrophobic interaction. The Gibbs energy of hydrophobic interaction (ΔG_{hp}) is expressed in terms of the molal surface tension increment of the salt (σ_s), as shown in equation (3), which is valid only in the absence of specific salt effects. In Equation (3) m_s is the salt molality, $\Delta G^0{}_{aq}$ represents the reduction in Gibbs energy due to other effects different form hydrophobic interactions, $\Delta A'$ is the difference between the molecular surface area of the unbound macromolecule (A_M) and the molecular surface area of the macromolecule attached to the stationary phase (A_s). $\Delta A'$ corresponds to the surface contact area between the bound protein and the hydrophobic site of the matrix.

$$\Delta G_{hf}^0 = \Delta G_{aq}^0 - \Delta A' \cdot \sigma_s \cdot m_s \tag{3}$$

2.3.3 Combined electrostatic and hydrophobic interaction

The retention factor (k'), given in equations (4) and (5), is represented in terms of salt molality when both electrostatic and hydrophobic interactions are present. This is accomplished by combining equations (2) and (3) to give equation (6). In equation (5), K is the equilibrium constant and ϕ is the phase ratio (stationary phase mass / mobile phase mass). In equation (6) α is the phase volume ratio (stationary phase/mobile phase). Equation (6) can be written in a simplified form, as given by equation (7), where A is a constant determined by all the system characteristics, B the electrostatic interaction parameter and C the hydrophobic interaction parameter. In equation (7), the term C accounts for the hydrophobic surface contact area between the macromolecule and the stationary phase, and is given by equation (8).

$$\log K = \left(\frac{-\Delta G_{es}^0}{2.3 \cdot R \cdot T}\right) - \left(\frac{-\Delta G_{hf}^0}{2.3 \cdot R \cdot T}\right) \tag{4}$$

$$k' = \phi \cdot K \tag{5}$$

$$\log k' = \log\left(\frac{N_{Av} \cdot b^2 \cdot \delta_p}{1000 \cdot e}\right) + \frac{Z_p}{Z_s}\log\left(\frac{1000 \cdot e}{\left(N_{Av} \cdot b^2 \cdot \delta_s \cdot m_s\right)\cdot\left(1 - Z_s \cdot \xi\right)}\right) +$$

$$\frac{\Delta G_{aq}^0}{2.3 \cdot R \cdot T} + \frac{\Delta A' \cdot \sigma_s \cdot m_s}{2.3 \cdot R \cdot T} + \log \alpha \qquad (6)$$

$$\log k' = A - B \cdot \log m_s + C \cdot m_s \qquad (7)$$

$$C = \frac{\Delta A' \cdot \sigma_s}{2.3 \cdot R \cdot T} \qquad (8)$$

Equation (7) corresponds to the Simplified Thermodynamic Model for Electrostatic and Hydrophobic Interactions. This model is of practical usefulness, since its parameters can be obtained from experimental runs in a relatively simple manner, depending on the salt concentration present in the macromolecule solution. At low salt concentration, up to 0.5 molal, hydrophobic interactions can be neglected, and therefore the parameters A and B in equation (7) can be estimated by means of a linear regression between isocratic retention factors obtained at different salt molalities. At high salt concentration, electrostatic interactions are negligible, and hence the parameters A and C can be obtained in a similar way, considering the isocratic retention factors. The hydrophobic contact area ($\Delta A'$ in equation (8)) can easily be obtained from the slope of the limiting plot of log k' versus molal salt concentration.

The simplified thermodynamic model has been used to investigate the effect of surface hydrophobicity distribution of proteins on retention in HIC (Mahn et al., 2004). The applicability of the model to predict protein retention time in HIC was demonstrated, and for the first time it was experimentally proven that surface hydrophobicity distribution has an important effect on protein retention in HIC. Furthermore, it was shown that the parameter $\Delta A'$ that comes from equations (7) and (8) was able to represent the protein retention in HIC with salt gradient elution. However, the methodology proposed by Mahn et al. (2004) requires the generation of a considerable amount of experimental data, thus limiting its application.

3. Hydrophobic Interaction Chromatography process

The HIC process consists of injecting a macromolecule solution in a column packed with a stationary phase specifically designed to promote hydrophobic interaction with macromolecules such as proteins (solute). Usually retention is accomplished under high salt concentration conditions. Elution is achieved by decreasing the ionic strength in the mobile phase, building a decreasing salt gradient. At a microscopic level, the macromolecule enters in contact with the hydrophobic ligands at the pores surface of the resin, suffering a spatial reorientation. The hydrophobic ligands of the stationary phase interact with the hydrophobic zones of the macromolecule exposed to the solvent (usually aqueous solution), and thus the protein is reversibly attached to the resin.

Figure 4 shows a schematic representation of a HIC process. Here, A and B represent the vessels that contain the buffers used to manage the chemical environment in order to promote adsorption and desorption of the macromolecules present in the sample. The solution in A corresponds to a buffer with a low concentration of a neutral salt (usually 0.1 M), aiming to stabilize the macromolecular three-dimensional structure. The solution in B corresponds to buffer "A" added with a high salt concentration (usually higher than 1 M). Adsorption is promoted by using buffer "B", while desorption is induced by mixing both A and B forming a decreasing gradient salt concentration.

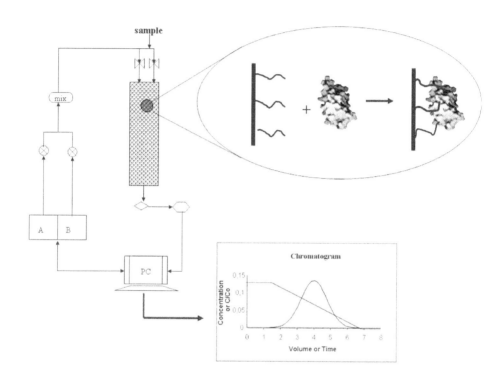

Fig. 4. Schematic representation of the HIC process. The HIC process consists of injecting a protein sample in a hydrophobic column under high salt concentration conditions such that hydrophobic interaction between the protein and the resin is promoted. Elution is achieved by decreasing the ionic strength in the mobile phase, building a decreasing salt gradient. In a microscopic level, the hydrophobic patches on the protein surface interact with the hydrophobic ligands of the resin, being reversibly attached to it. The protein concentration in the outlet is recorded as a function of time, and then a chromatogram is obtained.

The macromolecule concentration in the outlet solution is continuously determined through absorbance at 280 nm, and finally the elution curve or "chromatogram" is obtained. The chromatographic behavior in HIC can be characterized by several parameters, including the elution curve (most commonly by using the theoretical plate theory), the retention time or volume, or other parameters based on the preceding ones. To predict the behavior of proteins in HIC, the preferred parameter is the "Dimensionless Retention Time" (DRT), given by equation (9), where t_R is the time corresponding to the peak maximum, t_0 is the time at the beginning of the elution gradient, and t_f the time at the end of the gradient. In

HIC, the exploited property is hydrophobicity (Eriksson, 1998), and accordingly retention time is highly influenced by this property. Therefore, knowing macromolecule hydrophobicity allows predicting its behavior in HIC. Currently there is no universally agreed definition of protein hydrophobicity, but there is consensus in that it is determined by the hydrophobic contribution of the amino acids that compose the protein (Tanford, 1962).

$$DRT = \frac{t_R - t_0}{t_f - t_0} \qquad (9)$$

On the other hand, protein retention in HIC is significantly affected by the operating conditions, which influence the resolution and selectivity of purification processes that include a HIC step (Ladiwala et al., 2006). From a process point of view, it is essential to count on methodologies and mathematical models to describe and to predict a protein behavior in HIC, ideally under varying operating conditions. Many efforts have been carried out to develop theories to explain this behavior based on protein properties, mainly protein hydrophobicity. At this point, controversial approaches have been proposed to theoretically estimate or experimentally determine protein hydrophobicity. These approaches include different amino acid hydrophobicity scales as well as diverse methodologies to perform calculations that use some scale to describe and predict protein retention time in HIC.

3.1 Protein hydrophobicity
3.1.1 Amino acid hydrophobicity scales
As stated above, protein hydrophobicity is determined by the hydrophobicity of the amino acids that compose it. Hence, it becomes necessary to quantify in any way the hydrophobic contribution of each amino acid. For this purpose, different approaches have been proposed to assign a hydrophobicity value to each one of the standard amino acids (Biswas et al., 2003; Kovacs et al., 2006). These methods are based on theoretical calculations and/or experimental determinations. Besides, the amino acid hydrophobicity scales differ in the hydrophobicity value assigned to each amino acid as well as in the relative position occupied by each one. These scales have been classified into several categories by different authors (Lienqueo et al., 2002; Mahn et al., 2009), based on their underlying principles.
Despite the differences between the hydrophobicity assigned to each residue by the different scales, it is clear a global tendency. Isoleucine shows the highest hydrophobicity in most scales, followed by Tryptophan. Glycine usually has an intermediate hydrophobicity level, i.e. neutral hydrophobicity, and the lowest level is mostly assigned to Aspartic acid (Lienqueo et al., 2007), i.e., this is the most hydrophilic amino acid. The suitability of the hydrophobicity scale depends on the use that will be given to the estimation of the protein or peptide hydrophobicity, as well as on the way to estimate this property. The scales proposed by Miyazawa & Jernigan (1996) and by Cowan & Whittaker (1990) are the most adequate to estimate protein hydrophobicity based on its three-dimensional structure (Lienqueo et al., 2007), regarding its behavior in HIC. Additionally, Salgado et al. (2005) proposed that the scale developed by Wertz & Scheraga (1978) is the most adequate to estimate protein hydrophobicity based on the amino acid composition of that protein.

	Cowan & Whittaker (1990)		Miyazawa & Jernigan (1996)		Wertz & Scheraga (1978)	
	Original	Normalized	Original	Normalized	Original	Normalized
ALA	0.420	0.660	5.330	0.391	0.520	0.375
ARG	-1.560	0.176	4.180	0.202	0.490	0.321
ASN	-1.030	0.306	3.710	0.125	0.420	0.196
ASP	-0.510	0.433	3.590	0.105	0.370	0.107
CYS	0.840	0.763	7.930	0.819	0.830	0.929
GLN	-0.960	0.323	3.870	0.151	0.350	0.071
GLU	-0.370	0.467	3.650	0.115	0.380	0.125
GLY	0.000	0.557	4.480	0.252	0.410	0.179
HIS	-2.280	0.000	5.100	0.354	0.700	0.696
ILE	1.810	1.000	8.830	0.967	0.790	0.857
LEU	1.800	0.998	8.470	0.908	0.770	0.821
LYS	-2.030	0.061	2.950	0.000	0.310	0.000
MET	1.180	0.846	8.950	0.987	0.760	0.804
PHE	1.740	0.983	9.030	1.000	0.870	1.000
PRO	0.860	0.768	3.870	0.151	0.350	0.071
SER	-0.640	0.401	4.090	0.188	0.490	0.321
THR	-0.260	0.494	4.490	0.253	0.380	0.125
TRP	1.460	0.914	7.660	0.775	0.860	0.982
TYR	0.510	0.682	5.890	0.484	0.640	0.589
VAL	1.340	0.885	7.630	0.770	0.720	0.732

Table 1. Amino acid hydrophobicity scales useful in HIC.

The Miyazawa & Jernigan (1996) scale is based on the three-dimensional structure of proteins, and it represents the contact energy between adjacent amino acids in folded protein. The Wertz & Scheraga (1978) scale is also based on knowledge of the folded protein structure, and it estimates the amino acid hydrophobicity as the ratio between the number of buried residues and the number of residues exposed to the solvent, for each type of standard amino acid. Both scales are based on clusters composed by a significant number of proteins whose three-dimensional structure had been elucidated through experimental methods. Both scales have been classified as indirect scales (Mahn et al., 2009). On the other hand, the Cowan & Whittaker (1990) scale, which has been considered a direct scale, assigned a hydrophobicity value to each standard amino acid based on the retention time of z-derivatives of each amino acid in HPLC. The scales mentioned above are presented in Table 1.

3.1.2 Estimation of protein hydrophobicity

There are different approaches to estimate protein hydrophobicity, which are based on different principles. The classical approach consists of estimating the "average surface hydrophobicity" ($\phi_{surface}$) based on the three-dimensional structure of the macromolecule in its native conformation (Lienqueo et al., 2002; Berggren et al., 2002). This approach considers only the amino acid residues that are accessible to the solvent at the protein surface, by using three-dimensional structural data. This method considers that each amino acid on the protein surface has a hydrophobic contribution proportional to its solvent accessible area, and the hydrophobicity of each residue is given by the amino acid hydrophobicity scale developed by Miyazawa & Jernigan (1996) or Cowan & Whittaker (1990), in their normalized form (see Table 1), as shown by equation (10).

$$\phi_{surface} = \frac{\sum (s_{aai} \cdot \phi_{aai})}{s_p} \tag{10}$$

Here, $\phi_{surface}$ is the calculated value of the surface hydrophobicity for a given protein, i (i =1, . . ., 20; different i-values indicate different standard amino acids), s_{aai} is the solvent accessible area occupied by the amino acid i, ϕ_{aai} is the hydrophobicity value assigned to amino acid i by the hydrophobicity scale, and s_p is the total solvent accessible area of the entire protein. It has to be noted that for proteins with a prosthetic group s_p is bigger than the sum of the solvent accessible area occupied by the amino acids; and for proteins without prosthetic group, these values are equal. Table 2 shows the average surface hydrophobicity for a group of proteins using the amino acid hydrophobicity scales given in Table 1, and calculated by equation (10). This method for estimating protein hydrophobicity has proven to be valid in several cases (Lienqueo et al., 2002; Lienqueo et al., 2003; Lienqueo et al., 2007); however, this methodology is not valid for proteins that exhibit a highly heterogeneous distribution of the hydrophobic patches on their surfaces (Mahn et al., 2004).

Protein	Cowan & Whittaker	Miyazawa & Jernigan	Wetz & Scheraga
α-amylase	0.447	0.282	0.319
Citochrome c	0.362	0.185	0.171
Conalbumin	0.421	0.233	0.242
Concanavalin A	0.448	0.273	0.308
α-lactalbumin	0.491	0.318	0.304
β-lactoglobulin	0.468	0.279	0.284
Lysozyme	0.425	0.274	0.307
Myoglobin	0.392	0.214	0.220
Ovalbumin	0.457	0.257	0.270
Chymotrypsin	0.474	0.306	0.313
Chymotrypsinogen	0.468	0.298	0.305
Ribonuclease A	0.406	0.230	0.255
Thaumatin	0.464	0.269	0.279

Table 2. Surface hydrophobicity of proteins estimated by equation (9).

Genetic engineering is often used to improve the performance of separation and purification methods. Specifically in HIC, its performance has been improved by the fusion of short hydrophobic peptide tags such as T3, (TP)3, T3P2, T4, (TP)4, T6, T6P2, T8, (WP)2, (WP)4 to a protein of interest (Brandmann et al., 2000; Rodenbrock et al., 2000; Fexby & Bülow, 2004), thus increasing its original hydrophobicity. This genetic engineering strategy has the advantage that the structure/function changes are minimized in relation to the original properties of the native protein. Furthermore, the use of hydrophobic polypeptide tags allows investigating simple and less expensive stationary phases (in comparison with affinity chromatography supports), such as those used in HIC.

As a consequence, methods to calculate the surface hydrophobicity of tagged proteins have been proposed. One of those methods is the one proposed by Simeonidis et al. (2005) that allows computing the "tagged surface hydrophobicity" (ϕ_{tagged}), by equation (11). The surface hydrophobicity of the tagged protein is estimated as the average surface hydrophobicity of the original protein (without the tag) plus the hydrophobicity of the peptide tag. In this case, a fully exposed surface of the amino acids in the tag is assumed. In equation (11), n_k is the number of amino acids of "k" type (usually hydrophobic amino acids, such as tryptophan, leucine and isoleucine) in the tag, and s_{tag_aak} is the fully exposed surface of amino acid "k" in the tag.

$$\phi_{tagged} = \frac{\sum \left(s_{aai} \cdot \phi_{aai} \right)}{s_p} + \sum \left(\frac{\left(s_{tag\_aak} \cdot n_k \right)}{s_p + \sum \left(s_{tag\_aak} \cdot n_k \right)} \cdot \phi_{aak} \right) \tag{11}$$

Despite the remarkable results reached by the methods described above to estimate protein hydrophobicity, the need of knowing the three-dimensional structure appears as a serious disadvantage. This is especially clear from the ratio between the number of proteins of known three-dimensional structure available in the PDB database (Bermann et al., 2000) and the number of proteins sequenced in the UniProtKB/Swiss-Prot database (Bairoch et al., 2005). Currently (January 2011) this number is closer to 0.13 (70695/534420). This situation points out the need of a procedure based on low level information, such as the amino acidic composition. Salgado et al. (2005) developed a mathematical model to predict the average surface hydrophobicity of a protein based only on its amino acidic composition and, therefore, avoiding the use of its three-dimensional structure.

Equation (12) shows the basic structure of the model. In this equation, ASH represents the average surface hydrophobicity, n_i is the number of amino acids of class i in the protein, $\hat{l}$ is the normalized length of the protein sequence, and c_i correspond to adjustable parameters. The function f accounts for a correction of the amino acid composition of the protein according to different assumptions about the amino acids trend to be exposed to the solvent. The simplest form of f considers all the amino acids completely exposed. Parameters for building the function f were determined in a large set of non-redundant proteins by Salgado et al. (2005).

$$ASH = c_0 + \sum_{i=1}^{20} c_i \cdot f\left(n_i \right) + c_{21}\hat{l} \tag{12}$$

3.2 Methods for predicting retention time in HIC
The approaches discussed above to calculate protein hydrophobicity have been used to predict protein retention time by different methods. The simplest methodology uses straightforward quadratic models, whose parameters depend on the chromatographic conditions used in the HIC run (Lienqueo et al., 2007), and whose variables are DRT and the average surface hydrophobicity of the protein to be separated ($\phi_{surface}$). The most appropriate hydrophobicity scale was found to be that proposed by Miyazawa & Jernigan (1996), in its normalized form. The general model is given by equation (13), where A', B' and C' are the model parameters that depend on the chromatographic conditions, such as type and concentration of salt and type of stationary phase. These parameters have been obtained

from adjusting experimental data to the quadratic model. Table 3 shows the values of A′, B′ and C′ obtained for different operating conditions. The model given by equation (13) is useful for predicting retention times of structurally stable proteins that have a relatively homogeneous distribution of the surface hydrophobicity, such as ribonuclease A.

$$DRT = A' \cdot \phi_{surface}^2 + B' \cdot \phi_{surface} + C' \tag{13}$$

Figure 5 shows a scheme of the methodology to predict DRT based on protein hydrophobicity. The procedure begins with the calculation of the protein surface accessible to the solvent, and the fraction of that surface occupied by each kind of amino acid. To calculate this, it is necessary to count on a PDB file, i.e. to know the spatial coordinates of each atom composing the macromolecule, preferably determined experimentally through X-ray crystallography or nuclear magnetic resonance (NMR). Experimentally determined structures can be obtained in The Protein Data Bank (PDB; www.rcsb.org/pdb) database (Bermann et al., 2000). Additionally, three-dimensional models can be found in other databases such as ModBase (http://modbase.compbio.ucsf.edu/modbase-cgi/search_form.cgi) (Pieper et al., 2009). Also it is required using a computational program or suit to perform the calculation, such as the software GRASP (Nicholls et al., 1991). With this information, the average surface hydrophobicity is calculated by means of equation (10) and using the Miyazawa & Jernigan hydrophobicity scale, in its normalized form. Finally, through a quadratic model like equation (13) the retention time of the protein can be estimated as DRT.

Resin	Salt	Initial Salt molarity	A′	B′	C′
Phenyl Sepharose	Ammonium sulfate	1	11.79	-0.29	0.35
Phenyl Sepharose	Ammonium sulfate	2	-12.14	12.7	-1.14
Phenyl Sepharose	Sodium chloride	2	-77.10	42.33	-5.13
Phenyl Sepharose	Sodium chloride	4	-65.01	37.55	-4.71
Butyl Sepharose	Ammonium sulfate	1	36.76	-16.07	1.73
Butyl Sepharose	Ammonium sulfate	2	10.02	0.45	-0.38
Butyl Sepharose	Sodium chloride	2	-12.05	6.51	-0.80
Butyl Sepharose	Sodium chloride	4	-1.74	5.55	-1.01

Table 3. Parameters of equation (12) for different operating conditions.

The surface hydrophobicity of tagged proteins (ϕ_{tagged}) has been used by Lienqueo et al. (2007) for predicting the DRT of cutinases tagged with hydrophobic peptides in different matrices for HIC, by means of equation (13) and the methodology represented in Figure 3. The coefficients of the linear model are constants for each set of operating conditions. This approach has proven to be effective in predicting the behavior of tagged proteins in HIC, since it showed a low deviation between predicted and experimental DRT (in the order of 2%), for the tagged cutinases that were studied. Finally, the ASH obtained from equation

(11) based on amino acidic composition was used to predict chromatographic behavior in HIC, resulting in a performance 5% better than that observed in the model based on the three-dimensional structure of proteins (equation (10)) (Salgado et al., 2008).

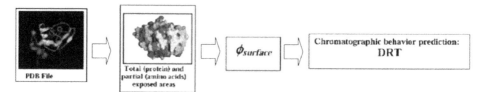

Fig. 5. Methodology for predicting protein retention time in HIC based on surface hydrophobicity. Using a PDB file as input to the program GRASP, the total and partial accessible areas of the exposed amino acids is determined. Using an amino acid hydrophobicity scale and equation (12), the average surface hydrophobicity can be obtained. Then, through simple mathematical correlations the DRT of the protein can be estimated.

4. Applications in biomedical engineering
4.1 General applications

Currently, many proteins of pharmacological and industrial interest are obtained through highly optimized purification processes, typically consisting of two or three chromatographic separation stages. Usually these processes involve one or two IEC steps followed by a HIC step (Asenjo & Andrews, 2004). In addition, most recombinant proteins can be obtained at therapeutic grade of purity, by processes of the same structure (Asenjo & Andrews, 2008). Then, HIC often forms part of processes to yield a purified macromolecule of biomedical interest, such as therapeutic proteins (Seely & Richey, 2001), DNA vaccines (Diogo et al., 2000), and enzymes (Teng et al., 2010), among others. Besides, the use of HIC to purify protein complexes (McCue et al., 2008), as well as to study protein folding from a thermodynamic point of view (Geng & Wang, 2007), have been reported. Some applications of HIC for purifying enzymes and protein complexes, and to studying protein folding are described below.

4.1.1 Purification of proteins and enzymes by HIC

Recently, many strategies that involve a HIC step to purify proteins and enzymes of industrial and/or biomedical interest have been reported. For instance, Liu et al. (2010) developed a purification process to isolate and characterize an antifungal protein from *Bacillus subtilis*, which can be used as a bio-control agent. The process consisted of a preliminary precipitation with ammonium sulfate at 30-70% saturation, followed by HIC (using Phenyl Sepharose as stationary phase) and finally an IEC step. The process gave an overall recovery of 1.2% of total protein in the cell extract. The antifungal protein showed ribonuclease, protease and hemagglutinating activities.

On the other hand, Teng et al. (2010) purified and characterized an endo-β-1,4-glucanase from the giant snail (*Achatina fulica frussac*) by means of a process consisting of three chromatographic steps: size exclusion chromatography (SEC), anion exchange chromatography (AEC), and finally hydrophobic interaction chromatography. A 29-fold purity increase was achieved, and an overall recovery of 14.7% was reached. In addition, this novel enzyme has a particularly high stability at a broad pH range, acidic pH optimum, and a very high thermostability, and therefore it would have a great potential use in industry.

Lavery et al. (2010) reported the purification of a peroxidase from horseradish roots (*Armoracia rusticana*) by means of a three-step strategy, consisting of ultrasonication, ammonium sulfate precipitation, and HIC (using Phenyl Sepharose). In this strategy, the only high-resolution purification step corresponded to HIC. An overall yield of 71% and a 291-fold purification were achieved, thus demonstrating the high efficiency of this technique. The purified peroxidase was extremely stable in different media, and therefore its commercialization seems promising. Bhuvanesh et al. (2010) used a single-step method to purify a filarial protein (expressed heterologously in *E. coli*) with great potential as a vaccine for preventing human lymphatic filariasis. The purification method consisted of a HIC step. An overall recovery of 60% and 100% purity were achieved.

4.1.2 Purification of protein aggregates by HIC

The use of HIC to separate product-related impurities in the biopharmaceutical industry is well documented (Queiroz et al., 2001). This method is also used to separate multimers from monomeric forms of proteins of biomedical interest, since these conformations often differ in surface hydrophobicity. This difference owes to the fact that the stabilization of quaternary structures occurs due to hydrophobic interaction between the monomers, resulting in protein aggregation. In this way, the hydrophobic patches of a multimer are somewhat hidden, and therefore less accessible to the hydrophobic ligands of a HIC support, unlike the monomer whose hydrophobic patches are exposed to the solvent and, accordingly, accessible to the HIC ligands. The adsorption mechanism of protein aggregates in HIC is complex and not fully understood so far.

Mc Cue et al. (2010) developed a chromatography model to predict the separation of monomer and aggregate species. Equation (14) shows the Langmuir isotherm that describes equilibrium between the protein adsorbed to the resin and the protein that remains in solution. Here, C is the protein concentration in the mobile phase, q is the protein concentration in the stationary phase, q_m is the resin maximum capacity and k is the equilibrium constant. Equation (15) shows the mass balance used to describe the protein concentration profiles. Mass conservation was assumed and the intra-particle mass transfer was considered to be driven by homogeneous diffusion. In equation (15), D_{eff} is the effective diffusivity of the protein from the mobile phase bulk to the inner of the porous resin bead, t is time and r is the radial coordinate. The validity of the model was assessed by experimental determinations. A fraction of the aggregate proteins bound irreversibly to the HIC resin, becoming the major factor governing the process. This phenomenon was adequately described by the model.

$$C = \frac{q}{k \cdot (q_m - q)} \tag{14}$$

$$\frac{\partial q}{\partial t} = D_{eff} \cdot \left(\frac{\partial^2 q}{\partial r^2} + 2 \cdot \frac{\partial q}{r \cdot \partial r} \right) \tag{15}$$

4.1.3 Protein folding in HIC

Protein folding is relevant from a process point of view, since most recombinant proteins produced in bacteria such as *E. coli* are accumulated as inclusion bodies, and therefore protein refolding constitutes an additional stage in the production and purification process in order to yield a "functional" product (especially in the case of enzymes). Hydrophobic interaction chromatography has been used to study thermodynamics aspects of protein folding. For instance, Geng et al. (2005) performed calorimetric determinations on the enthalpy change ($\Delta H_{folding}$) of denatured lysozyme during its adsorption to a hydrophobic surface, with the simultaneous protein refolding. The surface consisted of PEG-600 made of a silica base HP-HIC (High Performance- Hydrophobic Interaction Chromatography) packing. At 25°C, $\Delta H_{folding}$ was found to be - 34 439 KJ/mol, involving adsorption, dehydration and molecular conformation enthalpies changes.

Later, Geng & Wang (2007) used the concept of "Protein Folding Liquid Chromatography" (PFLC), to describe a chromatographic process aiming to either raise the efficiency, or shortening the time of protein folding. Besides, an optimal PFLC should be able to simultaneously remove denaturant substances, separate contaminant proteins, promote refolding of the target protein, and ease denaturant recovery. Any type of chromatography can be used in PFLC, mainly Size Exclusion Chromatography, Ion Exchange Chromatography, Affinity Chromatography, and Hydrophobic Interaction Chromatography.

In HIC, the process is governed by thermodynamic equilibrium and so does the protein folding. PFLC provides the chemical equilibrium that favors the conversion from aggregate to desorbed protein, resulting in a higher refolding efficiency and shorter refolding time. The unfolded proteins, at a high ionic strength, are driven by hydrophobic interactions from the mobile phase to the HIC stationary phase, and the hydrophobic patches on the proteins surface get attached to the hydrophobic ligands, while the hydrophilic zones of the unfolded molecules remain in contact with the solvent. As a consequence, unfolded molecules are not able to aggregate. The unfolded molecules desorb from the HIC support as ionic strength in the mobile phase decreases. Protein molecules with incorrectly folded domains would be corrected by the spontaneous disappearance of the domains in the mobile phase due to their thermodynamic instability. After many HIC runs, the incorrectly folded domains will decrease, while the correctly folded molecules will predominate, resulting in protein refolding at high efficiency.

4.2 Applications in biomedical engineering

Biomedical applications of HIC are broad, since this technique offers some advantages over other chromatographic techniques, such as Affinity Chromatography (AC) and Reverse-Phase Chromatography (RPC). The use of AC depends on the availability of a specific ligand for the protein or group of proteins to be separated, thus limiting their applicability and raising its cost. The main disadvantage of RPC relies on the nature of the solvent in which the purified protein is recovered, usually an organic solvent not suitable for human or animal use. Then, HIC constitutes a purification tool suitable for biomedical applications, such as vaccines, therapeutic proteins, plasmids and mainly antibodies. In addition, the use of chromatography in high-throughput studies, such as proteomics and protein interactions, is increasing. Some of these Biomedical Engineering applications of HIC are discussed below.

4.2.1 Antibodies purification

At the beginning of the antibody industry, purification was performed through AC. For instance, protein A - AC was used for purifying monoclonal antibodies (MAbs), due to the extremely low MAb concentration in the initial solution (fermentation broth), and the high amount of contaminant proteins. Therefore, affinity chromatography was the most suitable technique, given its high selectivity and resolution. Unfortunately, this purification technique has a serious disadvantage given by the high affinity of the MAb for the ligand (such as protein A), making it difficult to release the MAb from the ligand, with the consequent economical detriment. Moreover, MAbs are highly hydrophobic macromolecules, and then the use of HIC has been suggested (Asenjo & Andrews, 2008). At the present time molecular biology advances have enabled reaching high concentrations of MAbs in the fermentation broth, making it possible to use less selective but cheaper purification techniques, such as HIC. Figure 6 depicts a monoclonal antibody (A) and the antibody attached to a HIC stationary phase (B).

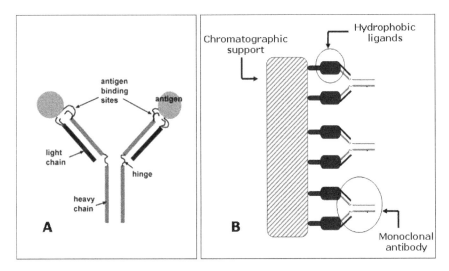

Fig. 6. (A) Schematic representation of a MAb. The antigen binding sites of the MAb are highlighted. Since this zone is characterized by an extremely high hydrophobicity, MAbs exhibit a high attraction for the hydrophobic ligands used in HIC resins. (B) Schematic representation of MAbs attached to a HIC resin. The antigen binding site interacts directly with the hydrophobic ligands of the HIC resin.

HIC is used as a polishing step in the purification processes of immunoglobulin-related products, since it has the ability to remove aggregated forms of the antibody (Rinderknecht & Zapata, 2006). Despite the high resolution offered by HIC, there are some drawbacks for its use in MAbs purification, given by the relatively low binding capacity of HIC supports and the consequent low yield in MAb recovery, compared to AC. Besides, MAb elution is usually achieved at a relatively high salt concentration, which implies that the solution containing the purified MAb also contains a high amount of salt that hinders sample manipulation and transitions during large-scale production.

This has encouraged research on HIC optimization, mainly regarding chromatographic supports. Recently, Chen et al., (2008) showed that the optimization of pore size of a HIC support significantly improved Immunoglobulin G binding capacity and also increased HIC efficiency, maintaining the MAb stability. Optimizing pore size facilitates mass transfer from mobile phase bulk towards the hydrophobic ligand. Kostareva et al. (2008) purified a heteropolymer (a kind of MAb consisting of a dual antibody conjugate) by HIC. They found that using a Propyl-HIC resin the heteropolymer was efficiently separated from free MAbs, thus confirming the ability of HIC for separating aggregates from monomers, and also its suitability for purifying MAbs.

4.2.2 Proteomics

Proteomics can be defined as the study of all the proteins codified by a genome, in a given tissue of a given organism at a given time. It involves studying how the concentration or "relative abundance" of the proteins change under a certain stimulus, protein conformational changes, protein – protein interactions (or "interactomics"), among others, as well as the use and development of experimental and bioinformatics technologies necessary to perform these studies. In this regard, protein separation techniques are essential. The fundamental separation methods used in proteomics are Sodium Dodecyl Sulfate- Polyacrylamide Gel Electrophoresis (SDS-PAGE) and/or Two-Dimensional Gel Electrophoresis (2DGE) and mass spectrometry (MS); the latter is used as separation but also as identification tool. Figure 7 depicts a classical proteomics experiment, starting from a biological sample, followed by preliminary fractionation by liquid chromatography and after that separation by 2DGE, and finally identification of protein spots by MS.

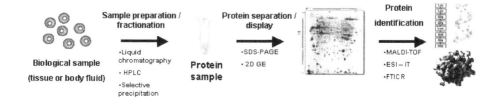

Fig. 7. Simplified representation of a gel-based proteomics experiment. Starting from a biological sample, a protein extract is obtained using different biochemical techniques to fractionate the sample. These fractionation steps allow the enrichment of protein fractions in low abundance proteins and to reduce the complexity of the sample. The protein fractions are then resolved by SDS-PAGE or 2D GE, and finally protein spots are excised form the gel and then analyzed by mass spectrometry in order to determine their identity and structural properties.

From a Biomedical point of view, proteomics is an important field in the task of discovering new biomarkers that reflect the health/disease status of living organisms. The use of proteomics with this purpose has been somewhat limited due to technical hurdles related to the high complexity of the biological samples to be analyzed, usually blood serum or plasma, but also cerebrospinal fluid, urine and tears. These samples show a wide dynamic range of protein concentration, exceeding 10^{10}. This means that the most abundant protein in plasma (albumin), for example, has a concentration 10^{10} times higher than that of the less abundant protein (such as transcription factors).

Two-dimensional gel electrophoresis can resolve a concentration range of up to 10^4, and therefore 2DGE images or "maps" of blood plasma are dominated by the highly abundant proteins, namely albumin, immunoglobulin, fibrinogen, among others, thus preventing the detection of low abundance proteins (Hoffmann et al., 2007). Mass spectrometry can resolve a range of 10^3 in a single spectrum, but combined with separation steps it can resolve a range of up to 10^6 (Jacobs et al., 2005). This range is still wide, and thus many proteins cannot be detected. Then, chromatographic separation steps should be used before 2DGE in order to reduce the dynamic range of proteins concentration, and consequently increase resolution.

The most abundant proteins in blood plasma are albumin, immunoglobulin, transferrin, haptoglobin, fibrinogen and α-1-antitrypsin, which amount to 90% of total protein mass. Then, total or partial depletion of these proteins allows detecting low abundance proteins. Different methods can be used to deplete these proteins, being liquid chromatography the most popular one (Nakamura et al., 2008). Different chromatographic strategies are available for this purpose, including affinity dye-based chromatography for albumin depletion, affinity to protein A and G for immunoglobulin depletion, specific antibody-affinity columns (Linke et al., 2007), and affinity columns containing lectins, peptides or inorganic ligands (Salih, 2005). Liquid chromatography has the advantage of being easy to use and to scale-up, but are relatively expensive, especially those involving affinity columns. Another drawback of affinity chromatography is the non-specific interactions that lead to the loss of some proteins, with the consequent loss of information (Altintas & Denizli, 2006). In order to overcome the disadvantages of affinity chromatography for its use in blood plasma proteomics, several complementary strategies have been examined, such as sequential anion and cation exchange chromatography followed by 2DGE; and strong cation exchange chromatography followed by liquid-phase isoelectric focusing (Ottens et al., 2005; Barnea et al., 2005). Since these approaches considerably improve the capacity to detect low abundance proteins, it was suggested that the optimization of combinatorial processes by coupling immuno-affinity depletion with other conventional separation methods such as hydrophobic interaction chromatography will probably lead to significant advances in proteomics (Mahn et al., 2010). Despite the research conducted in this area, there is still a lack of optimized processes that ensure detection of the complete proteome of a tissue or cell.

4.2.2.1 Plasma fractionation by HIC

The applicability of HIC as a plasma fractionation method has been recently proposed. Geng et al. (2009) developed a two-dimensional liquid chromatography resin having two types of ligands, and hence that functions in two retention modes: cation exchange and hydrophobic interaction. This method could be applied to the fast fractionation of intact proteins before mass spectrometry analysis. The results obtained by HIC were similar to those obtained by ion exchange chromatography. On the other hand, a HIC matrix consisting of highly acetylated agarose has been used for the isolation of immunoglobulin from porcine serum, with a relative success (Ramos-Clamont et al., 2006).

Recently, Mahn et al. (2010) investigated if the performance of 2DGE could be improved by fractionating blood plasma through a HIC step, thus reducing the relative concentration of some highly abundant proteins in plasma. First, the hydrophobicity of the main 56 proteins present in blood plasma was determined. To do this, the amino acidic composition of the proteins was considered, and hydrophobicity was calculated by equation (16) based on the methodology proposed by Salgado et al. (2005). In equation (16), ϕ_{aai} is given by the Cowan-Whittaker hydrophobicity scale in its normalized form (see Table 1), n_i is the number of amino acids of type i in the protein, $s_{i,max}$ is the maximum solvent accessible area that an amino acid X can have when forming part of the G–X–G tripeptide in extended conformation (Miller et al., 1987).

$$ASH = \sum_{i=1}^{20} \left(\phi_{aai} \cdot \frac{n_i \cdot s_{i,\max}}{\sum_{j \in A} n_j \cdot s_{j,\max}} \right) \tag{16}$$

After that, a cluster analysis was performed in order to classify them as low, medium or high hydrophobicity proteins. This analysis showed that the highly abundant proteins, i.e. albumin, immunoglobulins, fibrinogen and haptoglobin, exhibited a medium hydrophobicity, and thus they fell in the same cluster. With this information, a HIC step was designed to deplete highly abundant proteins from rat plasma samples. The HIC step consisted of stepwise elution to separate the three groups of proteins (low, medium and high hydrophobicity) using a maximum concentration of 2 M ammonium sulfate, and concentration for elution of 0.6 M (to desorb low hydrophobicity proteins), 0.5 M (to desorb medium hydrophobicity proteins), and 0.0 M (to desorb the highly hydrophobic proteins).

Finally, the depleted samples were analyzed by 2DGE and the performance of the HIC pre-fractionation step was compared with that exhibited by a commercial immuno-affinity column. The reproducibility of 2DGE was similar to that obtained from immuno-affinity depleted plasma. However, HIC was more successful in depleting albumin and α-1-antitrypsin. Besides, HIC resulted in a much lower increment of immunoglobulin and haptoglobin abundances than the immuno-affinity column. Then, HIC depletion allowed detecting twice the number of protein spots than immuno-affinity depletion did. Therefore, HIC could be used as a depletion method complementary to affinity columns. The operating conditions in HIC could be optimized in order to maintain the high number of spots that are detected if HIC is used as the sole depletion method. Finally, given the relatively low cost of HIC supports and HIC operation, its use could be proposed as a convenient choice for depleting highly abundant proteins in plasma samples prior to 2DGE-based proteomics.

4.2.2.2 Analysis of protein interaction networks by HIC

Protein–protein interactions are essential in biological processes. All the interactions in a cellular system are known as protein interaction network or 'interactome'. In Biomedicine there is great interest in recognizing these interactions, aiming to establish the role they play in certain diseases. The traditional approaches to study protein-protein interactions are the antibody pull-down method (APD) and the yeast two-hybrid method (YTH). Despite their popularity, these methods have some disadvantages. It is very likely that a protein forms part of different complexes; then, in an APD experiment, antibodies targeting such a protein will pull down together all the complexes where the protein participates, making them appear to be part of a single large complex, confusing the biological interpretation of the results. The YTH is an "*in vivo*" method that allows detecting only binary interactions. It tends to give false positives and is limited to binary interactions. Therefore it is not useful in studying the dynamics of complex formation triggered by different stimuli (Corvey et al., 2005).

Liu et al. (2008) investigated the potential of chromatography to allow the simultaneous examination of multiple protein complexes along with comparing and validating results from the traditional methods. Since protein complexes remain intact during mild forms of elution in AC, a similar behavior should be expected in other chromatographic supports, such as IEC and HIC. They studied the extent to which protein interaction partners from yeast (*S. cerevisiae*) lysate remain associated during IEC, SEC and HIC. Most protein complexes remained intact, and all the proteins forming part of the complex migrated as a single unit. Protein complexes exhibited a chromatographic behavior different from that shown by the individual proteins that compose the complex. Accordingly, studying protein complexes could be easily performed by multidimensional chromatographic methods when at least one of the fractionation dimensions included SEC of native proteins. This method enables the study and recognition of several protein complexes simultaneously, avoiding the use of genetic engineering.

5. Conclusion

HIC is a powerful tool for purifying macromolecules of biomedical interest whose potential has been relatively under-exploited so far. Its applications are diverse, including industrial processes as well as analytical methods. The performance of HIC can be improved by optimizing the supports and the operation mode considering the hydrophobicity of the macromolecule to be separated. Research on optimization of HIC for biomedical applications should be encouraged, since this method allows reducing production cost of biopharmaceuticals such as antibodies and therapeutic proteins.

6. Acknowledgement

Fondecyt Programme.

7. References

Altintas, E.B., Denizli, A. (2006). Efficient removal of albumin from human serum by monosize dye-affinity beads. *Journal of Chromatography B-Analytical Technologies in the Biomedical and Life Sciences*, 832, 2, 216-223.

Asenjo, J.A., Andrews, B.A. (2008). Challenges and trends in bioseparations. *Journal of Chemical Technology and Biotechnology*, 83, 117-120.

Asenjo, J.A., Andrews, B.A. (2004). Is there a rational method to purify proteins? From expert systems to proteomics. *Journal of Molecular Recognition*, 17, 236-247.

Bai, Q., Wei, Y.M., Geng, M.H., Geng X.D. (1997). High performance hydrophobic interaction chromatography A new approach to separate intermediates of protein folding .1. Separation of intermediates of urea-unfolded alpha-amylase. *Chinese Chemical Letters*, 8, 67-70.

Bairoch, A., Apweiler, R., Wu, C.H., Barker, W.C., Boeckmann, B., Ferro, S., Gasteiger, E., Huang, H., Lopez, R., Magrane, M., Martin, M.J., Natale, D.A., O'Donovan, C., Redaschi, N., Yeh, L.S. (2005). The universal protein resource (UniProt). *Nucleic Acids Research*, 33, D154-D159.

Bandmann, N., Collet, E., Leijen, J., Uhlen, M., Veide, A., Nygren, P.A. (2000). Genetic engineering of the Fusarium solani pisi lipase cutinase for enhanced partitioning in PEG-phosphate aqueous two-phase systems. *Journal of Biotechnology*, 79, 161-172.

Barnea, E., Sorkin, R., Ziv, T., Beer, I., Admon, A. (2005). Evaluation of prefractionation methods as a preparatory step for multidimensional based chromatography of serum proteins. *Proteomics*, 5, 3367-3375.

Berggren, K., Wolf, A., Asenjo, J.A., Andrews, B.A., Tjerneld, F. (2002). The surface exposed amino acid residues of monomeric proteins determine the partitioning in aqueous two-phase systems. *Biochimica et Biophysica Acta-Protein Structure and Molecular Enzymology*, 1596, 253-268.

Berman, H.M., Westbrook, J., Feng, Z., Gilliland, G., Bhat, T.N., Weissig, H., Shindyalov, I.N., Bourne, P.E. (2000). The Protein Data Bank. *Nucleic Acids Research*, 28, 235-242.

Bhuvanesh, S., Arunkumar, C., Kaliraj, P., Ramalingam, S. (2010). Production and single-step purification of Brugia malayi abundant larval transcript (ALT-2) using hydrophobic interaction chromatography. *Journal of Industrial Microbiology & Biotechnology*, 37, 1053-1059.

Biswas, K.M., DeVido, D.R., Dorsey, I.G. (2003). Evaluation of methods for measuring amino acid hydrophobicities and interactions. *Journal of Chromatography A*, 1000, 637-655.

Chaturvedi, R., Bhakuni, V., Tuli, R. (2000). The delta-endotoxin proteins accumulate in Escherichia coli as a protein-DNA complex that can be dissociated by hydrophobic interaction chromatography. *Protein Expression and Purification*, 20, 21-26.

Chen, J., Tetrault, J., Ley, A. (2008). Comparison of standard and new generation hydrophobic interaction chromatography resins in the monoclonal antibody purification process. *Journal of Chromatography A*, 1177, 272-281.

Corvey, C., Koetter, P., Beckhaus, T., Hack, J., Hofmann, S., Hampel, M., Stein, T., Karas, M., Entian, K.-D. (2005). Carbon source-dependent assembly of the Snf1p kinase complex in Candida albicans. *Journal of Biological Chemistry*, 280, 25323-25330.

Cowan, R., Whittaker, R.G. (1990). Hydrophobicity indices for amino acid residues as determined by high-performance liquid chromatography. *Peptide Research*, 3, 75-80.

Diogo, M.M., Queiroz, J.A., Prazeres, D.M. (2001). Studies on the retention of plasmid DNA and Escherichia coli nucleic acids by hydrophobic interaction chromatography. *Bioseparation*, 10, 211-220.

Diogo, M.M., Queiroz, J.A., Monteiro, G.A., Martins, S.A., Ferreira, G.N., Prazeres, D.M. (2000). Purification of a cystic fibrosis plasmid vector for gene therapy using hydrophobic interaction chromatography. *Biotechnology and Bioengineering*, 68, 576-583.

Eriksson, K., in: J.-C. Janson, L. Ryden (Eds.), Protein Purification: Principles, High-resolution Methods and Applications, 2nd ed., Wiley-Liss, New York, 1998, p. 151.

Fausnaugh, J.L., Regnier, F.E. (1986). Solute and mobile phase contributions to retention in hydrophobic interaction chromatography of proteins. *Journal of Chromatography*, 359, 131-146.

Fausnaugh, J.L., Kennedy, L.A., Regnier, F.E. (1984). Comparison of hydrophobic-interaction and reversed-phase chromatography of proteins. *Journal of Chromatography*, 31, 141-155.

Fexby, S., Bülow, L. (2004). Hydrophobic peptide tags as tools in bioseparation. *Trends in Biotechnology*, 22, 511-516.

Geng, X., Ke, C. , Chen, G., Liu, P., Wang, F., Zhang, H., Sun, X. (2009). On-line separation of native proteins by two-dimensional liquid chromatography using a single column. *Journal of Chromatography A*, 1216, 3553-3562.

Geng, X., Wang, C. (2007). Protein folding liquid chromatography and its recent developments. *Journal of Chromatography B-Analytical Technologies in the Biomedical and Life Sciences*, 849, 69-80.

Haimer, E., Tscheliessnig, A., Hahn, R., Jungbauer, A. (2007). Hydrophobic interaction chromatography of proteins IV - Kinetics of protein spreading. *Journal of Chromatography A*, 1139, 84-94.

Hjerten, S. (1973). Some general aspects of hydrophobic interaction chromatography. *Journal of Chromatography*, 87, 325-331.

Hoffman, S.A., Joo, W.A., Echan, L.A., Speicher, D.W. (2007). Higher dimensional (Hi-D) separation strategies dramatically improve the potential for cancer biomarker detection in serum and plasma. *Journal of Chromatography B-Analytical Technologies in the Biomedical and Life Sciences*, 849, 43-52.

Jacobs, J.M., Adkins, J.N., Qian, W.J., Liu, T., Shen, Y., Camp 2nd, D.G., Smith, R.D. (2005). Utilizing human blood plasma for proteomic biomarker discovery. *Journal of Proteome Research*, 4, 1073-1085.

Jennissen, H.P. (2000). Hydrophobic interaction chromatography. *International Journal of Bio-Chromatography*, 5, 131-138.

Kostareva, I., Hung, Campbell, F.C. (2008). Purification of antibody heteropolymers using hydrophobic interaction chromatography. *Journal of Chromatography A*, 1177, 254-264.

Kovacs, J.M., Mant, C.T., Hodges, R.S. (2006). Determination of intrinsic hydrophilicity/hydrophobicity of amino acid side chains in peptides in the absence of nearest-neighbor or conformational effects. *Biopolymers*, 84, 283-297.

Ladiwala, A., Xia, F., Luo, Q., Breneman, C.M., Cramer, S.M. (2006). Investigation of protein retention and selectivity in HIC systems using quantitative structure retention relationship models *Biotechnology and Bioengineering*, 93, 836-859.

Lavery, C. B., MacInnis, M.C., MacDonald, M.J., Williams, J.B., Spencer, C.A., Burke, A.A., Irwin, D.J., D'Cunha, G.B. (2010). Purification of Peroxidase from Horseradish (Armoracia rusticana) Roots. *Journal of Agricultural and Food Chemistry*, 58, 8471-8476.

Lienqueo, M.E., Mahn, A. V., Asenjo, J.A. (2002). Mathematical correlations for predicting protein retention times in hydrophobic interaction chromatography. *Journal of Chromatography A*, 978, 71- 79.

Lienqueo, M.E., Mahn, A., Vasquez, L., Asenjo, J.A. (2003). Methodology for predicting the separation of proteins by hydrophobic interaction chromatography and its application to a cell extract. *Journal of Chromatography A*, 1009, 189-196.

Lienqueo, M.E., Salazar, O., Henriquez, K., Calado, C.R.C., Fonseca, L.P., Cabral, J.M. (2007). Prediction of retention time of cutinases tagged with hydrophobic peptides in hydrophobic interaction chromatography. *Journal of Chromatograohy A*, 1154, 460-463.

Linke, T, Doraiswamy, S., Harrison, E.H. (2007). Rat plasma proteomics: Effects of abundant protein depletion on proteomic analysis. *Journal of Chromatography B-Analytical Technologies in the Biomedical and Life Sciences*, 849, 273-281.

Liu, B., Huang, L. Buchenauer, H., Kang, Z. (2010). Isolation and partial characterization of an antifungal protein from the endophytic Bacillus subtilis strain EDR4. *Pesticide Biochemistry and Physiology*, 98, 305-311.

Liu, X., Yang, W., Gao, Q., Regnier, F. (2008). Toward chromatographic analysis of interacting protein networks. *Journal of Chromatography A*, 1178, 24-32.

Mahn, A. V., Lienqueo, M.E., Asenjo, J.A. (2004). Effect of surface hydrophobicity distribution on retention of ribonucleases in hydrophobic interaction chromatography. *Journal of Chromatography A*, 1043, 47-55.

Mahn, A., Lienqueo, M.E., Salgado, J.C. (2009). Methods of calculating protein hydrophobicity and their application in developing correlations to predict hydrophobic interaction chromatography retention. *Journal of Chromatography B-Analytical Technologies in the Biomedical and Life Sciences*, 1216, 1838-1844.

Mahn, A., Reyes, A., Zamorano, M., Cifuentes, W., Ismail. M. (2010). Depletion of highly abundant proteins in blood plasma by hydrophobic interaction chromatography for proteomic analysis. *Journal of Chromatography B-Analytical Technologies in the Biomedical and Life Sciences*, 878, 1038-1044.

McCue, J.T., Engel, P., Ng, A., Macniven, R., Thömmes, J. (2008). Modeling of protein monomer/aggregate purification and separation using hydrophobic interaction chromatography. *Bioprocess and Biosystems Engineering*, 31, 261-275.

Melander, W.R., el Rassi, S., Horvath, Cs. (1989). Interplay of Hydrophobic and Electrostatic Interactions in Bio-polymer Chromatography - Effect of Salts on the Retention of Proteins. *Journal of Chromatography*, 469, 3-27.

Miller, S., Janin, J., Lesk, A.M., Chothia, C. (1987). Interior and surface of monomeric proteins. *Journal of Molecular Biology*, 196, 641-656.

Miyazawa, S., Jernigan, R. (1996). Residue-residue potentials with a favorable contact pair term and an unfavorable high packing density term, for simulation and threading. *Journal of Molecular Biology*, 256, 623-644.

Nakamura, T., Kuromitsu, J., Oda, Y. (2008). Evaluation of comprehensive multidimensional separations using reversed-phase, reversed-phase liquid chromatography/mass spectrometry for shotgun proteomics. *Journal of Proteome Research*, 7, 1007-1011.

Nicholls, A., Sharp, K.A., Honig, B. (1991). Protein Folding and Association - Insights from the Interfacial and Thermodynamic Properties of Hydrocarbons. *Proteins-Structure Function and Genetics*, 11, 281-291.

Ottens, A.K., Kobeissy, F.H., Wolper, R.A., Haskins, W.E., Hayes, R.L., Denslow, N.D., Wang, K.K. (2005). A multidimensional differential proteomic platform using dual-phase ion-exchange chromatography-polyacrylamide gel electrophoresis/reversed-phase liquid chromatography tandem mass spectrometry. *Analytical Chemistry*, 77, 4836-4845.

Pieper, U., Eswar, N., Webb, B.M., Eramian, D., Kelly, L., Barkan, D.T., Carter, H., Mankoo, P., Karchin, R., Marti-Renom, M.A., Davis, F.P., Sali, A. (2009). MODBASE, a database of annotated comparative protein structure models and associated resources. *Nucleic Acids Research*, 37, D347-D354.

Porath, J., Sundberg, L., Fornstedt, N., Olsson I. (1973). Salting-out in amphiphilic gels as a new approach to hydrophobic adsorption. *Nature*, 245, 465-466.

Queiroz J.A., Tomaz, C.T., Cabral, J.M.S. (2001). Hydrophobic interaction chromatography of proteins. *Journal of Biotechnology*, 87, 143-159.

Ramos-Clamont, G., Candia-Plata, M.C., Guzman, R., Vazquez-Moreno, L. (2006). Novel hydrophobic interaction chromatography matrix for specific isolation and simple elution of immunoglobulins (A, G, and M) from porcine serum. *Journal of Chromatography A*, 1122, 28-34.

Rinderknecht, E.H., Zapata, G.A. (2006). US Patent 7,038,017.

Rodenbrock, A., Selber, K., Egmond, M.R., Kula, M.R. (2000). Extraction of peptide tagged cutinase in detergent-based aqueous two-phase systems. *Bioseparation*, 9, 269-276.

Salgado, J.C., Andrews, B.A., Ortúzar, M.F., Asenjo, J.A. (2008). Prediction of the partitioning behaviour of proteins in aqueous two-phase systems using only their amino acid composition. *Journal of Chromatography A*, 1178, 134-144.

Salgado, J.C., Rapaport, I., Asenjo, J.A. (2005). Prediction of retention times of proteins in hydrophobic interaction chromatography using only their amino acid composition. *Journal of Chromatography A*, 1098, 44-54.

Salih, E. (2005). Phosphoproteomics by mass spectrometry and classical protein chemistry approaches. *Mass Spectrometry Reviews*, 24, 828-846.

Savard, J.M., Schneider, J.W. (2007). Sequence-specific purification of DNA oligomers in hydrophobic interaction chromatography using peptide nucleic acid amphiphiles: Extended dynamic range. *Biotechnology and Bioengineering*, 97, 367-376.

Seely, J.E., Richey, C.W. (2001). Use of ion-exchange chromatography and hydrophobic interaction chromatography in the preparation and recovery of polyethylene glycol-linked proteins. *Journal of Chromatography A*, 908, 235-241.

Shaltiel, S., Er-el, Z. (1973). Hydrophobic chromatography: use for purification of glycogen synthetase. *Proceedings of the National Academy of Sciences U.S.A.*, 70, 778-781.

Shepard, C.C., Tiselius, A. (1949). In "Chromatographic Analysis" p. 275. *Discussions of the Faraday Society*, 7. Hazell, Watson and Winey. London.

Simeonidis, E., Pinto, J.M., Lienqueo, M.E., Tsoka, S., Papageorgiou, L.G. (2005). MINLP models for the synthesis of optimal peptide tags and downstream protein processing. *Biotechnology Progress*, 21, 875-884.

Tanford, C. (1962). Contribution of hydrophobic interactions to the stability of globular conformation of proteins. *Journal of the American Chemical Society*, 84, 4240-4247.

Teng, Y., Yin, Q., Ding, M., Zhao, F. (2010). Purification and characterization of a novel endo-beta-1,4-glucanase, AfEG22, from the giant snail, Achatina fulica frussac. *Acta Biochimica et Biophysica Sinica*, 42, 729-734.

Wertz, D.H., Scheraga, H.A. (1978). Influence of water on protein structure. An analysis of the preferences of amino acid residues for the inside or outside and for specific conformations in a protein molecule. *Macromolecules*, 11, 9-15.

Zhang, Y., Martínez, T., Woodruff, B., Goetze, A., Bailey, R., Pettit, D., Balland, A. (2008). Hydrophobic interaction chromatography of soluble Interleukin 1 receptor type II to reveal chemical degradations resulting in loss of potency. *Analytical Chemistry*, 80, 7022-7028.

2

Poly (L-glutamic acid)-Paclitaxel Conjugates for Cancer Treatment

Shuang-Qing Zhang
National Center for Safety Evaluation of Drugs,
National Institutes for Food and Drug Control,
China

1. Introduction

One of the effective approaches to develop new anticancer drugs is to prepare polymer-anticancer drug conjugates. The polymer-anticancer drug conjugates include polymer-protein conjugates, polymer-drug conjugates and supramolecular drug-delivery systems. In 1975, the concept of a polymer-drug conjugate was first proposed by Ringsdorf (Ringsdorf, 1975). In his model, a bioactive anticancer agent was attached to a suitable polymeric carrier, directly or through a biodegradable linker. Currently developed delivery systems for anticancer agents encompass colloidal systems (liposomes, emulsions, nanoparticles and micelles), polymer implants and polymer conjugates. These delivery systems are able to provide enhanced therapeutic efficacy and reduce toxicity of anticancer agents mainly by altering the pharmacokinetics and biodistribution of the drugs (Kim & Lim, 2002). The idea is attractive and could form the basis of a new generation of anticancer agents (Sugahara et al., 2007).

Many polymers have been investigated as carriers for conjugates, including poly(glutamic acid), poly(L-lysine), poly(malic acid), poly(aspartamides), poly((N-hydroxyethyl)-L glutamine), poly(ethylene glycol), poly(styrene-co-maleic acid/anhydride), poly(N-(2-hydroxypropyl) methacrylamide) copolymer, poly(ethyleneimine), poly(acroloylmorpholine), poly(vinylpyrrolidone), poly(vinylalcohol), poly(amidoamines), divinylethermaleic anhydride/acid copolymer, dextran, pullulan, mannan, dextrin, chitosan, hyaluronic acid and proteins *etc.*. Coupling low molecular weight anticancer drugs to high or low molecular weight polymers is an effective method for improving the therapeutic index of clinically used agents.

Several candidates have been evaluated in clinical trials, such as N-(2-hydroxypropyl) methacrylamide conjugates of doxorubicin, camptothecin, paclitaxel, and platinum (II) complexes (Haag & Kratz, 2006). The conjugation of cytotoxic agents to a hydrophilic polymer may convey several advantages, (1) increased water solubility; (2) protection from hydrolysis and proteolysis; (3) prolonged half-life and enhanced bioavailability of drug; (4) reduction of toxicity, immunogenicity and antigenicity; (5) controlled release or specific targeting through an enhanced permeability and retention (EPR) effect. In the chapter, we will focus on poly (L-glutamic acid)-paclitaxel (PG-PTX), which can improve the therapeutic index, pharmacokinetic properties, safety and efficacy of paclitaxel (PTX).

2. The anticancer agent PTX

PTX, an anticancer agent isolated from the trunk bark of the Pacific Yew tree, Taxus brevifolia, shows a wide spectrum of anticancer activity for a variety of human cancers, including breast, ovarian, non-small-cell lung, prostate, head and neck, colon cancers and so on (Rowinsky, 1997). PTX can induce mitotic arrest and apoptosis in proliferating cells by targeting tubulin, which is a component of the mitotic spindle. PTX binds to the N-terminal 31 amino acids of the β-tubulin subunit and prevents depolymerization. As a result, the mitotic spindle is disabled, cell division can not be completed, and the cell replication in the late G2 or M phase of the cell cycle is inhibited. The cancer cells are killed by disrupting the dynamics necessary for cell division (Bhalla, 2003). The anticancer mechanism of PTX is shown in Fig. 1.

The clinical use of PTX is limited by its high hydrophobicity, low solubility, high systemic exposure, poor pharmacokinetic characteristics, and the lack of selective tumor uptake (Parveen & Sahoo, 2008; ten Tije et al., 2003). The clinical use of PTX also leads to many side effects. Side effects of PTX include nausea, vomiting, diarrhea, mucositis, myelosuppression, cardiotoxicity, neurotoxicity and hypersensitivity reactions, and the latter two are mainly owing to polyoxyethylated castor oil (Cremophor® EL) and ethanol used for solubilizing PTX (Rogers, 1993; Sugahara et al., 2007). PTX for injection is supplied in 50% Cremophor® EL and 50% dehydrated ethanol.

Despite premedication with corticosteroids and antihistamines, PTX still induces minor reactions (e.g., flushing and rash) in approximately 40% of patients and major potentially life-threatening reactions in 1.5%–3% of patients (Lemieux et al., 2008; Price & Castells, 2002). Hydrophobicity of PTX is also associated with unfavorable kinetics, high levels of protein binding, and high volumes of distribution often greatly exceeding total body water. Together, all these factors have a negative impact on the therapeutic index because only a small proportion of the drug administered actually reaches the tumor site (Singer, 2005). As mentioned above, the efficacy and tolerability of PTX are limited by its low solubility, high systemic exposure, and the lack of selective tumor uptake.

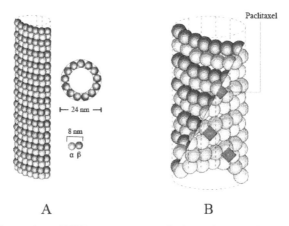

A   B

Fig. 1. Schematic illustration of PTX anticancer mode: heterdimers of α- and β-tubulin assemble to form a highly dynamic microtubule which plays an extremely important role in the process of mitosis (A); microtubule-targeted paclitaxel binds along the interior surface of the microtubule, suppressing its dynamics mitosis (B).

3. The EPR effect and endocytosis of polymer-drug conjugates

Anticancer polymer-drug conjugates can be divided into two targeting modalities: passive and active. While clinical anticancer activity has been achieved by passive macromolecular drug delivery systems, further selectivity is possible by active targeting (Luo & Prestwich, 2002). Polymer-drug conjugates can promote passive tumor targeting by EPR effect and allow for lysosomotropic drug delivery following endocytic capture of the drugs (Greco & Vicent, 2008).

The view that the polymer-drug conjugates passively accumulate in tumor tissues because EPR effect is clearly supported by the electron microscopic observation that the peripheral tumor vascular endothelium has quantitatively more fenestrations and open junctions than normal vessels (Li, 2002; Roberts & Palade, 1997). Tumor vasculature is more permeable to macromolecules than normal vasculature because the structures between the neovasculature in tumors and the mature vasculature in normal organs are different (Roberts & Palade, 1997; Singer, 2005). The paucity of lymphatic vessels in tumor tissues allows the retention of these macromolecules in the interstitial space, which leads to 10 to 100-fold increase in intratumoral drug concentrations for a prolonged time when compared with an equivalent dose of the anticancer drug given according to the conventional methods.

The phenomenon of EPR effect is applicable for almost all rapidly growing solid tumors and it has been widely used in cancer-targeting drug design (Iyer et al., 2007; Maeda et al., 2000; Reddy, 2005). The EPR effect is molecular weight (MW) - and size-dependent and is most effective with agents whose MWs are 50 000 or greater, which is above the threshold for renal excretion. Due to the different pathways to enter the cells between the small molecule drugs and the macromolecule drugs, multi-drug resistance (MDR) can be minimized at the same time (Boddy et al., 2005; Greish et al., 2003; Shaffer et al., 2007). Because of the stronger metabolic activity of the cancer cells, in addition to the EPR effect, cancer cells show a higher degree of uptake of macromolecules by endocytosis than normal cells (Li, 2002). The process of EPR and endocytosis is illustrated in Fig. 2.

4. PG-PTX conjugate

4.1 Modified formulations of PTX

Because of the unfavorable properties of PTX, it is urgent to develop a more effective strategy to improve its water solubility and selectivity towards tumor tissues (Maeda et al., 2009). Several approaches have been utilized to increase the therapeutic index of PTX. More water-soluble formulations of PTX have been investigated, including a nanoparticulate formulation (ABI-007), a polymeric micellar formulation (Genexol-PM), and a liposomal formulation and covalent linkage to macromolecule polymers that alter the pharmacokinetics of the parent drug (Ibrahim et al., 2002; Kim et al., 2004; Soepenberg et al., 2004).

4.2 The formation and the anticancer mechanism of PG-PTX

PG-PTX (paclitaxel poliglumex, CT-2103, Xyotax®, Opaxio®) is a water-soluble macromolecular conjugate that links PTX with PG (Li et al., 1996). PTX is conjugated by ester linkage to the γ-carboxylic acid side chains, leading to a relatively stable conjugate (Li et al., 1998a). Because the conjugation site is the 2 hydroxyl group of PTX, which is a crucial site for tubulin binding, the conjugate does not interact with β-tubulin and is inactive (Li, 2002; Rogers, 1993). The median MW of PG-PTX is 38.5 kDa, with a PTX content of approximately 36% on a w/w basis, equivalent to about one PTX ester linkage per 11 PG units of the polymer (Fig. 3) (Bonomi, 2007; Rogers, 1993).

Morphological analysis and biochemical characterizations demonstrate that both PTX and PG-PTX are able to induce apoptosis in cells expressing wild-type p53 or mutant p53, to arrest cells in the G2/M phase of the cell cycle, and to down-regulate HER-2/neu expression. Furthermore, when PG-PTX is compared with other water-soluble derivatives of PTX, including small-molecular-weight sodium pentetic acid-PTX and polyethylene glycol-PTX conjugate (MW 5 kDa), they all show the same effects on telomeric association, mitotic index chromatin condensation, and formation of apoptotic bodies (Multani et al., 1999). These results indicate that PG-PTX has the same mechanisms of action as PTX.

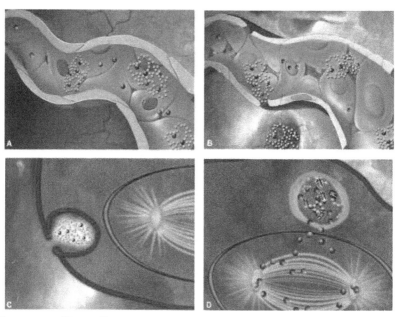

Fig. 2. Illustration of EPR effect and endocytosis. Different from blood vessels in normal tissue (A), those in tumor tissue (B) have porous openings, through which large-size conjugate leaks and is preferentially trapped and distributed to the tumor tissue. Once in the tumor tissue, the conjugate is taken up by the tumor cells through a cellular process called endocytosis (C). The conjugate releases active agent (D) via metabolism by lysosomal enzymes inside the lysosome of the tumor cell.

4.3 PG as the carrier of PG-PTX

Compared with other synthetic polymers that have been tested in clinical studies, PG is unique because it is composed of natural L-glutamic acid linked together through amide bonds rather than the nondegradable C-C backbone. The free γ-carboxyl group in each repeating unit of L-glutamic acid is negatively charged under a neutral pH condition, which makes the polymer water-soluble. The carboxyl groups can also provide functionality for drug attachment. PG is not only water-soluble and biodegradable, it is also nontoxic. All these characteristics make PG a unique candidate as the carrier of polymer-drug conjugates for selective delivery of chemotherapeutic agents, especially for PTX (Parveen & Sahoo, 2008).

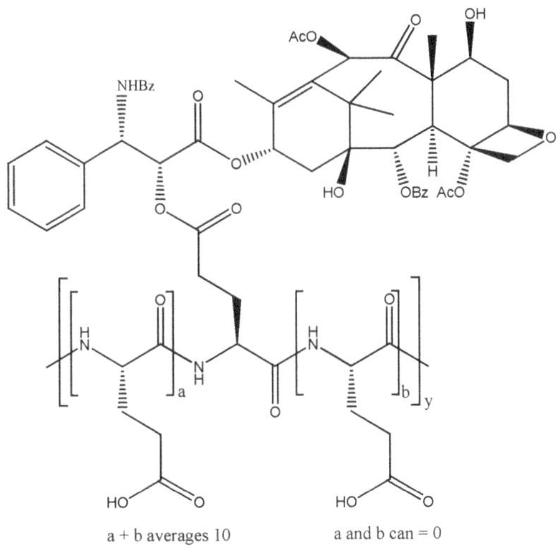

Fig. 3. Schematic representation of PG-PTX structure. The structure shown is illustration of a fragment of the molecule. On average there are approximately 10 non-conjugated monomer glutamic acid units (a + b) for every molecule conjugated to a PTX molecule (y).

4.4 The metabolism of PG-PTX

PG-PTX conjugate is stable in the systemic circulation but can be broken down by intracellular lysosomal enzymes to release the drug after entering cells by endocytosis. The proposed mechanism by which PG-PTX is metabolized includes endocytosis of the polymer-drug conjugate followed by intracellular release of active PTX by proteolytic activity of the lysosomal enzyme cathepsin B, an exocarboxydipeptidase, and diffusion of PTX into the nucleus (Turk et al., 2001). The process is presented in Fig. 4. This finding may have biological relevance as expression of cathepsin B is upregulated in malignant cells, particularly during tumor progression period (Podgorski & Sloane, 2003). These data support a model in which PG-PTX accumulates in tumor tissues through the EPR effect, followed by the cathepsin B-mediated release of PTX. The kinetics of intracellular formation of several PG-PTX metabolites have been quantified in vitro and have been found to be largely dependent on cathepsin B. Metabolites that have been detected in vivo include diglutamyl-PTX and monoglutamyl-PTX. Monoglutamyl-PTX is an unstable compound that can be nonenzymatically degraded to release free PTX. Specific enzyme inhibitors such as CA-074 methyl ester, a cell-permeable irreversible inhibitor of cathepsin B, and EST, a cell-permeable irreversible inhibitor of cysteine proteases, dramatically decrease the formation of monoglutamate PTX and unconjugated PTX in tumor cells that have been incubated with PG-PTX (Rogers, 1993).

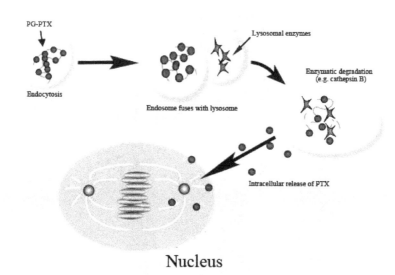

Fig. 4. Metabolism of PG-PTX by endocytosis and enzymatic degradation.

4.5 The advantages of PG-PTX
4.5.1 PG-PTX reduces the side effects of PTX in clinical application

Compared with PTX, the solubility, uptake, tumor retention, and anticancer efficacy of PG-PTX are increased. Many clinical studies so far have confirmed several advantages of PG-PTX in the treatment of cancer patients. This macromolecular conjugate PG-PTX eliminates the need for Cremophor® EL and, therefore, decreases infusion time and the risk of hypersensitivity. Compared to standard taxanes, PG-PTX in phase I and phase II studies shows encouraging outcomes with reduced neutropenia and alopecia, and allows a more convenient administration schedule without the need for routine premedications (Rogers, 1993). PG-PTX induces hypersensitivity reactions in less than 1% of patients, without premedication, and only rare severe reactions have been reported. Furthermore, patients undergoing PG-PTX therapy have a better quality of life, because there is no significant hair loss, nerve damage, or neutropenia at the current dose. PG-PTX is water-soluble and can be administrated rapidly as a 10 -20 min infusion rather than hours when administrating PTX. The recommended phase II dose of PG-PTX is 235 mg/m^2 every 3-week administered over a 10 min infusion without premedication (Sabbatini et al., 2004). Twenty-six patients were treated with PG-PTX in the Phase I study of PG-PTX administered weekly for patients with advanced solid malignancies. The recommended dose of PG-PTX for subsequent disease-directed studies is 70 mg/m^2 weekly. Most patients experienced at least one drug-related adverse event during the study (Table 1) (Mita et al., 2009). Ninety-nine patients were treated with PG-PTX in a multi-center phase II study of PG-PTX as an intravenous (i.v.) infusion (approximately 10 min) at a dose of 175 mg/m^2 on day 1 of each 3-week cycle. And the treatment-related adverse events including non-laboratory-based and laboratory-based maximum CTC toxicities in all cycles are listed in Table 2.

	Grade 1	Grade 2	Grade 3	Total
Adverse event -70 mg/m^2				
Neutropenia		1(13%)	1(13%)	2(25%)
Anemia	1(13%)	1(13%)		2(25%)
Peripheral neuropathy	2(25%)	2(25%)		2(50%)
Anorexia	1(13%)			1(13%)
Diarrhea			1(13%)	1(13%)
Nausea	1(13%)			1(13%)
Pain in limb	1(13%)			1(13%)
Fatigue		1(13%)		1(13%)
Adverse event-all cohorts				
Neutropenia		2(8%)	3(12%)	5(19%)
Anemia	1(4%)	1(4%)	2(8%)	4(15%)
Peripheral neuropathy	5(16%)	4(8%)	1(4%)	10(38%)
Anorexia	1(4%)	1(4%)		2(8%)
Diarrhea	3(12%)	1(4%)	1(4%)	5(19%)
Nausea	3(12%)	1(4%)		4(15%)
Rash	1(4%)	2(8%)		3(12%)
Myalgia	3(12%)	2(8%)		5(19%)
Pain in limb	1(4%)	1(4%)		2(8%)
Fatigue	4(15%)	3(12%)		7(27%)
Hypersensitivity	1(4%)	1(4%)		2(8%)

Table 1. Adverse events related to PG-PTX administration (n = 26).

	Grade 1	Grade 2	Grade 3	Grade 4
Non-laboratory-based maximum CTC toxicity, all cycles				
Neuropathy	20	15	15	0
Fatigue	10	19	5	0
Musculoskeletal	11	10	1	0
Myalgia/arthralgia	7	5	1	0
Nausea	21	4	1	0
Stomatitis/mucositis	7	1	1	0
Hoarseness	0	0	1	0
Drug hypersensitivity	5	8	1	0
Laboratory-based maximum CTC toxicity, all cycles				
Neutropenia	20	14	15	9
Leukopenia	13	18	18	0
Anemia	46	18	6	0

Abbreviation: CTC, National Cancer Institute Common Toxicity Criteria.

Table 2. Treatment-related adverse events (n = 99).

4.5.2 PG-PTX promotes anticancer efficacy and reduces toxicity by prolonging tumor exposure and minimizing systemic exposure to active drug

A single i.v. injection of PG-PTX at its maximum tolerated dose (MTD) equivalent to 60 mg of PTX/kg and at a lower dose equivalent to 40 mg of PTX/kg results in the disappearance of an established implanted 13762F mammary adenocarcinoma (mean size, 2000 mm³) in rats. Similarly, mice bearing syngeneic OCa-1 ovarian carcinoma (mean size, 500 mm³) are tumor-free within 2 weeks after a single i.v. injection of PG-PTX at a dose equivalent to 160 mg of PTX/kg (Li et al., 1998b). MTD of PTX in rats and mice are 20 mg/kg and 60 mg/kg, respectively. In contrast, MTD of a single i.v. injection of PG-PTX in rats and mice are 60 mg/kg and 160 mg/kg, respectively. MTD of PG-PTX was approximately 160 mg/kg to 200 mg/kg in immunocompetent mice and 120 mg/kg to 150 mg/kg in immunodeficient animals. At their respective MTDs, single-dose PG-PTX is more efficacious than PTX in Cremophor® EL/ethanol (Li et al., 1999). PG-PTX has shown anticancer activity in preclinical studies with human tumor xenografts and in early phase I trials. MTD of PG-PTX as a single agent, based on the first cycle toxicity of patients, is 235 mg/m² (Verschraegen et al., 2009). Biodistribution in mice bearing OCa-1 tumor treated with i.v. injections of tritium-labeled PG-[³H] PTX shows a five times greater distribution of PTX to tumor tissues than those treated with PTX (Li et al., 2000c), which was demonstrated by whole-body autoradiograph (Fig. 5).

A   B

Fig. 5. Whole-body autoradiographs of mice killed 1 day (A) and 6 days (B) after tail vein injection of PG-[³H]PTX. Most radioactivity was localized to tumor periphery at 1 day after injection, but by day 6, radioactivity had diffused into the center of the tumor. L: liver; M: muscle; Arrow head: tumor.

Preclinical studies in animal tumor models demonstrate that PG-PTX is more effective than PTX and it is associated with prolonged tumor exposure but minimized systemic exposure to the active drug. The slow release of the active drug from a well-designed polymer carrier results in sustained high intratumoral drug levels and lower plasma concentrations of the active drug. To accomplish this, the polymer conjugate should release the active drug in tumor tissues rather than in the plasma during circulation. As a result, exposure of normal tissues will be limited, which is potentially associated with a more favorable toxicity profile (Li et al., 2000c). Thus, enhanced tumor uptake and sustained release of PTX from PG-PTX in tumor tissues are major factors contributing to its markedly improved in vivo anticancer activities.

4.5.3 The favorable pharmacokinetic properties of PG-PTX

The superior anticancer activity of PG-PTX in preclinical studies suggests that PG-PTX might have favorable pharmacokinetic properties. Many studies suggest that PG-PTX exerts the anticancer activity by the continuous release of free PTX, and that the favorable pharmacokinetics of PG-PTX conjugate in vivo is likely the main cause contributing to its advanced anticancer activity (Oldham et al., 2000). Female mice with subcutaneous B16 murine melanomas are given PG-[^{3}H] PTX at the equivalent dose 40 mg/kg of [^{3}H] PTX i.v. infusion. Tumor samples are collected at regular intervals up to 144 h after infusion, and the concentrations of PG-PTX and PTX are determined by LC/MS analysis. Tumor exposure to total taxanes is increased by a factor of 3 (C_{max}) or a factor of 12 (AUC) in mice treated with PG-[^{3}H] PTX compared with the mice treated with [^{3}H] PTX (Table 3) (Chipman et al., 2006). PG-[^{3}H] PTX has a much longer half-life in plasma than [^{3}H] PTX (Fig. 6).

Whereas PTX has an extremely short half-life in the plasma of mice ($t_{1/2}$ = 29 min), the apparent half-life of PG-PTX is prolonged ($t_{1/2}$ = 317 min) (Li et al., 1998b). In clinical trials, PG-PTX is given as a 30-min infusion every 3 weeks. Patients were treated at dose levels ranging from 30 mg/m^2 to 720 mg/m^2. PG-PTX is detectable in plasma of all patients and has a long plasma half-life of up to 185 h, and the results are consistent with preclinical findings. Furthermore, concentrations of free PTX released from PG-[^{3}H] PTX remain relatively constant up to 6 d after infusing. Moreover, peak plasma concentrations of free PTX are less than 0.1 µM 24 h after PG-PTX administration at doses up to 480 mg/m^2 (176 mg/m^2 PTX equivalents) (Boddy et al., 2001).

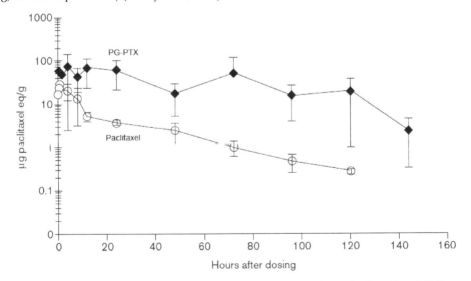

Fig. 6. Tumor concentration of [^{3}H]paclitaxel after treatment with PG-[^{3}H] PTX and [^{3}H] paclitaxel in female mice with s.c. B16 melanomas at a dose of 40 mg paclitaxel.

In another phase I study, PG-PTX is administrated at 70 mg/m^2. The mean maximal concentration (C_{max}) is 41.2 ± 8.60 µg/mL and the C_{max} is reached right after the end of the infusion. The plasma concentration declines with a mean terminal half-life of 15.7 ± 3.17 h. The mean AUC at the MTD is 455 ± 112 µg/h/mL and the mean average systemic plasma clearance is 0.16 ± 0.04 L/h/m^2 (Sabbatini et al., 2004). At the MTD, the mean volume of distribution at steady state and during the terminal phase are 1.41 ± 0.28 L/m^2 and 3.62 ± 1.13 L/m^2, the mean C_{max} of unconjugated PTX is 0.21 ± 0.07 µg/mL, the mean T_{max} is 0.56 ± 0.18 h, the mean terminal half life is 16.6 ± 7.85 h, and the mean AUC is 3.15 ± 1.16 µg/h/mL. The ratio of the free PTX AUC to the conjugated PTX AUC is 0.7%. The plasma concentration of PTX released from PG-PTX increases largely in proportion to the dose and remains similar after repeated administration (Sabbatini et al., 2004).

	C_{max} (µg/g)	T_{max} (h)	AUC (µg/h/g)	MRT (h)
PG-[^{3}H] PTX				
Total taxanes	72.0	4	4547	51
PTX	4.0	72	345	66
[^{3}H] PTX				
Total taxanes	26.7	1.5	384	23
PTX	22.4	1.5	261	17

Table 3. Preclinical tumor pharmacokinetics.

4.5.4 PG-PTX facilitates the radiotherapy and chemotherapy

Preclinical studies in animal tumor models demonstrate the enhanced safety and efficacy of PG-PTX relative to PTX when administered as a single agent or in conjunction with radiation. Studies show that PG-PTX given 24 h before or after radiotherapy enhances tumor growth delay significantly more than PTX. PG-PTX dramatically potentiates tumor radiocurability after single-dose or fractionated irradiation without affecting acute normal tissue injury. PG-PTX increases the therapeutic ratio of radiotherapy more than that previously reported for other taxanes (Milas et al., 2003). PG-PTX not only can produce a much stronger radiopotentiating effect than PTX, the kinetics of its radiopotentiating effect is also different from that of PTX. Delays in the growth of syngeneic murine ovarian OCa-1 tumors grown intramuscularly in C3Hf/Kam mice are used as the treatment end point.

PG-PTX given 24 h before tumor irradiation increases the efficacy of tumor radiation by a factor of more than 4 (Li et al., 2000a). Furthermore, the combination of radiation and PG-PTX can produce a significantly greater tumor growth delay than treatment with radiation and PTX when both drugs are given at the same equivalent PTX dose of 60 mg/kg 24 h after tumor irradiation (enhancement factors, 4.44 versus 1.50) (Li et al., 2000b). When the treatment end point is tumor cure, the enhancement factors are 8.4 and 7.2 of fractionated and single dose radiation, respectively. These values are greater than those produced by other taxanes or by any other chemotherapeutic drugs or radiosensizer tested so far. PG-PTX may exert its radiopotentiation activity through increased tumor uptake of PG-PTX and sustained release of PTX in the tumor (Li et al., 2000a). To determine whether prior irradiation affects tumor uptake of PG-PTX, PG-[^{3}H] PTX is injected into mice with OCa-1 tumors 24 h after 15 Gy local irradiation (Li et al., 2000b). The uptake of PG-[³H] PTX in irradiated tumors is 28%–38% higher than that in nonirradiated tumors at different times after PG-[³H] PTX injection, indicating that tumor irradiation can increase the accumulation of PG-PTX in the tumors (Fig. 7).

Anticancer activity in patients who have failed previous chemotherapy, including PTX treatment, is observed with PG-PTX (Sabbatini et al., 2004). Ninety-nine patients in a multi-center phase II study treated with PG-PTX as an i.v. infusion approximately 10 min at a dose of 175 mg/m² on day 1 of each 3-week cycle have received at least one cycle of treatment. Response rates categorized by platinum sensitivity are shown in Table 4.

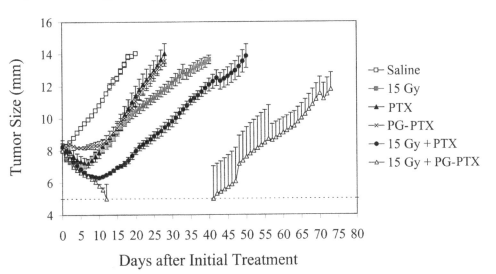

Fig. 7. Effects of combined radiation and PG-PTX and radiation and PTX on the growth of OCa-1 tumors in mice.

No. of Regimens	Response					
	PR		SD		PD	
	No. of Patients	(%)	No. of Patients	(%)	No. of Patients	(%)
Platinum-sensitive, n = 42						
1 or 2 prior regimens	5/18	28	6/18	33	7/18	39
≥3 prior regimens	1/24	4	11/24	46	12/24	50
Total	6/42	14	17/42	40	19/42	45
Platinum-resistant/refractory, n = 57						
1 or 2 prior regimens	2/21	10	4/21	19	15/21	71
≥3 prior regimens	2/36	6	11/36	31	23/36	64
Total	4/57	7	15/57	26	38/57	67

Abbreviations: PR, partial response; SD, stable disease; PD, progressive disease.

Table 4. Response by Platinum Sensitivity.

5. Summary

Clinical proof of concept for polymer conjugates has already been achieved over the last three decades, with a family of polymer-protein conjugates reaching the market and a growing list of polymer-drug conjugates currently in clinical studies. The application of polymer-anticancer drug conjugates in anticancer treatment is a promising field with growing opportunities to achieve medical treatments with highly improved therapeutic value (Vicent et al., 2008). PG-PTX conjugate can improve the anticancer activity, enhance the safety and efficacy, ameliorate the pharmacokinetic properties and so on. Therefore, the application of PG-PTX facilitates the clinical therapy of a variety of human cancers.

However, many challenges still exist, providing opportunities to improve this platform technology further. The clinical development of anticancer agents utilizing various delivery systems is actively undergoing. New technologies and multidisciplinary approaches to develop advanced drug delivery systems, applicable to a wide range of anticancer agents, may eventually lead to an effective cancer therapy in the future.

6. Acknowledgement

This work was supported by the National Key Technologies R & D Program for Drug Safety Evaluation (Grant No. 2008ZX09305-002) and the Scientific Research Foundation for the Returned Overseas Chinese Scholars, State Education Ministry.

7. References

Bhalla, K. N. (2003). Microtubule-targeted anticancer agents and apoptosis. *Oncogene*, Vol. 22, No. 56, (December 2003), pp. 9075-9086, ISSN 0950-9232

Boddy, A. V.; Griffin M. J.; Sludden J.; Thomas H. D.; Fishwick K.; Wright J. G.; Plumner E. R.; Highley M. & Calvert A. H. (2001). Pharmacological study of paclitaxel duration of infusion combined with GFR-based carboplatin in the treatment of ovarian cancer. *Cancer chemotherapy and pharmacology*, Vol. 48, No. 1, (July 2001), pp. 15-21, ISSN 0344-5704

Boddy, A. V.; Plummer E. R.; Todd R.; Sludden J.; Griffin M.; Robson L.; Cassidy J.; Bissett D.; Bernareggi A.; Verrill M. W. and others. (2005). A phase I and pharmacokinetic study of paclitaxel poliglumex (XYOTAX), investigating both 3-weekly and 2-weekly schedules. *Clinical Cancer Research*, Vol. 11, No. 21, (November 2005), pp. 7834-7840, ISSN 1078-0432

Bonomi, P. (2007). Paclitaxel poliglumex (PPX, CT-2103): macromolecular medicine for advanced non-small-cell lung cancer. *Expert Review of Anticancer Therapy*, Vol. 7, No. 4, (April 2007), pp. 415-422, ISSN 1744-8328

Chipman, S. D.; Oldham F. B.; Pezzoni G. & Singer J. W. (2006). Biological and clinical characterization of paclitaxel poliglumex (PPX, CT-2103), a macromolecular polymer-drug conjugate. *International Journal of Nanomedicine*, Vol. 1, No. 4, 2006), pp. 375-383, ISSN 1176-9114

Greco, F. & Vicent M. J. (2008). Polymer-drug conjugates: current status and future trends. *Frontiers in Bioscience*, Vol. 13, No., 2008), pp. 2744-2756, ISSN 1093-4715

Greish, K.; Fang J.; Inutsuka T.; Nagamitsu A. & Maeda H. (2003). Macromolecular therapeutics: advantages and prospects with special emphasis on solid tumour targeting. *Clinical pharmacokinetics*, Vol. 42, No. 13, 2003), pp. 1089-1105, ISSN 0312-5963

Haag, R. & Kratz F. (2006). Polymer therapeutics: concepts and applications. *Angewandte Chemie*, Vol. 45, No. 8, (February 2006), pp. 1198-1215, ISSN 1433-7851

Ibrahim, N. K.; Desai N.; Legha S.; Soon-Shiong P.; Theriault R. L.; Rivera E.; Esmaeli B.; Ring S. E.; Bedikian A.; Hortobagyi G. N. and others. (2002). Phase I and pharmacokinetic study of ABI-007, a Cremophor-free, protein-stabilized, nanoparticle formulation of paclitaxel. *Clinical Cancer Research*, Vol. 8, No. 5, (May 2002), pp. 1038-1044, ISSN 1078-0432

Iyer, A. K.; Greish K.; Seki T.; Okazaki S.; Fang J.; Takeshita K. & Maeda H. (2007). Polymeric micelles of zinc protoporphyrin for tumor targeted delivery based on EPR effect and singlet oxygen generation. *Journal of drug targeting*, Vol. 15, No. 7-8, (August-September 2007), pp. 496-506, ISSN 1061-186X

Kim, C. K. & Lim S. J. (2002). Recent progress in drug delivery systems for anticancer agents. *Archives of pharmacal research*, Vol. 25, No. 3, (June 2002), pp. 229-239, ISSN 0253-6269

Kim, T. Y.; Kim D. W.; Chung J. Y.; Shin S. G.; Kim S. C.; Heo D. S.; Kim N. K. & Bang Y. J. (2004). Phase I and pharmacokinetic study of Genexol-PM, a cremophor-free, polymeric micelle-formulated paclitaxel, in patients with advanced malignancies. *Clinical Cancer Research*, Vol. 10, No. 11, (June 2004), pp. 3708-3716, ISSN 1078-0432

Lemieux, J.; Maunsell E. & Provencher L. (2008). Chemotherapy-induced alopecia and effects on quality of life among women with breast cancer: a literature review. *Psychooncology*, Vol. 17, No. 4, (April 2008), pp. 317-328, ISSN 1099-1611

Li, C. (2002). Poly(L-glutamic acid)--anticancer drug conjugates. *Advanced Drug Delivery Reviews*, Vol. 54, No. 5, (September 2002), pp. 695-713, ISSN 0169-409X

Li, C.; Ke S.; Wu Q. P.; Tansey W.; Hunter N.; Buchmiller L. M.; Milas L.; Charnsangavej C. & Wallace S. (2000a). Potentiation of ovarian OCa-1 tumor radioresponse by poly (L-glutamic acid)-paclitaxel conjugate. *International Journal of Radiation Oncology, Biology, Physics*, Vol. 48, No. 4, (November 2000a), pp. 1119-1126, ISSN 0360-3016

Li, C.; Ke S.; Wu Q. P.; Tansey W.; Hunter N.; Buchmiller L. M.; Milas L.; Charnsangavej C. & Wallace S. (2000b). Tumor irradiation enhances the tumor-specific distribution of poly(L-glutamic acid)-conjugated paclitaxel and its antitumor efficacy. *Clinical Cancer Research*, Vol. 6, No. 7, (July 2000b), pp. 2829-2834, ISSN 1078-0432

Li, C.; Newman R. A.; Wu Q. P.; Ke S.; Chen W.; Hutto T.; Kan Z.; Brannan M. D.; Charnsangavej C. & Wallace S. (2000c). Biodistribution of paclitaxel and poly(L-glutamic acid)-paclitaxel conjugate in mice with ovarian OCa-1 tumor. *Cancer chemotherapy and pharmacology*, Vol. 46, No. 5, 2000c), pp. 416-422, ISSN 0344-5704

Li, C.; Price J. E.; Milas L.; Hunter N. R.; Ke S.; Yu D. F.; Charnsangavej C. & Wallace S. (1999). Antitumor activity of poly(L-glutamic acid)-paclitaxel on syngeneic and xenografted tumors. *Clinical Cancer Research*, Vol. 5, No. 4, (April 1999), pp. 891-897, ISSN 1078-0432

Li, C.; Yu D.; Inoue T.; Yang D. J.; Milas L.; Hunter N. R.; Kim E. E. & Wallace S. (1996). Synthesis and evaluation of water-soluble polyethylene glycol-paclitaxel conjugate as a paclitaxel prodrug. *Anticancer Drugs*, Vol. 7, No. 6, (August 1996), pp. 642-648, ISSN 0959-4973

Li, C.; Yu D. F.; Newman R. A.; Cabral F.; Stephens L. C.; Hunter N.; Milas L. & Wallace S. (1998a). Complete regression of well-established tumors using a novel water-soluble poly(L-glutamic acid)-paclitaxel conjugate. *Cancer Res*, Vol. 58, No. 11, (Jun 1998a), pp. 2404-2409, ISSN 0008-5472

Li, C.; Yu D. F.; Newman R. A.; Cabral F.; Stephens L. C.; Hunter N.; Milas L. & Wallace S. (1998b). Complete regression of well-established tumors using a novel water-soluble poly(L-glutamic acid)-paclitaxel conjugate. *Cancer Research*, Vol. 58, No. 11, (June 1998b), pp. 2404-2409, ISSN 0008-5472

Luo, Y. & Prestwich G. D. (2002). Cancer-targeted polymeric drugs. *Current Cancer Drug Targets*, Vol. 2, No. 3, (September 2002), pp. 209-226, ISSN 1568-0096

Maeda, H.; Bharate G. Y. & Daruwalla J. (2009). Polymeric drugs for efficient tumor-targeted drug delivery based on EPR-effect. *European Journal of Pharmaceutics and Biopharmaceutics:*, Vol. 71, No. 3, (March 2009), pp. 409-419, ISSN 1873-3441

Maeda, H.; Wu J.; Sawa T.; Matsumura Y. & Hori K. (2000). Tumor vascular permeability and the EPR effect in macromolecular therapeutics: a review. *Journal of Controlled Release*, Vol. 65, No. 1-2, (March 2000), pp. 271-284, ISSN 0168-3659

Milas, L.; Mason K. A.; Hunter N.; Li C. & Wallace S. (2003). Poly(L-glutamic acid)-paclitaxel conjugate is a potent enhancer of tumor radiocurability. *International Journal of Radiation Oncology, Biology, Physics*, Vol. 55, No. 3, (March 2003), pp. 707-712, ISSN 0360-3016

Mita, M.; Mita A.; Sarantopoulos J.; Takimoto C. H.; Rowinsky E. K.; Romero O.; Angiuli P.; Allievi C.; Eisenfeld A. & Verschraegen C. F. (2009). Phase I study of paclitaxel poliglumex administered weekly for patients with advanced solid malignancies. *Cancer chemotherapy and pharmacology*, Vol. 64, No. 2, (July 2009), pp. 287-295, ISSN 1432-0843

Multani, A. S.; Li C.; Ozen M.; Imam A. S.; Wallace S. & Pathak S. (1999). Cell-killing by paclitaxel in a metastatic murine melanoma cell line is mediated by extensive telomere erosion with no decrease in telomerase activity. *Oncology Reports*, Vol. 6, No. 1, (January-February 1999), pp. 39-44, ISSN 1021-335X

Oldham, E. A.; Li C.; Ke S.; Wallace S. & Huang P. (2000). Comparison of action of paclitaxel and poly(L-glutamic acid)-paclitaxel conjugate in human breast cancer cells. *International Journal of Oncology*, Vol. 16, No. 1, (January 2000), pp. 125-132, ISSN 1019-6439

Parveen, S. & Sahoo S. K. (2008). Polymeric nanoparticles for cancer therapy. *Journal of drug targeting*, Vol. 16, No. 2, (February 2008), pp. 108-123, ISSN 1061-186X

Podgorski, I. & Sloane B. F. (2003). Cathepsin B and its role(s) in cancer progression. *Biochemical Society Symposium*, Vol. No. 70, 2003), pp. 263-276, ISSN 0067-8694

Price, K. S. & Castells M. C. (2002). Taxol reactions. *Allergy and Asthma Proceedings*, Vol. 23, No. 3, (May-June 2002), pp. 205-208, ISSN 1088-5412

Reddy, L. H. (2005). Drug delivery to tumours: recent strategies. *Journal of Pharmacy and Pharmacology*, Vol. 57, No. 10, (October 2005), pp. 1231-1242, ISSN 0022-3573

Ringsdorf, H. (1975). Structure and properties of pharmacologically active polymers. *Journal of Polymer Science: Polymer Symposia*, Vol. 51, No. 1, 1975), pp. 135-153,

Roberts, W. G. & Palade G. E. (1997). Neovasculature induced by vascular endothelial growth factor is fenestrated. *Cancer Research*, Vol. 57, No. 4, (February 1997), pp. 765-772, ISSN 0008-5472

Rogers, B. B. (1993). Taxol: a promising new drug of the '90s. *Oncology nursing forum*, Vol. 20, No. 10, (November-December 1993), pp. 1483-1489, ISSN 0190-535X

Rowinsky, E. K. (1997). The development and clinical utility of the taxane class of antimicrotubule chemotherapy agents. *Annual review of medicine*, Vol. 48, No., 1997), pp. 353-374, ISSN 0066-4219

Sabbatini, P.; Aghajanian C.; Dizon D.; Anderson S.; Dupont J.; Brown J. V.; Peters W. A.; Jacobs A.; Mehdi A.; Rivkin S. and others. (2004). Phase II study of CT-2103 in patients with recurrent epithelial ovarian, fallopian tube, or primary peritoneal carcinoma. *Journal of Clinical Oncology*, Vol. 22, No. 22, (November 2004), pp. 4523-4531, ISSN 0732-183X

Shaffer, S. A.; Baker-Lee C.; Kennedy J.; Lai M. S.; de Vries P.; Buhler K. & Singer J. W. (2007). In vitro and in vivo metabolism of paclitaxel poliglumex: identification of metabolites and active proteases. *Cancer chemotherapy and pharmacology*, Vol. 59, No. 4, (March 2007), pp. 537-548, ISSN 0344-5704

Singer, J. W. (2005). Paclitaxel poliglumex (XYOTAX, CT-2103): a macromolecular taxane. *Journal of Controlled Release*, Vol. 109, No. 1-3, (December 2005), pp. 120-126, ISSN 0168-3659

Soepenberg, O.; Sparreboom A.; de Jonge M. J.; Planting A. S.; de Heus G.; Loos W. J.; Hartman C. M.; Bowden C. & Verweij J. (2004). Real-time pharmacokinetics guiding clinical decisions; phase I study of a weekly schedule of liposome encapsulated paclitaxel in patients with solid tumours. *European Journal of Cancer*, Vol. 40, No. 5, (March 2004), pp. 681-688, ISSN 0959-8049

Sugahara, S.; Kajiki M.; Kuriyama H. & Kobayashi T. R. (2007). Complete regression of xenografted human carcinomas by a paclitaxel-carboxymethyl dextran conjugate (AZ10992?) *Journal of Controlled Release,* Vol. 117, No. 1, (January 2007), pp. 40-50, ISSN 0168-3659

ten Tije, A. J.; Verweij J.; Loos W. J. & Sparreboom A. (2003). Pharmacological effects of formulation vehicles: implications for cancer chemotherapy. *Clinical pharmacokinetics*, Vol. 42, No. 7, 2003), pp. 665-685, ISSN 0312-5963

Turk, V.; Turk B. & Turk D. (2001). Lysosomal cysteine proteases: facts and opportunities. *EMBO Journal*, Vol. 20, No. 17, (September 2001), pp. 4629-4633, ISSN 0261-4189

Verschraegen, C. F.; Skubitz K.; Daud A.; Kudelka A. P.; Rabinowitz I.; Allievi C.; Eisenfeld A.; Singer J. W. & Oldham F. B. (2009). A phase I and pharmacokinetic study of paclitaxel poliglumex and cisplatin in patients with advanced solid tumors. *Cancer chemotherapy and pharmacology*, Vol. 63, No. 5, (April 2009), pp. 903-910, ISSN 1432-0843

Vicent, M. J.; Dieudonne L.; Carbajo R. J. & Pineda-Lucena A. (2008). Polymer conjugates as therapeutics: future trends, challenges and opportunities. *Expert Opinion on Drug Delivery*, Vol. 5, No. 5, (May 2008), pp. 593-614, ISSN 1742-5247

3

Development and Engineering of CSαβ Motif for Biomedical Application

Ying-Fang Yang
Biomedical Technology and Device Research Laboratories,
Industrial Technology Research Institute,
Republic of China

1. Introduction

Protein engineering is a process that modifies/creates functions or increases stabilities of proteins through artificial selection and evolution (Angeletti, 1998). For the fast development of molecular biology techniques, numbers of proteins have been successfully engineered to equip noval functions in the past several decades (Alahuhta et al., 2008;Bottcher & Bornscheuer, ;Ehren et al., 2008;Huston et al., 1988;Leta Aboye et al., 2008). These engineered proteins are developed either for academic research purposes or biomedical applications. The core of protein engineering is an appropriate protein scaffold (Hey et al., 2005;Pessi et al., 1993;Skerra, 2007). An excellent protein scaffold not only provides a platform for developing noval functions but also has many benefits, such as cost reduction during development/production or lasting efficacy of protein. An suitable protein scaffold should equip several characteristics (Hey et al., 2005;Pessi et al., 1993;Skerra, 2007). To introduce tailored functions into protein scaffolds is a major challenge and has a unique significance in protein design (Hey et al., 2005;Pessi et al., 1993).

A protein scaffold is a peptide framework that exhibits a high tolerance of its fold for modifications (Hey et al., 2005). Candidates for suitable protein scaffolds should exhibit a compact and structurally rigid core. The folding properties of the protein scaffolds should not be significantly changed when the side chains in a contiguous surface region are replaced or loops of varying sequence and length are presented (Skerra, 2007). Several additional priorities have to be considered if the scaffold is used in biomedical applications. The scaffold must display extra stabilities to environments, such as low pH, high concentration of chaotropic agents and high temperatures. Molecular weight of the scaffold should be low and small molecules have advantages in passing tissue barrier (Baines & Colas, 2006). The scaffold should highly resist protease degradation and this ensures the engineered proteins can be safe in the gastrointestinal tract and not degraded (Aharoni et al., 2005). Low immunogenecity is important to reduce unexpected side effects and damage of healthy tissue (Van Walle et al., 2007). Post-translational modification of protein is an important issue, too. Most of eukarytic proteins require post-translationally modified then gain their functions. These modifications includ glycosylation, cleavage of pre-peptide, formation of disulphide bridge and association of multiple peptides. Appropriate glycosylation also could reduce immunogenicity of proteins (Kosloski et al., 2009;Wang et al., 2010;Wu et al.). Currently, bacterial cells are the most conventional and convenious tools for mass production of recombinant proteins, but it is not easy for bacterial cells to undergo post-translational modication of eukaryotic proteins (Jacobs & Callewaert, 2009;Muir et al., 2009). Even expressed in eukaryotic cells, glycosylation of proteins is not exactly the same among difference specises (Perego et al., 2010). Unappropriatly modified recombinant proteins might lead to unexpected immune respones, if the proteins are used for medical purpose (Kosloski et al., 2009; Li & d'Anjou, 2009).

Protein designing challenges our understanding of the principles underlying protein structure and is also a good method to access our understanding of sequence-structure and structure-function relationship (Nikkhah et al., 2006). Rational design of proteins requires detailed knowledge of protein folding, structure, function, and dynamics (Chen et al., 2005). To build expression libraries, it is necessary to understand a protein scaffold in detail to amino acids usage on each residue position. This would reveal the key elements that affect functions and stabilities of a protein scaffold.

The appearance of new intellectual property, the breakthroughs in technology, or the increase in a market need are three major impacts to biopharma industry. Protein engineering is a branch of the biopharm industry, where intellectual property rights apparently play the most important role in the development and commercialization of final products. The intellectual property strategy to protect inventions of biopharma industry is to patent and to license them on an exclusive basis. The intellectual property also must be included in the linkage of research/development and business to ensure the commercial viability of biopharma products and to cope with the rapid changes of market. Currently, the intellectual property situation of biopharma industry is too complex and hampers the generation and production of recombinant protein/peptide drugs. Pantent analysis and patent map are necessary and helpful while planing the research and development and marketing. A well planed intellectual property strategy can not only protect the output of research and development but also defend market. A protein scaffold with simple legal situation will avoid knotty lawsuits. In recent years, shouts for alternative protein scaffolds is urgent and alternative protein scaffolds usually provide a favourable intellectual property situation. In this chapter, we will focus on a protein scaffold, cysteine-stabilized α/β (CS$\alpha\beta$) motif, and discuss protein engineering based on the scaffold.

2. Cysteine-stabilized α/β motif

2.1 Specific pattern and structural feature

CS$\alpha\beta$ motif is a very unique protien scaffold. Proteins with a CS$\alpha\beta$ motif express a high diversity in their protein primary structures (Figure 1) but share a common core sturcture that consists of a α-helix and an anti-parallel triple-stranded β-sheet (Figure 2 a to d). This is especially interesting not only for academic research but also very useful for applied utilities. From the protein sequence alignment analysis, six cysteines and one glycine (-C-X_l-CXXXC-X_m-GXC-X_n-CXC-) are exzactly conserved in all proteins containing a CS$\alpha\beta$ motif (Carvalho & Gomes, 2009;Lay & Anderson, 2005). The cysteines form a framework and tightly connect the intramolecular structures. The most significant feature of the framework is two disulphide bonds which are formed with a patten of $(i, i+4)$ and $(j, j+1)$. The cysteines of $(i, i+4)$ and $(j, j+1)$ are located in the α-helix and the $\beta3$ strand of the β sheet; respectively (Figure 1). The helix and the sheet are connected by the disulphide bonds pairing with the pattern (*i*, *i+4*) and (*j*, *j+1*) (Assadi-Porter et al., 2000;Fant et al., 1999;Sun et al., 2002;Zasloff, 2002;Zhu et al., 2005). The third disulphide bond connects loop 1 and and β2 strand. In the β2 strand one glycin is conserved and locates in the central region of the motif. In plant defensins, there is usually a forth disulphide bond that seal the N- and C- terminee. The positively charged α-helix is another character. The positively charged residues are especially concentrated in the helix. β-sheet of the motif is formed by a hydrophobic amino acid clust and the sheet is composed by two or three anti-parallal strands. Loop regions of the mofit are highly diversed in lenght among proteins. There is a small turn that connects the helix and β2 strand.

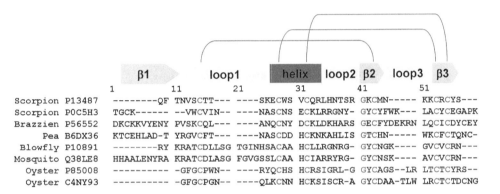

Fig. 1. The sequence alignment of proteins containing CSαβ motif. Sequences of two scorpion toxins (UniProt P13487, P0C5H3), two plant defensins (UniProt P56552, B6DX36), two insect defensins (UniProt P10891, Q38LE8) and two mollusk defensins (UniProt P85008, C4NY93) are aligned. The restrictly conserved residues, include six cysteins and one glycin, are high lighted in red. The green arrow box, red box and cyan boxs above the alignment represent β strands, α-helix and loops; respectively. The blue lines above on the top indcate the disulphide bond bwteen cysteines.

2.2 Biological distribution and functions

To date, there are at least four hundred proteins with a CSαβ motif have been discovered and deposited in databases. Proteins with the motif are widely distributed among plants, insects, arachnidia and mollusca (Sun et al., 2002,Zhu et al., 2005). They exhibit a wide spectrum of biological activities, including antimicrobial activity, enzyme inhibitory function, inhibition of protein translation and sweet taste (Assadi-Porter et al., 2000;Chen et al., 2005;Clauss & Mitchell-Olds, 2004;Spelbrink et al., 2004;Stec., 2006;Wong & Ng, 2005;Zhu et al., 2002). Proteins with the CSαβ motif usually serve a common function as defenders of their hosts (Lobo et al., 2007;Song et al., 2005;Zasloff, 2002).

Before designing a unique function into a protein scaffold, it is important to understand the relationship between each part of the scaffold and its biological function. Based on the three-dimensional structure, protein scaffold containing a CSαβ motif can be devided into three parts: one α-helix, one β-sheet and three loops. It is well known that the three parts bearing different biological functions (Liu et al., 2006;Thevissen et al., 1996;Wong & Ng, 2005;Zhao et al., 2002).

Core structure   2NZ3

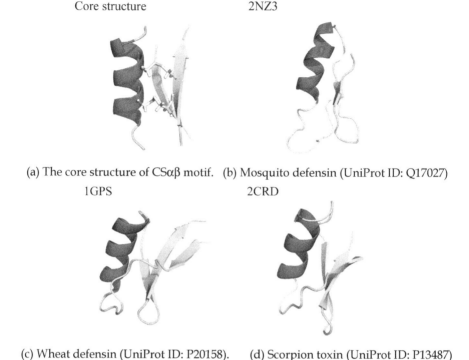

(a) The core structure of CSαβ motif.   (b) Mosquito defensin (UniProt ID: Q17027)

1GPS   2CRD

(c) Wheat defensin (UniProt ID: P20158).   (d) Scorpion toxin (UniProt ID: P13487)

Fig. 2. The core structure and three dimensional structrues of different specices. The structures are presented in color ribbon. Red: α-helix, green: β-sheet, and cyan: loop. Protein structures are retreved from Protein Data Bank and visualized with PyMol 0.99rc6.

The α-helix is related to antimicrobial ability. As described previously, the helix forms a positively charged cluster. When the positively charged residues were replaced with null or negatively charged amino acids, the anti-microbial ability of the proten is also changed. For its net positive charge, it is believed that proteins with the motif could interact with negatively charged cell membrane (Thevissen et al., 1996;Thomma et al., 2003). Several studies have demostrated that plant defensins are able to pass artificial membranes and lead to ions leaking from one side of the membrane. The mechanism about how plant defensin passing membrane is not revealed, yet.

Role of the β-sheet is less studied and disscussed. The direct mutagensis studies showed when the hydrophobic residues in the β-sheet are alanine substituted, biological function of mutated proteins are dromatically reduced (Walters et al., 2009;Yang et al., 2009). The maximal effects of mutated proteins are only 30-40% maximal effects of the wild type, even at a high protein concentration. Protein-protein docking model showed that the β-sheet could form interaction interface with their counterpart. These data imply that, the distribution of the hydrophobilc residues in the β-sheet plays a role in protein-protein interactions and β-sheet could relate to the protein-protein interaction specificity to their targets.

The length of loop regions are different from protein to protein and the loops connecting CSαβ motif can serve as functional epitopes (Figure 3) (Lay & Anderson, 2005;Wijaya et al., 2000;Zhao et al., 2002). For example, loop 1 of the *Arabidopsis thaliana* trypsin inhibitor (ATT$_p$) and loop 2 of cowpea thionine are the functional epitopes for trypsin inhibition (Wijaya et al., 2000;Zhao et al., 2002). The loop 3 of *Vigna radiate* defensin 1 (VrD1) is the functional loop for insect α-amylase inhibition (Lin et al., 2007;Liu et al., 2006).

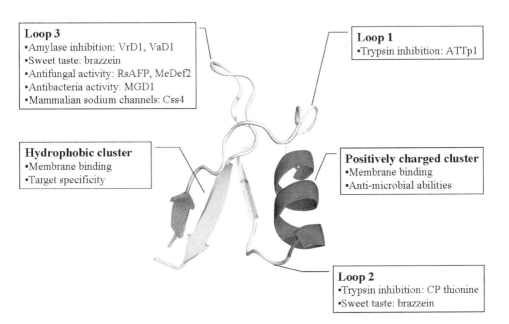

Fig. 3. Structure-function relationship of peptide with a CSαβ motif. Red: α-helix, green: β strands, and cyan: loops.

2.3 Structural ultra-stability

A protein scaffold with ultra high stability can maintain its three dimensional structure in extreme environments and its functions can be preserved. Therefore, when a protein scaffold applied in biomedical applications, its structural stabilities must be considered. The scaffold should be able to pass a low pH environment in the stomach, resist protease digestion, endure chaotropic agents at high concentrations and so on. These criteria are to ensure the engineered proteins could reach their target sites and perform functions inside body, and are not destroyed. It has been reported that proteins with a CSαβ motif equip ultra stabilities to several extreme environments such as urea at a concentration of 6M, a temperature over than 95°C, and resist protease digestion without changing its structure (Malavasic et al., 1996;Yang et al., 2009). In Table 1, the properties of CSαβ motif and single-domain antibody are listed and compared (Holt et al., 2003;Skerra, 2007;Yang et al., 2009;Yang & Lyu, 2008). For their advantages, such as high specificity and affinity, antibodies still are the most popular protein scaffold for engineering. Antibodies are widely used both in routine laboratorial experiments and clinical diagnosis. In spite of their significant clinical success, several disadvantages, including high cost in manufacturing, large in size, undesired effector functions and complex intellectual property situations, obstruct their development and applications (Jones et al., 2008). Single-domain antibody is only constituted of antibody variable regions and does not have constant regions (Holt et al., 2003). Compared with single-domain antibodies, the proteins with a CSαβ motif have smaller molecular weights and higher structural stabilities (Holt et al., 2003;Skerra, 2007;Yang et al., 2009). As previously described, proteins with a CSαβ motif share low sequence identity but high structural similarity (Lin et al., 2007). Their functions are highly varied and sturctures are ultra-stable. There is a possibility to utilize CSαβ motif as an engineering scaffold for biomedical applications (Yang & Lyu, 2008).

Item	Molecule	
	Single-domain antibody	CSαβ motif protein
Molecular weight	11-15 kDa	5-7 kDa
Generation of expression library	B cell mRNA	Synthetic
Water solubility	Less	High
Ultra stability	Unfolded 6M Gdn-HCl T_m 60-78°C	Ultra stable 6M Gdn-HCl $T_m > 96$°C
Post-translational modification	Glycosylation	Disulfide bridge
High functional diversity	Yes	Yes
Tolerate to amino acid substitution	Loop regions	Structural and loop regions
Enzyme inhibition	Not certain	Direct inhibition
Membrane binding	Not certain	Direct binding
Legal problem	Very complex	Simple

Table 1. Comparison of properties bwtween CSαβ motif and single-domain antibody.

2.4 Amino acid usage of CSαβ motif

To understand the amino acid usage of a protein scaffold can reveal relationships among structures, functions and sequence residues (Kristensen et al., 1997;Yang et al., 2009). To completely understand the relationships could be an approach through extensive amino acid substitution and analysis of protein sequences (Corzo et al., 2007;Wang et al., 2006). Amino acid substitution have been performed in plant defensins, brazzein of *Pentadiplandra brazzeana* and VrD1 of *Vigna radiate*, and some key residue positions are discovered (Assadi-Porter et al., 2010;Yang et al., 2009). In both cases, amino acid substitution does not lead to the structure significantly being changed in all positions along the sequence but the replacement in some positions have effects on biochemical function (Assadi-Porter et al., 2010;Yang et al., 2009).

It has been noted that certain amino acids have preference to fold into a given secondary structure (Chan et al., 1995;Zhong & Johnson, 1992). Comprehending preference of amino acids usage will be really helpful to protein engineering and can be as a fundment for designing innovative peptides. The two major classes of CSαβ motif protein are plant defensins and scorpion toxins (Zhu et al., 2005). Currently, there are at least 140 sequences of scorpion toxin and 180 sequences of plant defensin deposited in the SwissProt database and the numbers are continuously increased. The peptide sequences of the scorpion toxin and plant defensin peptides deposited in the SwissProt database are retrieved and amino acid usage perferacnes are separately analyzed (Table 2 and Table 3).

In Table 2 and Table 3, the twenty amino acids are listed on the top row, the secondary structures and residue positions are listed on the left two columns. Use of amino acids in each position is counted and the usage frequency is calculated (form 0.00 to 1.00). Amino acids with a high usage frequency (> 0.20) are listed on the right column. The results are interesting and different between scorpion toxins and plant defensins. In scorpion toxins, peptides tend to employ similar amino acids in the same position. Therefore, sequences of scorpion toxins are less diverse and more uniformed. In plant defensins, sequences of peptides are more diverse. The major targets of scorpion toxins are ion channels on the neuron cell surface but the targets of plant defensins are various form peptide to peptide (Zhu et al., 2005). This might be the reason for the amino acid usage difference between the two classes of CSαβ motif peptides. In the helix region, scorpion toxins prefer using charged residues and plant defensins are more inclusive of different type of amino acids. The most interesting is the two residues flanking the exzactly conserved glycine on $\beta_2 2$. In scorpion toxins, two tyrosins are prefered on $\beta_2 1$ and $\beta_2 3$, and the rates are 0.44 and 0.75; respectively. In plant defensins, a preceded glycine ($\beta_2 1$) is prefered to the glycine on $\beta_2 2$ and the frequency is 0.35. It is worth to clarify the role of the two tyrosins surrounding the glycine on $\beta_2 2$ of scorpion toxins. Although the frequnectly used amino acids are listed, it just provides a reference for protein designing and there is no necessarity that combination of these amino acids forms a universal sequence.

Region	Position	A	G	V	L	I	F	Y	W	C	S	T	D	E	N	Q	R	H	K	M	P	High frequency[a]
Strand 1	S1		0.81												0.12	0.03						G
	S2 (C)									1.00												C
	S3	0.05	0.03	0.01						0.03							0.02		0.85			K
	S4			0.34	0.13	0.10	0.03	0.22					0.02	0.01	0.02		0.10			0.03		V Y
	S5	0.05	0.07	0.02	0.00	0.06			0.04		0.51	0.06	0.03	0.03	0.02		0.02	0.05			0.01	S
Helix	H1	0.04	0.16	0.08							0.04	0.03	0.05	0.20	0.01	0.06	0.03	0.06	0.16		0.07	E
	H2		0.13		0.01	0.00	0.16	0.37	0.03		0.10		0.03	0.03	0.05	0.08	0.01					Y
	H3 (C)									1.00												C
	H4	0.09	0.05	0.03	0.05	0.10	0.01				0.01	0.25	0.06	0.22	0.03	0.03	0.03	0.01	0.01			D E
	H5	0.01	0.01	0.01	0.01	0.01					0.09	0.09	0.14	0.03	0.17	0.04	0.07	0.01	0.30	0.01		K
	H6	0.01		0.09	0.08	0.07	0.01			0.03	0.01	0.01	0.01	0.43	0.02	0.02	0.13	0.01	0.02		0.07	E
	H7 (C)									1.00												C
	H8	0.03	0.03	0.01	0.01						0.09	0.02		0.03	0.01	0.07	0.11	0.01	0.57			K
	H9	0.22	0.05	0.03	0.11	0.03	0.01	0.02		0.06	0.01	0.14	0.04	0.01			0.09		0.16	0.01		A
β_2 strand	$\beta_2 1$	0.01	0.04	0.01		0.01	0.09	0.44		0.13	0.02	0.15	0.02	0.01			0.01	0.05				Y
	$\beta_2 2$ (G)		1.00																			G
	$\beta_2 3$				0.01	0.02	0.75			0.01	0.03	0.01	0.02			0.01	0.01	0.01	0.11			Y
	$\beta_2 4$ (C)									1.00												C
	$\beta_2 5$	0.01		0.01		0.01	0.02	0.34	0.26	0.00	0.03	0.01	0.03	0.01	0.00	0.01			0.10	0.01	0.07	Y W
β_3 strand	$\beta_3 1$	0.49	0.05		0.03	0.00	0.01	0.01			0.05	0.01	0.01	0.02	0.00	0.04	0.02	0.01	0.25			A K
	$\beta_3 2$ (C)									1.00												C
	$\beta_3 3$	0.01			0.01	0.01	0.01	0.28	0.45				0.02			0.01	0.01	0.05	0.14	0.01		W Y
	$\beta_3 4$ (C)									1.00												C
	$\beta_3 5$	0.04	0.07		0.01	0.03	0.01	0.22	0.00		0.02	0.07	0.01	0.42	0.01	0.05		0.05	0.01			E Y

Table 2. The high frequent amino acids in the structural regions of scorpion β toxin.

Position	A	G	V	L	I	F	Y	W	C	S	T	D	E	N	Q	R	H	K	M	P	High frequency[a]
Strand 1 S1	0.02	0.08	0.20	0.07	0.01	0.03			0.01	0.27	0.02	0.08	0.01	0.02	0.07	0.01	0.04	0.04	0.02		T L
S2 (C)									1.00												C
S3	0.01		0.04	0.07	0.02	0.02	0.02	0.01	0.05	0.03	0.05	0.34	0.01	0.04	0.08	0.04	0.11	0.07	0.01		E
S4	0.06	0.02	0.07	0.05	0.03				0.18	0.10	0.04	0.07	0.08	0.02	0.13	0.03	0.12		0.01		
S5	0.05	0.04	0.05	0.12	0.05	0.01	0.04		0.06	0.05	0.03	0.07	0.00	0.12	0.08	0.01	0.08	0.01	0.13		
Helix H1	0.07	0.10	0.02	0.01	0.02	0.01	0.01		0.15	0.15	0.05	0.03	0.14	0.05	0.06	0.04	0.08		0.04		
H2	0.12	0.04	0.01	0.08	0.02	0.02	0.00	0.02	0.07	0.05	0.11	0.05	0.27	0.02	0.02	0.03	0.06	0.01	0.03		N
H3 (C)									1.00												C
H4	0.24	0.03	0.04	0.02	0.02	0.00	0.03		0.06	0.02	0.17	0.03	0.08	0.02	0.09		0.19				A
H5	0.06	0.01	0.02	0.04	0.02	0.03		0.01	0.11	0.08	0.09	0.06	0.23	0.09	0.03	0.04	0.10		0.01		N
H6	0.05	0.01	0.25	0.06	0.01	0.01	0.04		0.02	0.08	0.05	0.01	0.05	0.05	0.16	0.01	0.11	0.02	0.03	0.02	V
H7 (C)									1.00												C
H8	0.03	0.01	0.04	0.11	0.18	0.02	0.03	0.01	0.01	0.04	0.01	0.01	0.02	0.09	0.18	0.05	0.16	0.03	0.02		
H9	0.06	0.02		0.01		0.01	0.01	0.01	0.09	0.21	0.02	0.07	0.19	0.11	0.08	0.01	0.10	0.01	0.02		T
β₂ strand β₂1	0.05	0.35	0.05	0.01	0.02	0.01	0.01		0.01	0.11	0.03	0.08	0.02	0.05	0.01	0.05	0.14		0.05		G
β₂2 (G)		1.00																			G
β₂3	0.04	0.02	0.05	0.01	0.01	0.05	0.06		0.18	0.07	0.05	0.06	0.07	0.03	0.11	0.06	0.17	0.01			
β₂4 (C)									1.00												C
β₂5	0.02	0.01	0.07	0.05	0.06	0.04	0.01	0.01	0.06	0.02	0.14	0.04	0.17	0.03	0.16	0.09	0.06	0.01	0.01		
β₃ strand β₃1	0.05		0.02	0.02	0.07	0.02	0.02		0.04	0.05	0.01	0.02	0.04	0.06	0.30	0.04	0.23	0.05			K
β₃2 (C)									1.00												C
β₃3	0.01	0.02	0.07	0.12	0.15	0.19	0.13	0.07		0.02	0.02	0.03	0.01	0.01	0.03	0.05	0.05	0.05			
β₃4 (C)									1.00												C
β₃5	0.02	0.02	0.03	0.04	0.06	0.02	0.22		0.04	0.07	0.30	0.03	0.02	0.03	0.04	0.04	0.01	0.04	0.02		Y

Table 3. The high frequent amino acids in the structural regions of plant defensin.

3. Applications of CSαβ motif peptides

Human beings have been using natural products containing CSαβ motif peptides for several hundreds years. The most well known natural products are sweet peptides of plants and scorpion venom. For their anti-microbial and pesticide abilities, in many cases, peptides with CSαβ motif are transfected into industrial crops to protect these transgenic plant from pathogens. In recent years, some of plant defensins are also screened and selected as antibiotics for therapeutic utilities. In this section, some peptides with therapeutic potential are selected and discussed.

3.1 Sweet peptide: Brazzein

Brazzein is isolated from *Pentadiplandra brazzeana*, a climbing plant plant of the West Africa. Berry fruits of *Pentadiplandra brazzeana* have sweetness and are traditional foods of African natives (Assadi-Porter et al., 2008;Yang & Lyu, 2008). For a long time, the sweetness of the berry fruits is a secret. Ding and Hellekant at University Wisconsin Madison isolated a small peptide with sweetness from *Pentadiplandra brazzeana* and named it as brazzein 1994 (Ming & Hellekant, 1994). Peptide sequence analysis shows that brazzein contains a set of (i, $i+4$) and (j, $j+1$) cysteine pattern. Its NMR structure is solved and confirms it as a CSαβ motif peptide 1998 (Caldwell et al., 1998). Brazzein is the smallest, most heat-stable and pH-stable protein known to have an intrinsic sweetness (Jin et al., 2006). Brazzein is 200 times sweeter than sugar and it can be used both in cold and hot drinks/foods without change its sweet taste (Jin et al., 2006). Brazzein is widely recruited as sugar substitute in low caloric dietary formulas or for diabetes patients. Brazzein is a food additive for several years and no side effects have been reported (Yang & Lyu, 2008).

3.2 Scorpion venom: Traditional remedies

Scorpion is therapeutically used in traditional Chinese medicine for hundreds of years. Figure 4 shows the therapeutic instruction of using scorpion described in the *Compendium of Materia Medica*, an ancient Chinese pharmacopoeia compiled in the 1500s AD. In traditional Chinese medicine, scorpion is recruited as one of the major treatments for stroke, rheumatism, epilepsy, hemiplegia, convulsions, cramps, twitchs, headaches, tetanus and scrofula. *Buthus martensii Karsch* is the most common species of scorpion that are processed as traditional Chinese medicine. Captured scorpions have to be preserved in salt. The preserved scorpions are usually boiled with herbs, such as licorice, to reduce toxicity after washing off the salt. Patients take the boiled extracts twice daily and the amount of scorpions should be controlled to under 10 grams for an adult daily. Most of proteins will be denatured and loss functions after being exposed to salt at a high concentration and water boiling temperature. The major effective composition of scorpion is scorpion venom. It has been demonstrated the venom can block cation channel of insects (Li et al., 2000;Wang et al., 2001). Several peptides containing CSαβ motif have been isolated from scorpion venom. The peptides can keep their structures and functions after be exposed to water boiling temperature and solvents of high salt concentration. BmK AEP, a peptide containing a CSαβ motif isolated from *Buthus martensii Karsch* venom, is the first anticonvulsant purified in the venom (Wang et al., 2001). BmK AEP has inhibitory effect to coriaria lactone-induced epilepsy model in rodents. It has a low toxicity to mice in doses up to 20 mg/animal (Rajendra et al., 2004). There is nearly no effect on heart rate, electroencephalogram, breathing rate and blood pressure (Rajendra et al., 2004;Wang et al., 2001). CII9, another peptide with a CSαβ motif isolated from *Centruroides limpidus Karsch*, can partially inhibit

Fig. 4. The therapeutic instruction of scorpion described in the ancient Chinese pharmacopoeia: *Compendium of Materia Medica*.

sodium ion currents from superior cervical ganglion neurons of rats (Dehesa-Davila et al., 1996;Possani et al., 1999). In the rat model, CII9 also exerts a depressant effect on the behavior and electroencephalographic activity and antagonizes the epileptiform activity induced by penicillin (Dehesa-Davila et al., 1996;Gazarian et al., 2005;Possani et al., 1999).

3.3 Peptide antibiotics: Plectasin

Peptides with a CSαβ motif act as primary defenders for their host. Microbe-killing abilities of plant defensins have been demonstrated and transgenic industrial crops have been bred to carry specific defensins to increase their tolerance to plant pathogens. For directly binding to membranes or cell walls, target pathogens do not easily resist peptides with a CSαβ motif as they do to protein-targeting antibiotics. Pathogenic microbes would escape attacking from conventional antibiotics by mutating proteins but it is never easy to change the properties of cell membranes or cell walls. Scaffold of a CSαβ motif could be as a suitable platform for developing peptide antibiotics. Some plant defensins are tested and show activities againt human pathogenic microbes. For example, Rs-AFP2, a defensin isolated from seeds of *Raphanus sativus*, has an inhibitory activity against *Candida albicans* (Landon et al., 2004;Terras et al., 1992;Thomma et al., 2003). Scientists of Novozymes, a Dnaish biotech company, isolated a novel plant defensin, plectasin, with excellent antibiotics ability.

Plectasin is a 40-amino acid residue peptide isolated from saprophytic ascomycete, *Pseudoplectania nigrella*, and might be the first plant defensin with a CSαβ motif isolated from fungus (Mygind et al., 2005). Plectasin expresses anti-bacterial activities against a broad Gram-positive bacteria and has an inhibitory effect at a low concentration of 0.25 μg/ml to growth of *Streptococcus pneumoniae* (Mygind et al., 2005). Plectasin also has a comparable killing rate to *Streptococcus pneumoniae* as do conventional antibiotics (Mygind et al., 2005). *Streptococcus pneumoniae* is a major pathogenic bacteria and the most common cause of hospital/community-acquired pneumonia, bacterial meningitis, bacteremia, sinusitis, septic arthritis, osteomyelitis, peritonitis, and endocarditis (Whitney et al., 2000). Currently, antibiotics are major treatments to patients with infection of *Streptococcus pneumoniae*. In recent years, *Streptococcus pneumoniae* is more and more resistant to antibiotics and the demand for new drugs to cure *Streptococcus pneumoniae* infection is urgent (Baquero et al., 1991;Whitney et al., 2000). Another challenge of theraping *Streptococcus* infection is that *Streptococcus* can be an intracellular pathogen and avoids targeting by the immune system or drugs (Gordon et al., 2000;Talbot et al., 1996). Different from the conventional antibiotics, plectasin can directly act on the cell wall precursor lipid II of *Streptococcus* (Mygind et al., 2005). Studies also showed that plectasin has an intracellular activity against *Streptococcus aureus* both in human monocytes and in mouse peritonitis model without effecting the cells viability or inducing IL-8 production (Brinch et al., 2009;Hara et al., 2008).

4. Protein engineering based on CSαβ motif scaffold

Based on the scaffold, some CSαβ proteins have been engineered to exhibit new functions or changing of antimicrobial activities (Lin et al., 2007;Vita et al., 1999;Vita et al., 1995). Different approaches, including minimal residue substitution, functional epitope exchange, structural based modification and combinatorial chemistry have been employed to engineer the scaffold to exhibit new functions (Lin et al., 2007;Thevissen et al., 2007;Van Gaal et al.,

2004;Vita et al., 1999;Vita et al., 1995;Zhao et al., 2004). Protein structure is important to the rational design of a protein scaffold. It does not only reveal the shape of a protein molecule but sidechain orientation of each residue. This information is crucial for investigating possible interactions between designed protein and its target.

4.1 Site-direct mutagenesis

In some studies, researchers focus on the relationships of structures, functions and each residue of peptides with a CSαβ motif. Extensive residue substitution is usually performed on the peptides and the changes of structures and functions are observed. The process is time-consuming and labor intensive but it is required for collecting basic information about the scaffold. Amino acid substitution has been extensively performed on two peptides with a CSαβ motif, VrD1 and brazzein (Assadi-Porter et al., 2010;Yang et al., 2009). In the two studies, molecular docking models are also established to investigate the interactions between receptors and ligands.

In both cases, similar positions along the structures are discovered that are crucial to functions of both peptides (Figure 5). These sites are widely distributed on β1 strand, loop 1, loop 2 and loop 3 (Assadi-Porter et al., 2000;Walters et al., 2009;Yang et al., 2009). From the molecular docking models, these sites either directly interact with their targets or play as functional epitopes and insert into the active site of their targets (Assadi-Porter et al., 2010;Liu et al., 2006;Yang et al., 2009). It is interesting that, when two negatively charged residues, D29 and E41, of brazzein are replaced by a positively charged amino acid, the sweetness of brazzein is greatly improved (Assadi-Porter et al., 2010). The mutants should be with high interests to industrial utilities. In the case of VrD1, there is no functional improvement observed on its mutants (Yang et al., 2009). Comparing conformations of the wild-type proteins and their mutations, there are only minimal shifts measured (Assadi-Porter et al., 2010). This implies that structure of peptide with a CSαβ motif is relatively stable and has high tolerance to amino acid substitution. Side chains of these residues on the interactive surface are crucial to biological functions of the peptides.

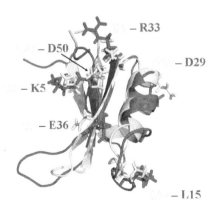

Fig. 5. Superposition of brazzein and VrD1. Some critical residues are overlapping between brazzein and VrD1. Red: brazzein, cyan: VrD1. Red and cyan labeled residues: critical residues of brazzein and VrD1; respectively. The two structures are aligned with SARST (http://sarst.life.nthu.edu.tw/iSARST/).

Multiple mutation also has been performed on the scaffold. Vital *et al* performed minimal residue substitutions on the charybodotxin (Chtx), an scorpion toxin containing CSαβ motif, to equip metal ion binding ability (Vita et al., 1995). Three residues, K27, M28 and R34, on the β sheet are substituted with histidines. The modified protein exhibts a chelatine property but has the same circular dichroism spectrum profile as the native Chtx does.

4.2 Functional epitope exchange

Grafting epitopes with known function is a straightforward stratage for a protein to gain new function. Functional epitopes could be exchanged between proteins have a high sequence identity or share a high structural similarity. Based on the structural homologous, the β2-β3 hairpin of scyllatoxin, a scorpion toxin containing the CSαβ motif, is replaced by the CDR2-like loop of human CD4. The chimeric protein can bind to the HIV-1 envelope glycoprotein and the affinity is increased 100 fold as compared with the native scyllatoxin (Vita et al., 1999). In another case, VrD1 and VrD2, two defensins of *Vigna radiate*, share 80% sequence identities. The major difference of sequences concentrats on the loop 3. When the loop 3 of VrD1 is grafted to VrD2, the chimeric peptide exhibits the enzyme inhibitory function as VrD1 does (Lin et al., 2007).

4.3 Combinatorial chemistry approach

Combinatorial chemistry approach are powerful in screening and selecting binders with high affinity and high specificity (Hosse et al., 2006). It has been widely applied in antibody scaffold and thousands of antibodies are generated through the technology. Combinatorial chemistry approach also has been recruited to develop and isolate artificial proteins with new functions (Zhao et al., 2004). The approach could accelerate protein engineering based on a CSαβ motif to develop novel peptides with biomedical interesting (Thevissen et al., 2007;Van Gaal et al., 2004). Based on scaffold of insect defensin A, an expression library of peptides with 29 residues is constructed and used in screeening novel binders to targets. The expression library is artificially synthesized and amino acid of seven positions on the loops are randomized (Zhao et al., 2004). Tumor necrosis factor α (TNF-α), TNF receptor 1, TNF receptor 2 and antibody against BMP-2 are selected as targets and the screening results show significant enrichment in all cases.

5. The challenges of engineering of CSαβ motif

Although several successful engineered peptides based on the CSαβ motif scaffold have been reported, to design new functions into the scaffold does perplex us. In this section, we would like to discuss challenges in engineering a CSαβ motif to equip desired functions.

5.1 Lacking co-structural model

To introduce tailored functions into a protein scaffold is never an easy task. For its ultra stability and high diversity of function, there is no doubt that peptide with a CSαβ motif is a suitable scaffold to be engineered for biomedical applications. To date, there are only a few of successful cases are reported. Many structures of peptides with CSαβ motif have been solved and deposited in database, but knowledge about the motif is poor. In spite of hundreds of peptides with a CSαβ motif deposited in the database, it is hard to make conclusions relating to structures and biochemical functions of the motif. Key residues are revealed in some peptides with a CSαβ motif and molecular docking models provide reasonable explainations (Figure 6). The real interaction is not clarified, for lacking of a co-structural model. A co-structural model of the motif and its target will provide information about the dynamic interactions of the two protein molecules (Dumas et al., 2004;Thioulouse & Lobry, 1995). Structures of the complexes of peptides with a CSαβ motif and its counterparts will provide the situation of protein-protein interaction and have a great benefit to engineering of the scaffold.

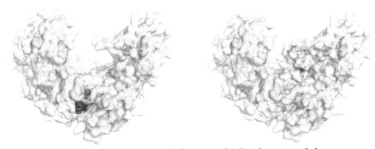

(a) HIV reverse trascriptase with inhibitor   (b) Docking model

Fig. 6. Docking model of plant defensin to HIV reverse transcriptase. Structures of HIV reverse transcriptase (3DRP, HIV RTase) and VrD2 (2GL1) are retrieved from the Protein data bank. Molecular docking is performed with PatchDock (http://bioinfo3d.cs.tau.ac.il/PatchDock/). (a) The crystal structure of HIV RTase with inhibitor. (b) Molecular docking model of VrD2 and HIV RTase. Cyan: HIV RTase, yellow: VrD2 and red: HIV RTase inhibitor.

5.2 Generation of gene library and construction of expression system

The advantages of combinatorial chemistry approach to protein engineering is its ability to associate every protein with its genetic material (Weng & DeLisi, 2002). There are two core parts about applying combinatorial chemistry approach in protein engineering: display techniques and gene expression library. These two core parts are equally important and affect each other. To date, several display techniques have been developed (Table 4) (Daugherty, 2007;Gronwall & Stahl, 2009). Natural properties of peptides with a CSαβ motif are as defenders for their native hosts and most of the peptides express anti-bacterial abilities, virus inhibitory abilities and inhibition of protein translation. How to design a expression library based on a CSαβ motif scaffold is a tough task.

Technique	Library limitation	Expression system	Disulphide bond
Bacterial display	10^{11}	Circular	No
Phage display	10^{11}	Circular	No
Ribosome display	10^{14}	Circular/Linear	Yes

Table 4. Comparison of three display techniques.

Peptides with a CSαβ motif are tightly held by disulphide bridges and may be not easily folded into appropriate structures inside bacterial cells (Villemagne et al., 2006). Ribosomal display technique can overcome the problems (Figure 7), but it requires an ultra clean

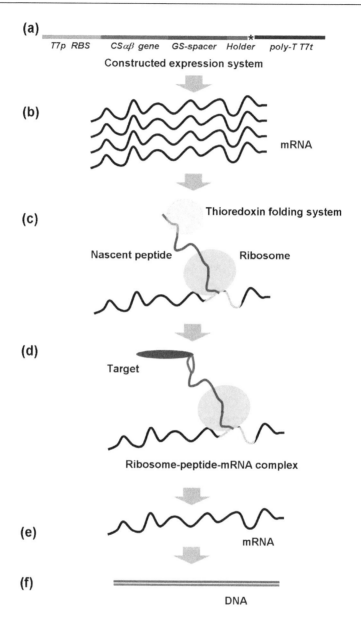

Fig. 7. Concept of ribosome display techniques. (a) expression system, (b) & (c) transcription, translation and folding while translation, (d) binding and selection, (e) isolation of mRNA from the complex, (f) reverse transcription and construction of expression system for next cycle of selection. A specific holder gene usually is constructed to the expression system. The holder can arrest protein translation and the nascent peptide can form a stable complex with ribomsome.

environment to protect mRNA from degradation (Fennell et al., 2010;Villemagne et al., 2006). The gene library of a CSαβ motif has to be artificially synthesized then constructed to the expression system. There are two key issues of the synthesized gene library have to be considered, one is the size of library the other one is how many random residue positions it should have (Gronwall & Stahl, 2009;Michel-Reydellet et al., 2005). These two questions are hard to answer. A large library with more randomized residue positions could have more genotypes/phenotypes but it also has a high possibility to generate nonsense genes for the unvoided stop codon on unexpected positions. A library with a randomized loop 3 of a CSαβ motif scaffold has been constructed and successfully used in a noval binder to human TNF-α (Zhao et al., 2004). It also has been reported that all three loops of peptides with a CSαβ motif can equip unique functions and the structural regions also fine regulate the functions of the peptides (Assadi-Porter et al., 2010;Liu et al., 2006;Yang et al., 2009;Zhao et al., 2002). For a library of long length artificial genes, several cycles of PCR overlapping extensions are required and may reduce quality of the library. A circular or a linear expression systems can be recruited (Fennell et al., 2010;Shimizu et al., 2001;Villemagne et al., 2006). To construct the linear library, all required DNA elements, as promotor, ribosomal binding site and poly T tail, are synthesized or copied and constructed with the desired gene library through PCR (Katzen et al., 2005;Shimizu et al., 2001) . In the linear expression system, the library can have a maximal size up to 10^{13} (Table 4) (Gronwall & Stahl, 2009).When a linear expression system is recruited, the qualities of final PCR products should be well controlled. It is better for the library to be freshly constructed everytime the protein display is performed. The size of a library constructed with a circular expression system is much smaller for the genotype lost during construction, however, the library can be stably stored in a good condition after construction.

5.3 Mass production and folding of proteins

To screening possible funtions of a protein, it requires a large amount of the protein. The amount of proteins with a CSαβ motif in their native host is rare and to obtain eounght of the native proteins for assays requires a lot of raw materials. To produce recombinant proteins of a CSαβ motif in bacteria cells, the proteins have to be fused with a large tag to reduce their toxicity to bacteria. Protein folding is also another problem for apporpriately forming the disulphide bonds of a CSαβ motif. To overcome the problems, peptides with CSαβ motif can be fused with a thioredoxin tag and a cleavagable site is placed between the peptide and the tag. The recombinant proteins are expressed in *E. Coli* and purified. After the thioredoxin tag is cut off, the desired proteins are further purified. A simple method to cut off the thioredoxin tag is to place a acidohydrolysis site, -Asp-Pro-, between the peptide and the tag (Liu et al., 2006). A large amount of reconbinant proteins can be obtained by following the procedures and the native properties and functions of the proteins are preserved. The acidohydrolysis process is not suitable for proteins with a CSαβ motif containing proline for the proteins would be unstable in a heat and acidic environment (Cunningham & O'Connor, 1997;LeBlanc & London, 1997;Wilce et al., 1998). To resolve the problem, an alternative enzyme cleavage site can be introduced to replace the acidohydrolysis site.

5.4 Unpredictable biochemical function

Enzyme inhibitory function and microbial killing ability are the two greatest advantages of peptide with a CSαβ motif. The conventional bacterial/fungal killing assay is time consuming and it is possible for a performer to be exposed to pathogens during the operation. The peptides with a CSαβ motif have been shown to form pores on an artificial membrane, but the artificial membrane does not represent the membrane of different organisms. A new high throughput screening recruiting membrane of different pathogens will be a great help in resolving the problem (Yang & Lyu, 2008).

Not simple as binding assay, enzymes act on specific substrates and it is extremely difficult to unify experimental conditions for a broad spectrum of enzymes. *In vitro* enzyme inhibitory assays are costly and unable to represent the actually physiological conditions. Computer aided drug screening methods have been established to screen drugs with small molecule weights that can fit into active sites of target proteins (Barrons, 2004;Weideman et al., 1999). Drug dynamics also can be *in silico* simulated (Sinek et al., 2009;Zunino et al., 2009). An antibody can be as a simple binder and its binding targets usually are short fragments exposed to the surface of the proteins. The binding fragments are predicable for a huge amount of knowledge accumulated in the last several decades (Blythe & Flower, 2005;Kulkarni-Kale et al., ;Odorico & Pellequer, 2003). Currently, we do not well understand the interface between peptides with a CSαβ motif and their targets. How functional loops access active sites of the targets is needed to be decrypted. We know too little about the scaffold and it is to difficult to engineer it. It is hard to develop an algorithm to predict the biochemical functions of the peptides with a CSαβ motif. For the complexity of protein-protein interaction, it is a long way to screen peptide drugs *in silico*.

6. Business issue

To biopharma companies, protein engineering/evolution platforms must provide solutions to improve the throughput, safety, biodistribution, activity and frequency of dosing of designed protein, evading intellectual property issues and reducing manufacturing costs, especially in the late process of development and production stages. Intellectual property rights are definite issues, the intellectual property situations of antibodies are obscure as previously described. An alternative scaffold equips the same abilities as antibody does is a valuable prospect and desired to avoid the complex intellectual property landscape of antibody. A protein scaffold can access to a cardinal target and circumvent intellectual property issues that does not mean that it is necessarily a platform worth investing in. The successfully designed protein must equip an improved performance and provide solutions to the concerned points described above.

7. Conclusion

The knowledge about the scaffold is not enough to establish rules to predict its biochemical functions. There are still some technical limitations and the development of the scaffold is confined. Peptides with a CSαβ motif bear many excellent properties and can serve as an alternative protein scaffold for biomedical application. Natural products containing peptides with a CSαβ motif have been therapeutically used for hundreds of years. Some of the purified peptides are tested and have low toxicities to mammalian species. Prudential is needed when recruiting new proteins/peptides as treatments for diseases. The intellectual property situation of peptides with a CSαβ motif is not as complex as antibody, especially in biomedical application (Yang & Lyu, 2008). For its original expression in plant, the engineered proteins with medicinal effect can be expressed in transgenic crops and diseases can be controlled through foods. To predict the biochemical functions of peptides with a CSαβ motif is difficult, some prediction tools have been developed to predict functions of protein from the primary structure, but how proteins interact with their counterparts has to be considered. Folding and post-translational modification of proteins has to be thought over. We believe that peptides with a CSαβ motif will be an alternative scaffold and can be widely applied in biomedical utilities.

8. Acknowledgment

We sincerely thank Professor Ping-Chiang Lyu at the College of Life Science, National Tsing-Hua University for his kind and generous suggestions. We also thank Dr. Chung-Chang Liu, the senior expert of Industrial Technology Research Institute (ITRI), for his supporting part of the research of this work during his term as the director of Biomedical Engineering Research Laboratories, ITRI.

9. References

Aharoni, A., Gaidukov, L., Khersonsky, O., Gould, S. M., Roodveldt, C.&Tawfik, D. S. (2005) The 'evolvability' of promiscuous protein functions *Nature Genetics* Vol. 37 No. 1 pp. 73-76

Alahuhta, M., Salin, M., Casteleijn, M. G., Kemmer, C., El-Sayed, I., Augustyns, K., Neubauer, P.&Wierenga, R. K. (2008) Structure-based protein engineering efforts with a monomeric TIM variant: the importance of a single point mutation for generating an active site with suitable binding properties *Protein Engineering, Design and Selection* Vol. 21 No. 4 pp. 257-266

Angeletti, R. H. (1998) *Proteins: Analysis and Design*, Academic Press, 0120587858 New York

Assadi-Porter, F., Tonelli, M., Radek James, T., Cornilescu Claudia, C.&Markley John, L. (2008) How Sweet It Is: Detailed Molecular and Functional Studies of Brazzein, a Sweet Protein and Its Analogs. In: *Sweetness and Sweeteners*, pp. (560-572), American Chemical Society, 9780841274327, New York

Assadi-Porter, F. M., Aceti, D. J.&Markley, J. L. (2000) Sweetness Determinant Sites of Brazzein, a Small, Heat-Stable, Sweet-Tasting Protein *Archives of Biochemistry and Biophysics* Vol. 376 No. 2 pp. 259-265

Assadi-Porter, F. M., Maillet, E. L., Radek, J. T., Quijada, J., Markley, J. L.&Max, M. (2010) Key Amino Acid Residues Involved in Multi-Point Binding Interactions between Brazzein, a Sweet Protein, and the T1R2-T1R3 Human Sweet Receptor *Journal of Molecular Biology* Vol. 398 No. 4 pp. 584-599

Baines, I. C.&Colas, P. (2006) Peptide aptamers as guides for small-molecule drug discovery *Drug Discovery Today* Vol. 11 No. 7-8 pp. 334-341

Baquero, F., Beltran, J. M.&Loza, E. (1991) A review of antibiotic resistance patterns of Streptococcus pneumoniae in Europe *Journal of Antimicrobial Chemotherapy* Vol. 28 No. Suppl C pp. 31-38

Barrons, R. (2004) Evaluation of personal digital assistant software for drug interactions *American Journal of Health-System Pharmacy* Vol. 61 No. 4 pp. 380-385

Blythe, M. J.&Flower, D. R. (2005) Benchmarking B cell epitope prediction: Underperformance of existing methods *Protein Science* Vol. 14 No. 1 pp. 246-248

Bottcher, D.&Bornscheuer, U. T. (2010) Protein engineering of microbial enzymes *Current Opinion in Microbiology* Vol. 13 No. 3 pp. 274-282

Brinch, K. S., Sandberg, A., Baudoux, P., Van Bambeke, F., Tulkens, P. M., Frimodt-Moller, N., Hoiby, N.&Kristensen, H.-H. (2009) Plectasin Shows Intracellular Activity against Staphylococcus aureus in Human THP-1 Monocytes and in a Mouse Peritonitis Model *Antimicrob Agents Chemother.* Vol. 53 No. 11 pp. 4801-4808

Caldwell, J. E., Abildgaard, F., Dzakula, Z., Ming, D., Hellekant, G.&Markley, J. L. (1998) Solution structure of the thermostable sweet-tasting protein brazzein *Nature Structural & Molecular Biology* Vol. 5 No. 6 pp. 427-431

Carvalho, A. d. O.&Gomes, V. M. (2009) Plant defensins--Prospects for the biological functions and biotechnological properties *Peptides* Vol. 30 No. 5 pp. 1007-1020

Chan, H. S., Bromberg, S.&Dill, K. A. (1995) Models of Cooperativity in Protein Folding *Philosophical Transactions of the Royal Society B: Biological Sciences* Vol. 348 No. 1323 pp. 61-70

Chen, G.-H., Hsu, M.-P., Tan, C.-H., Sung, H.-Y., Kuo, C. G., Fan, M.-J., Chen, H.-M., Chen, S.&Chen, C.-S. (2005) Cloning and Characterization of a Plant Defensin VaD1 from Azuki Bean *Journal of Agricultural and Food Chemistry* Vol. 53 No. 4 pp. 982-988

Clauss, M. J.&Mitchell-Olds, T. (2004) Functional Divergence in Tandemly Duplicated Arabidopsis thaliana Trypsin Inhibitor Genes *Genetics* Vol. 166 No. 3 pp. 1419-1436

Corzo, G., Sabo, J. K., Bosmans, F., Billen, B., Villegas, E., Tytgat, J.&Norton, R. S. (2007) Solution Structure and Alanine Scan of a Spider Toxin That Affects the Activation of Mammalian Voltage-gated Sodium Channels *Journal of Biological Chemistry* Vol. 282 No. 7 pp. 4643-4652

Cunningham, D. F.&O'Connor, B. (1997) Proline specific peptidases *Biochimica et Biophysica Acta (BBA) - Protein Structure and Molecular Enzymology* Vol. 1343 No. 2 pp. 160-186

Das, A., Wei, Y., Pelczer, I.&Hecht, M. H. (2011) Binding of small molecules to cavity forming mutants of a de novo designed protein *Protein Science* Vol. 20 No. 4 pp. 702-711

Daugherty, P. S. (2007) Protein engineering with bacterial display *Current Opinion in Structural Biology* Vol. 17 No. 4 pp. 474-480

Dehesa-Davila, M., Ramfrez, A. N., Zamudio, F. Z., Gurrola-Briones, G., Lievano, A., Darszon, A.&Possani, L. D. (1996) Structural and functional comparison of toxins from the venom of the scorpions Centruroides infamatus infamatus, centruroides limpidus limpidus and Centruroides noxius *Comparative Biochemistry and Physiology Part B: Biochemistry and Molecular Biology* Vol. 113 No. 2 pp. 331-339

Dumas, J. J., Kumar, R., McDonagh, T., Sullivan, F., Stahl, M. L., Somers, W. S.&Mosyak, L. (2004) Crystal Structure of the Wild-type von Willebrand Factor A1-Glycoprotein Ib Complex Reveals Conformation Differences with a Complex Bearing von Willebrand Disease Mutations *Journal of Biological Chemistry* Vol. 279 No. 22 pp. 23327-23334

Ehren, J., Govindarajan, S., Moron, B., Minshull, J.&Khosla, C. (2008) Protein engineering of improved prolyl endopeptidases for celiac sprue therapy *Protein Engineering, Design & Selection* Vol. 21 No. 12 pp. 699-707

Fant, F., Vranken, W. F.&Borremans, F. A. M. (1999) The three-dimensional solution structure of *Aesculus hippocastanum* antimicrobial protein 1 determined by ^{1}H nuclear magnetic resonance. *Proteins: Structure, Function, and Genetics* Vol. 37 No. 3 pp. 388-403

Fennell, B. J., Darmanin-Sheehan, A., Hufton, S. E., Calabro, V., Wu, L., Miller, M. R., Cao, W., Gill, D., Cunningham, O.&Finlay, W. J. J. (2010) Dissection of the IgNAR V Domain: Molecular Scanning and Orthologue Database Mining Define Novel IgNAR Hallmarks and Affinity Maturation Mechanisms *Journal of Molecular Biology* Vol. 400 No. 2 pp. 155-170

Fisher, M. A., McKinley, K. L., Bradley, L. H., Viola, S. R.&Hecht, M. H. (2011) *De Novo* Designed Proteins from a Library of Artificial Sequences Function in *Escherichia Coli* and Enable Cell Growth *PLoS ONE* Vol. 6 No. 1 pp. e15364

Gazarian, K. G., Gazarian, T., Hernodez, R.&Possani, L. D. (2005) Immunology of scorpion toxins and perspectives for generation of anti-venom vaccines *Vaccine* Vol. 23 No. 26 pp. 3357-3368

Gordon, S. B., Irving, G. R. B., Lawson, R. A., Lee, M. E.&Read, R. C. (2000) Intracellular Trafficking and Killing of Streptococcus pneumoniae by Human Alveolar Macrophages Are Influenced by Opsonins *Infection and Immunity* Vol. 68 No. 4 pp. 2286-2293

Gronwall, C.&Stahl, S. (2009) Engineered affinity proteins--Generation and applications *Journal of Biotechnology* Vol. 140 No. 3-4 pp. 254-269

Hara, S., Mukae, H., Sakamoto, N., Ishimoto, H., Amenomori, M., Fujita, H., Ishimatsu, Y., Yanagihara, K.&Kohno, S. (2008) Plectasin has antibacterial activity and no affect on cell viability or IL-8 production *Biochemical and Biophysical Research Communications* Vol. 374 No. 4 pp. 709-713

Hey, T., Fiedler, E., Rudolph, R.&Fiedler, M. (2005) Artificial, non-antibody binding proteins for pharmaceutical and industrial applications *Trends in Biotechnology* Vol. 23 No. 10 pp. 514-522

Holt, L. J., Herring, C., Jespers, L. S., Woolven, B. P.&Tomlinson, I. M. (2003) Domain antibodies: proteins for therapy *Trends in Biotechnology* Vol. 21 No. 11 pp. 484-490

Hosse, R. J., Rothe, A.&Power, B. E. (2006) A new generation of protein display scaffolds for molecular recognition *Protein Science* Vol. 15 No. 1 pp. 14-27

Huston, J. S., Levinson, D., Mudgett-Hunter, M., Tai, M. S., Novotna, J., Margolies, M. N., Ridge, R. J., Bruccoleri, R. E., Haber, E.&Crea, R. (1988) Protein engineering of antibody binding sites: recovery of specific activity in an anti-digoxin single-chain Fv analogue produced in Escherichia coli *Proceedings of the National Academy of Sciences of the United States of America* Vol. 85 No. 16 pp. 5879-5883

Jacobs, P. P.&Callewaert, N. (2009) N-glycosylation Engineering of Biopharmaceutical Expression Systems *Current Molecular Medicine* Vol. 9 No. 7 pp. 774-800

Jin, Z., Markley, J. L., Assadi-porter, F. M.&Hellekant, B. G. (2006) Protein sweetener. In., Wisconsin, Alumni Res Found (US), United States

Jones, D. S., Silverman, A. P.&Cochran, J. R. (2008) Developing therapeutic proteins by engineering ligand-receptor interactions *Trends in Biotechnology* Vol. 26 No. 9 pp. 498-505

Katzen, F., Chang, G.&Kudlicki, W. (2005) The past, present and future of cell-free protein synthesis *Trends in Biotechnology* Vol. 23 No. 3 pp. 150-156

Kosloski, M., Miclea, R.&Balu-Iyer, S. (2009) Role of Glycosylation in Conformational Stability, Activity, Macromolecular Interaction and Immunogenicity of Recombinant Human Factor VIII *The AAPS Journal* Vol. 11 No. 3 pp. 424-431

Kristensen, C., Kjeldsen, T., Wiberg, F. C., Schaffer, L., Hach, M., Havelund, S., Bass, J., Steiner, D. F.&Andersen, A. S. (1997) Alanine Scanning Mutagenesis of Insulin *Journal of Biological Chemistry* Vol. 272 No. 20 pp. 12978-12983

Kulkarni-Kale, U., Bhosle, S.&Kolaskar, A. S. (2005) CEP: a conformational epitope prediction server *Nucleic Acids Research* Vol. 33 No. suppl 2 pp. W168-W171

Landon, C., Barbault, F., Legrain, M., Menin, L., Guenneugues, M., Schott, V., Vovelle, F.&Dimarcq, J.-L. (2004) Lead optimization of antifungal peptides with 3D NMR structures analysis *Protein Science* Vol. 13 No. 3 pp. 703-713

Lay, F. T.&Anderson, M. A. (2005) Defensins - Components of the Innate Immune System in Plants *Current Protein & Peptide Science* Vol. 6 No. 1 pp. 85-101

LeBlanc, D. A.&London, R. E. (1997) Cleavage of the X-Pro Peptide Bond by Pepsin Is Specific for the trans Isomer? *Biochemistry* Vol. 36 No. 43 pp. 13232-13240

Leta Aboye, T., Clark, R. J., Craik, D. J.&Göransson, U. (2008) Ultra-Stable Peptide Scaffolds for Protein Engineering—Synthesis and Folding of the Circular Cystine Knotted Cyclotide Cycloviolacin O2 *ChemBioChem* Vol. 9 No. 1 pp. 103-113

Li, H.&d'Anjou, M. (2009) Pharmacological significance of glycosylation in therapeutic proteins *Current Opinion in Biotechnology* Vol. 20 No. 6 pp. 678-684

Li, Y.-J., Tan, Z.-Y.&Ji, Y.-H. (2000) The binding of BmK IT2, a depressant insect-selective scorpion toxin on mammal and insect sodium channels *Neuroscience Research* Vol. 38 No. 3 pp. 257-264

Lin, K. F., Lee, T. R., Tsai, P. H., Hsu, M. P., Chen, C. S.&Lyu, P. C. (2007) Structure-based protein engineering for α-amylase inhibitory activity of plant defensin *Proteins: Structure, Function, and Bioinformatics* Vol. 68 No. 2 pp. 530-540

Liu, Y. J., Cheng, C. S., Lai, S. M., Hsu, M. P., Chen, C. S.&Lyu, P. C. (2006) Solution structure of the plant defensin VrD1 from mung bean and its possible role in insecticidal activity against bruchids *Proteins: Structure, Function, and Bioinformatics* Vol. 63 No. 4 pp. 777-786 1097-0134

Lobo, D. S., Pereira, I. B., Fragel-Madeira, L., Medeiros, L. N., Cabral, L. M., Faria, J., Bellio, M., Campos, R. C., Linden, R.&Kurtenbach, E. (2007) Antifungal Pisum sativum Defensin 1 Interacts with Neurospora crassa Cyclin F Related to the Cell Cycle *Biochemistry* Vol. 46 No. 4 pp. 987-996

Malavasic, M., Poklar, N., Macek, P.&Vesnaver, G. (1996) Fluorescence studies of the effect of pH, guanidine hydrochloride and urea on equinatoxin II conformation *Biochimica et Biophysica Acta (BBA) - Biomembranes* Vol. 1280 No. 1 pp. 65-72

Michel-Reydellet, N., Woodrow, K.&Swartz, J. (2005) Increasing PCR Fragment Stability and Protein Yields in a Cell-Free System with Genetically Modified Escherichia coli Extracts *Journal of Molecular Microbiology and Biotechnology* Vol. 9 No. 1 pp. 26-34

Ming, D.&Hellekant, G. (1994) Brazzein, a new high-potency thermostable sweet protein from Pentadiplandra brazzeana B *FEBS Letters* Vol. 355 No. 1 pp. 106-108

Muir, E. M., Fyfe, I., Gardiner, S., Li, L., Warren, P., Fawcett, J. W., Keynes, R. J.&Rogers, J. H. (2009) Modification of N-glycosylation sites allows secretion of bacterial chondroitinase ABC from mammalian cells *Journal of Biotechnology* Vol. 145 No. 2 pp. 103-110

Mygind, P. H., Fischer, R. L., Schnorr, K. M., Hansen, M. T., Sonksen, C. P., Ludvigsen, S., Raventos, D., Buskov, S., Christensen, B., De Maria, L., Taboureau, O., Yaver, D., Elvig-Jorgensen, S. G., Sorensen, M. V., Christensen, B. E., Kjaerulff, S., Frimodt-Moller, N., Lehrer, R. I., Zasloff, M.&Kristensen, H.-H. (2005) Plectasin is a peptide antibiotic with therapeutic potential from a saprophytic fungus *Nature* Vol. 437 No. 7061 pp. 975-980

Nikkhah, M., Jawad-Alami, Z., Demydchuk, M., Ribbons, D.&Paoli, M. (2006) Engineering of β-propeller protein scaffolds by multiple gene duplication and fusion of an idealized WD repeat *Biomolecular Engineering* Vol. 23 No. 4 pp. 185-194

Odorico, M.&Pellequer, J.-L. (2003) BEPITOPE: predicting the location of continuous epitopes and patterns in proteins *Journal of Molecular Recognition* Vol. 16 No. 1 pp. 20-22

Perego, P., Gatti, L.&Beretta, G. L. (2010) The ABC of glycosylation *Nature Reviews Cancer* Vol. 10 No. 7 pp. 523-523

Pessi, A., Bianchi, E., Crameri, A., Venturini, S., Tramontano, A.&Sollazzo, M. (1993) A designed metal-binding protein with a novel fold *Nature* Vol. 362 No. 6418 pp. 367-369

Possani, L. D., Becerril, B., Delepierre, M.&Tytgat, J. (1999) Scorpion toxins specific for Na+-channels *European Journal of Biochemistry* Vol. 264 No. 2 pp. 287-300

Rajendra, W., Armugam, A.&Jeyaseelan, K. (2004) Neuroprotection and peptide toxins *Brain Research Reviews* Vol. 45 No. 2 pp. 125-141

Shimizu, Y., Inoue, A., Tomari, Y., Suzuki, T., Yokogawa, T., Nishikawa, K.&Ueda, T. (2001) Cell-free translation reconstituted with purified components *Nature Biotechnology* Vol. 19 No. 8 pp. 751-755

Sinek, J., Sanga, S., Zheng, X., Frieboes, H., Ferrari, M.&Cristini, V. (2009) Predicting drug pharmacokinetics and effect in vascularized tumors using computer simulation *Journal of Mathematical Biology* Vol. 58 No. 4 pp. 485-510

Skerra, A. (2007) Alternative non-antibody scaffolds for molecular recognition *Current Opinion in Biotechnology* Vol. 18 No. 4 pp. 295-304

Song, X., Wang, J., Wu, F., Li, X., Teng, M.&Gong, W. (2005) cDNA cloning, functional expression and antifungal activities of a dimeric plant defensin SPE10 from Pachyrrhizus erosus seeds *Plant Molecular Biology* Vol. 57 No. 1 pp. 13-20

Spelbrink, R. G., Dilmac, N., Allen, A., Smith, T. J., Shah, D. M.&Hockerman, G. H. (2004) Differential Antifungal and Calcium Channel-Blocking Activity among Structurally Related Plant Defensins *Plant Physiol.* Vol. 135 No. 4 pp. 2055-2067

Stec., B. (2006) Plant thionins-the structural perspective *Cellular and Molecular Life Sciences (CMLS)* Vol. 63 No. 12 pp. 1370-1385

Sun, Y. M., Liu, W., Zhu, R. H., Wang, D. C., Goudet, C.&Tytgat, J. (2002) Roles of disulfide bridges in scorpion toxin BmK M1 analyzed by mutagenesis *Journal of Peptide Research* Vol. 60 No. 5 pp. 247-256

Talbot, U. M., Paton, A. W.&Paton, J. C. (1996) Uptake of Streptococcus pneumoniae by respiratory epithelial cells *Infection and Immunity* Vol. 64 No. 9 pp. 3772-3777

Terras, F. R., Schoofs, H. M., De Bolle, M. F., Van Leuven, F., Rees, S. B., Vanderleyden, J., Cammue, B. P.&Broekaert, W. F. (1992) Analysis of two novel classes of plant antifungal proteins from radish (*Raphanus sativus L.*) seeds *Journal of Biological Chemistry* Vol. 267 No. 22 pp. 15301-15309

Thevissen, K., Ghazi, A., De Samblanx, G. W., Brownlee, C., Osborn, R. W.&Broekaert, W. F. (1996) Fungal Membrane Responses Induced by Plant Defensins and Thionins *Journal of Biological Chemistry* Vol. 271 No. 25 pp. 15018-15025

Thevissen, K., Kristensen, H.-H., Thomma, B. P. H. J., Cammue, B. P. A.&Francois, I. E. J. A. (2007) Therapeutic potential of antifungal plant and insect defensins *Drug Discovery Today* Vol. 12 No. 21-22 pp. 966-971

Thioulouse, J.&Lobry, J. R. (1995) Co-inertia analysis of amino-acid physico-chemical properties and protein composition with the ADE package *Computer Applications In The Biosciences* Vol. 11 No. 3 pp. 321-329

Thomma, B. P. H. J., Cammue, B. P. A.&Thevissen, K. (2003) Mode of Action of Plant Defensins Suggests Therapeutic Potential *Current Drug Targets - Infectious Disorsers* Vol. 3 No. pp. 1-8

Van Gaal, L., Mertens, I., Ballaux, D.&Verkade, H. J. (2004) Modern, new pharmacotherapy for obesity. A gastrointestinal approach *Best Practice & Research Clinical Gastroenterology* Vol. 18 No. 6 pp. 1049-1072

Van Walle, I., Gansemans, Y., Parren, P. W. H. I., Stas, P.&Lasters, I. (2007) Immunogenicity screening in protein drug development *Expert Opinion on Biological Therapy* Vol. 7 No. 3 pp. 405-418

Villemagne, D., Jackson, R.&Douthwaite, J. A. (2006) Highly efficient ribosome display selection by use of purified components for in vitro translation *Journal of Immunological Methods* Vol. 313 No. 1-2 pp. 140-148

Vita, C., Drakopoulou, E., Vizzavona, J., Rochette, S., Martin, L., Menez, A., Roumestand, C., Yang, Y.-S., Ylisastigui, L., Benjouad, A.&Gluckman, J. C. (1999) Rational engineering of a miniprotein that reproduces the core of the CD4 site interacting with HIV-1 envelope glycoprotein *Proceedings of the National Academy of Sciences of the United States of America* Vol. 96 No. 23 pp. 13091-13096

Vita, C., Roumestand, C., Toma, F.&Menez, A. (1995) Scorpion toxins as natural scaffolds for protein engineering *Proceedings of the National Academy of Sciences of the United States of America* Vol. 92 No. 14 pp. 6404-6408

Walters, D. E., Cragin, T., Jin, Z., Rumbley, J. N.&Hellekant, G. (2009) Design and Evaluation of New Analogs of the Sweet Protein Brazzein *Chemical Senses* Vol. 34 No. 8 pp. 679-683

Wang, C.-G., He, X.-L., Shao, F., Liu, W., Ling, M.-H., Wang, D.-C.&Chi, C.-W. (2001) Molecular characterization of an anti-epilepsy peptide from the scorpion Buthus martensi Karsch *European Journal of Biochemistry* Vol. 268 No. 8 pp. 2480-2485

Wang, L. K., Schwer, B.&Shuman, S. (2006) Structure-guided mutational analysis of T4 RNA ligase 1 *RNA* Vol. 12 No. 12 pp. 2126-2134

Wang, W., Lu, B., Zhou, H., Suguitan, A. L., Jr., Cheng, X., Subbarao, K., Kemble, G.&Jin, H. (2010) The Hemagglutin protein 158N glycosylation and receptor binding specificity synergistically affect antigenicity and immunogenicity of a live attenuated H5N1 A/Vietnam/1203/2004 vaccine virus in ferrets *Journal of Virology* Vol. 84 No. 13 pp. 6570-6577

Weideman, R. A., Bernstein, I. H.&McKinney, W. P. (1999) Pharmacist recognition of potential drug interactions *American Journal of Health-System Pharmacy* Vol. 56 No. 15 pp. 1524-1529

Weng, Z.&DeLisi, C. (2002) Protein therapeutics: promises and challenges for the 21st century *Trends in biotechnology* Vol. 20 No. 1 pp. 29-35

Whitney, C. G., Farley, M. M., Hadler, J., Harrison, L. H., Lexau, C., Reingold, A., Lefkowitz, L., Cieslak, P. R., Cetron, M., Zell, E. R., Jorgensen, J. H.&Anne Schuchat. (2000) Increasing Prevalence of Multidrug-Resistant Streptococcus pneumoniae in the United States *New England Journal of Medicine* Vol. 343 No. 26 pp. 1917-1924

Wijaya, R., Neumann, G. M., Condron, R., Hughes, A. B.&Polya, G. M. (2000) Defense proteins from seed of Cassia fistula include a lipid transfer protein homologue and a protease inhibitory plant defensin *Plant Science* Vol. 159 No. 2 pp. 243-255

Wilce, M. C. J., Bond, C. S., Dixon, N. E., Freeman, H. C., Guss, J. M., Lilley, P. E.&Wilce, J. A. (1998) Structure and mechanism of a proline-specific aminopeptidase from Escherichia coli *Proceedings of the National Academy of Sciences of the United States of America* Vol. 95 No. 7 pp. 3472-3477

Wong, J. H.&Ng, T. B. (2005) Sesquin, a potent defensin-like antimicrobial peptide from ground beans with inhibitory activities toward tumor cells and HIV-1 reverse transcriptase *Peptides* Vol. 26 No. 7 pp. 1120-1126

Wu, C., Zhang, X., Tian, Y., Song, J., Yang, D., Roggendorf, M., Lu, M.&Chen, X. (2010) Biological significance of amino acid substitutions in hepatitis B surface antigen (HBsAg) for glycosylation, secretion, antigenicity and immunogenicity of HBsAg and hepatitis B virus replication *Journal General Virology* Vol. 91 No. 2 pp. 483-492

Yang, Y.-F., Cheng, K.-C., Tsai, P.-H., Liu, C.-C., Lee, T.-R.&Lyu, P.-C. (2009) Alanine substitutions of noncysteine residues in the cysteine-stabilized αβ motif *Protein Science* Vol. 18 No. 7 pp. 1498-1506

Yang, Y.-F.&Lyu, P.-C. (2008) The Proteins of Plant Defensin Family and their Application Beyond Plant Disease Control *Recent Patents on DNA & Gene Sequences* Vol. 2 No. 3 pp. 214-218

Zasloff, M. (2002) Antimicrobial peptides of multicellular organisms *Nature* Vol. 415 No. 6870 pp. 389-395

Zhao, A., Xue, Y., Zhang, J., Gao, B., Feng, J., Mao, C., Zheng, L., Liu, N., Wang, F.&Wang, H. (2004) A conformation-constrained peptide library based on insect defensin A *Peptides* Vol. 25 No. 4 pp. 629-635

Zhao, Q., Chae, Y. K.&Markley, J. L. (2002) NMR Solution Structure of ATT$_p$, an *Arabidopsis thaliana* Trypsin Inhibitor *Biochemistry* Vol. 41 No. 41 pp. 12284-12296

Zhong, L.&Johnson, W. C. (1992) Environment affects amino acid preference for secondary structure *Proceedings of the National Academy of Sciences of the United States of America* Vol. 89 No. 10 pp. 4462-4465

Zhu, Q., Liang, S., Martin, L., Gasparini, S., Menez, A.&Vita, C. (2002) Role of Disulfide
    Bonds in Folding and Activity of Leiurotoxin I: Just Two Disulfides Suffice
    Biochemistry Vol. 41 No. 38 pp. 11488-11494
Zhu, S., Gao, B.&Tytgat, J. (2005) Phylogenetic distribution, functional epitopes and
    evolution of the CSαβ superfamily *Cellular and Molecular Life Sciences* Vol. 62 No. 19
    - 20 pp. 2257-2269
Zunino, P., D'Angelo, C., Petrini, L., Vergara, C., Capelli, C.&Migliavacca, F. (2009)
    Numerical simulation of drug eluting coronary stents: Mechanics, fluid dynamics
    and drug release *Computer Methods in Applied Mechanics and Engineering* Vol. 198
    No. 45-46 pp. 3633-3644

4

Application of Liposomes for Construction of Vaccines

Jaroslav Turánek[1], Josef Mašek[1], Milan Raška[2] and Miroslav Ledvina[3]
[1]Veterinary Research Institute,
[2]Palacky University,
[3]Institute of Organic Chemistry and Biochemistry,
Czech Republic

1. Introduction

Vaccinology as a scientific field is undergoing a dramatic development. Never before such sophisticated techniques and in-depth knowledge of immunological processes have been at hand to exploit fully the potential of protecting from as well as curing diseases through vaccination. In spite of great successes like eradication of smallpox in the 1970s and poliomyelitis elimination from all but six countries in the world (two important milestones in the medical history), new challenges have arisen to be faced. Rapidly changing ecosystems and human behaviour, an ever-increasing density of human and farmed animal populations, a high degree of mobility resulting in rapid spreading of pathogens in infected people and animals, new contacts between human and animals in endemic areas, poverty, and war conflicts in the third world, and many other factors contribute to the more frequent occurrence and rapid dissemination of new diseases. Three diseases that most heavily afflict global health are AIDS, tuberculosis, and malaria. As an example of new viral pathogens we can mention Ebola virus, SARS-coronavirus, or new strains of influenza virus (Wack & Rappuoli, 2005).

Among re-emerging diseases of the past few years, diphteria and cholera should be mentioned. Moreover, multi drug-resistant bacteria frequently occur as a result of overdosing on antibiotics. One of the most important future challenges will be to respond promptly to emerging diseases such as those mentioned above. Rapid sequencing of the genome of the pathogen implicated the speed of the development of diagnostic tools as well as the identification and expression of recombinant targets for vaccines and therapeutic agents development (Stadler et al., 2003). Immunotherapy of cancer represents a special field, where anticancer vaccines could be powerful weapons/tools for long-term effective treatment.

The progress in the vaccine development is closely related with the progress in immunology and molecular biology. A new term "Reverse vaccinology" was proposed by Rappuoli (Rappuoli, 2000) to specify a complex genome-based approach in the vaccine development. Unlike the conventional approach that requires a laborious process of a selection of individual components important for the induction of protective immune response, reverse vaccinology offers a possibility to use genomic information derived from *in silico* analysis of the sequenced organisms. This approach can significantly reduce the time necessary to identify the antigens for the development of a candidate vaccine and enables a systematic identification of all potential antigens of pathogens including those which are difficult or currently impossible to culture. Of course, this approach is limited to the identification of protein or glycoprotein antigens, omitting such important vaccine components as polysaccharides and glycolipids. Nevertheless, reverse vaccinology can enable scientists to systemically classify the potential protective antigens, thereby helping to improve the existing vaccines and to develop efficient preparations against virtually any pathogen that has had its genome sequence determined.

As regards the process of activation of the immune system to produce an adaptive immune response, it is generally observed that the antigen by itself may not be adequate as a stimulating agent. Many potential antigens have no apparent immunizing activity at all when tested alone. In general, seamy side of pure recombinant protein antigens and synthetic peptide antigens is their poor immunogenicity. Therefore, potent adjuvants are required for highly purified antigen-based vaccines to be effective.

2. Adjuvants

2.1 Toll-like receptors and pathogen-associated molecular patterns

The word *adjuvant* is derived from the Latin root of *adjuvare*, which means to help. Thus, an adjuvant can be defined as any product which increases or modulates the specific humoral or cellular immune response against an antigen. The interaction between the innate and the adaptive immune responses is paramount in generating an antigen-specific immune response. The initiation of innate immune responses begins with the interaction of pathogen-associated molecular patterns (PAMPs) on the pathogen side with pattern-recognizing receptors (PRR) such as Toll-like receptors (TLRs) on the host cells involved in the innate immunity (e.g., dendritic cells). A major functional criterion commonly used for the evaluation of various new adjuvants involves their ability to stimulate the innate immunity cells. This would include engaging and other PRRs and the co-receptors and intracellular adaptor signalling proteins with which they are associated. PAMPs and their derivatives are utilized by adjuvant developers to harness the power of innate immunity to channel the immune response in a desired direction.

Based on the identification of several TLRs and PAMPs recognized by them, various PAMP agonists were tested as adjuvans. Examples of TLR-PAMP specific interaction include bacterial or viral unmethylated immunostimulating CpG oligonucleotides interacting with TLR9, liposaccharide and its component monophosphoryl lipid A (MPLA) interacting with TRL4. These two types of adjuvants are in advanced stage of testing in clinical trials and some already licensed vaccines contain MPLA in liposomal form. Further liposomal or lipid-based particle formulations of both CpG and MPLA are under development and testing.

2.2 Muramyl dipeptide and other muropeptides

Very specific group of PAMPs is represented by peptidoglycans (PGN). Both Gram-positive and Gram-negative bacteria contain PGN which consists of numerous glycan chains that are cross-linked by oligopeptides. These glycan chains are composed of alternating N-acetylglucosamine (GlcNAc) and N-acetylmuramic acid (MurNAc) with the amino acids coupled to the muramic acid. Muropeptides are breakdown products of PGN that bear at least the MurNAc moity and one amino acid (Traub et al., 2006). One of the prominent muropeptides is muramyl dipeptide (MDP), which is known since the 1970s.

Recently, the molecular bases for MDP recognition and subsequent stimulation of the host immune system have been uncovered. Myeloid immune cells (monocytes, granulocytes, neutrophils, and also DCs) possess two types of intracellular receptor for MDP, namely NOD2 and Cryopyrin (inflammasome-NALP-3 complex) (Agostini et al., 2004; Girardin and Philpott, 2004; McDonald et al., 2005). These two receptors recognize MDP/MDP analogues minimal recognition motifs for bacterial cell wall peptidoglycans (Girardin & Philpott, 2004; McDonald et al., 2005). NOD2 is also expressed in specialised epithelial cells, Paneth cells, localised in crypts of Lieberkün, which are producers of antimicrobial peptides having direct antimicrobial activity together with signalling functions within the immune system. Induction of an innate immune response against *Cryptosporidium parvum* infection by the liposomal preparation of lipophilic norAbuGMDP was demonstrated by us in newborn goats (Turanek et al., 2005) and this data is supportive of the present view of the role of MDP recognition in inducing both specific and innate immune responses.

Here GMDP abbreviates N-acetylglucosaminylmuramyl dipeptide. Another recently reported sensor of MDP is Cryopyrin (also known as CIAS1 and NALP3), which is a member of the NOD-LRR family (Agostini et al., 2004). Cryopyrin is a part of the inflammasome complex that is responsible for processing caspase-1 to its active form. Caspase-1 cleaves the precursors of interleukin IL-1β and IL-18, thereby activating these proinflammatory cytokines and promoting their secretion. IL-1β is known to be a strong endogenous pyrogen induced by MDP. We showed that norAbu-MDP analogues were not pyrogenic even at a high concentration, much higher than the concentrations used for vaccination. We supposed that the modification introduced into the structure of MDP to get norAbuMDP analogues had not changed their affinity to NOD2 but had substantially decreased the affinity to cryopyrin. This hypothesis is in accordance with our data on pyrogenicity and is being currently tested in appropriate *in vitro* models. In addition, murabutide, another nonpyrogenic derivative of MDP was shown not to be able to induce detectable level of IL-1ß in sera of treated volunteers (Darcissac et al., 2001).

The expression of NOD2 in dendritic cells is of importance with respect to the application of MDP analogues as adjuvants. Nanoparticles like liposomes are able to provide a direct co-delivery of a danger signal (e.g., MDP) together with the recombinant antigen and therefore to induce an immune response instead of an immune tolerance. This is especially important for weak recombinant antigens or peptide antigens. Clearly, the recognition of MDP by DCs is crucial for the application of MDP analogues as adjuvants. Although the immuno-stimulatory effects of MDPs have been described for over three decades, the process of molecular recognition and binding of MDP/MDP analogues to NOD2 and cryopyrin receptors remains unclear. Within the cell, MDP/MDP analogues trigger intracellular signalling cascades that culminate in the transcriptional activation of inflammatory mediators such as the nuclear transcription factor NF-кB. The biological effects of muramyl peptides have been described for over three decades. The mechanism underlying their internalization of MDP to the cytosol, where it is sensed by NOD2 and cryopyrin, remains unclear. Liposomes probably play the role of efficient carriers for MDP and its analogues on the pathway from extracellular milieu into the cytosol, where they trigger intracellular signalling cascades that culminate in transcriptional activation of inflammatory mediators such as the nuclear transcription factor NF-кB pathway. In case of liposomal formulation of various MDP analogues, the relevant intracellular pharmacokinetics, molecular recognition and binding affinity towards NOD2 and Cryopyrin remain to be determined. Such differences found for various MDP analogues are responsible for their various biological activities (e.g., pyrogenicity, ability to induce the innate immune response etc.) and, therefore, could be utilised for a precise tuning of the intensity and type of immune response. Since the discovery and first synthesis of MDP, about one thousand various derivatives of MDP have been designed, synthesised, and tested to develop an appropriate drug for an immunotherapeutic application that would be free of the side effect exerted by MDP. The main side effects of MDP are pyrogenicity, rigor, headache, flue-like symptoms, hypertension etc. Only several preparations reached the stage of clinical testing and only Mifamurtide (Fig. 1) was approved for the treatment of osteosarcoma.

2.2.1 Mifamurtide (MTP-PE)

In pyrogenicity test in rabbits, pyrogenic activity of Mifamurtide i.v. was comparable to that of MDP. In several studies with cancer patients refractory to standard therapy, infused with liposomal Mifamurtide at a dose range of $0.01 - 1.8$ mg/m^2/dose, dose-dependent fever (in common about 70% of patients) and rigor (about 50% of patients) were the most prominent from a number of acute systemic toxicities (Creaven et al., 1990).

Mifamurtide was also assayed as an additional immunomodulator in an MF59-adjuvanted influenza virus vaccine (Keitel et al., 1993) and HIV-1 vaccine (Keefer et al., 1996); systemic symptoms including fever, chill, and nausea made these vaccines unsuitable for clinical use. Today, the main interest lies in clinical trials for liposomal Mifamurtide as a component of three-drug chemotherapy of osteosarcoma (Anderson P.M, 2006; Anderson et al., 2010).

2.2.2 Romurtide (Muroctasine)

In healthy volunteers, s.c. administration of Romurtide (Fig. 1) at a dose of 200 µg induced - besides local pain and redness - an approximately $1°$ C increase in body temperature with great individual variability in the course of pyrogenicity curves; normalization occurred within 48 hours (Ichihara et al., 1988). Fever accompanied by chill and headache was also the most common adverse reaction in cancer patients treated with Romurtide at a dose range of $100 - 400$ µg/dose s.c. for the restoraion of haemopoiesis after chemotherapy and/or radiotherapy (Tsubura et al., 1988) (Azuma & Seya, 2001; Tsubura et al., 1988).

2.3 nor-Muramyl glycopeptides

We found that a combination of structural modifications both in the saccharide and peptide moiety of MDP and GMDP molecules leads to significant suppression or elimination of pyrogenicity and potentiation of immune-stimulatory activity (Fig. 2). The substitution of muramic acid with normuramic acid and L-alanine with L-2-aminobutyric acid has lead, in the case of the norAbu-MDP molecule, to a decrease of pyrogenicity and, at the same time, to the potentiation of immunoadjuvant activity. If the same structural change is carried out in GMDP molecule, a non-pyrogenic and highly immunoadjuvant analog, norAbu-GMDP is obtained. Furthermore, it has been demonstrated that by the introduction of bulky lipophilic residues into the molecules of these analogs, immunomodulatory activity can be effectively profiled, while the favourable pharmacological parameters of the parent structures are retained. These facts motivated our aims to design and prepare the new groups of lipophilic analogs of norAbu-MDP and norAbu-GMDP, which differ in the character and topology of the lipophilic residue. We primarily aimed to modify their immunopharmacologic parameters. norAbu-MDP-Lys(L18), i.e. MT05, belongs to them. In accordance with our premise, all the new compounds were nonpyrogenic (rabbit test), and the character and topology of the lipophilic residue had a significant effect on their immunologic parameters. As an example, the structural differences between norAbu-MDP-Lys(18) (MT05) and Romurtide, which influenced the profile of the effects and lead to the elimination of pyrogenicity, are depicted in Fig. 3. (Ledvina M., Turánek J., Miller A.D., Hipler K.: Compound (Adjuvants): PCT appl., WO 2009/11582 A2, 2009.)

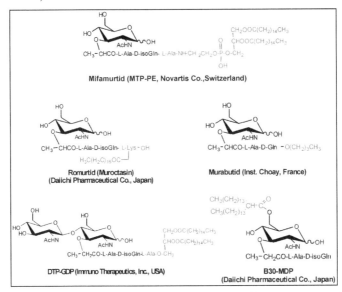

Fig. 1. MDP derivatives developed by various pharmaceutical companies as adjuvants and immunotherapeutics.

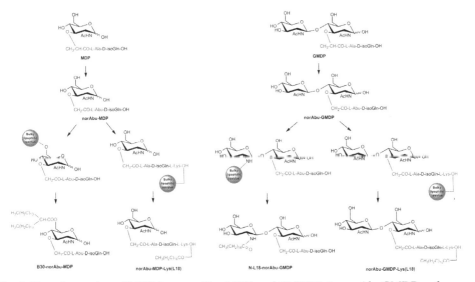

Fig. 2. Transformation of MDP into norAbu-MDP and GMDP into norAbuGMDP and formulae of their hydrophobised derivatives suitable for development of lipid-based adjuvants.

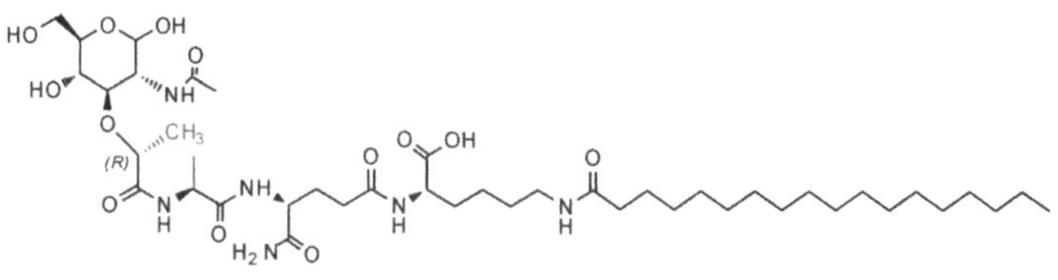

Romurtide (MDP-Lys(L18))

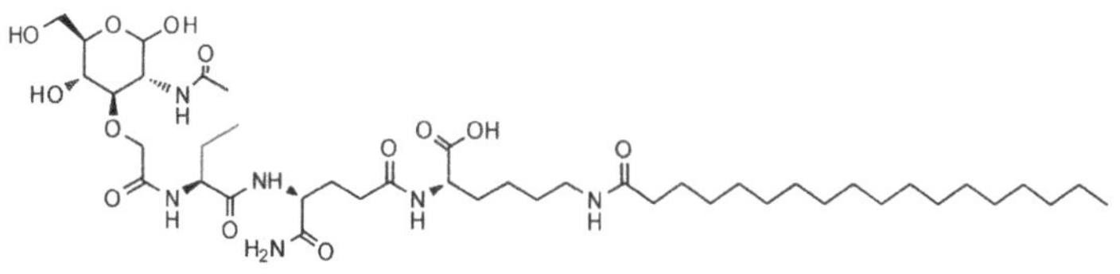

MT-05 (MDP-Lys(L18))

Fig. 3. Structural differences between Romurtide and MT05.

3. Liposomes

3.1 General characterisation of liposomes

Liposomes, membrane-like spherical structures consisting of one or more concentric lipid bilayers enclosing aqueous compartments were first formulated and described by Alex Bangham in 1965 (Bangham et al., 1965), and have become a useful tool and model in various areas of science. Liposomes represent the oldest and the most explored nano- and micro systems for biological studies on model membranes and for medical applications, especially for drug formulations, because they eliminate or suppress organ-specific toxic side-effects of various drugs (Allen, 1997). Through their 46-year history, liposomes have been approved as suitable delivery systems for applications ranging from cosmetics and dermatology, anti-infection and anticancer therapy and diagnostics up to human as well as veterinary vaccines (Gregoriadis, 1995).

Liposomes are classified in terms of number of bilayers enclosing the sequestered aqueous volume as follows: unilamellar, oligolamellar, and multilamellar. Unilamellar vesicles can be further divided into small unilamellar vesicles (SUVs) with a large curvature, and large unilamellar vesicles (LUVs) with a low curvature and hence, with properties similar to those of a flat surface. Multilamellar vesicles (MLVs) are liposomes that represent a heterogenous group in terms of size and morphology (Cullis et al., 1987). Lipid composition, size and morphology are variables determining the fate of liposomes in biological milieu; therefore, the selection of suitable method for the preparation of liposomal drugs and vaccines is of importance in respect to subsequent animal experiments and future successful marketing of the product. Schematic structures of various types of liposomes as well as realistic picture obtained by cryoelectron microscopy are presented in Fig. 4.

3.2 Liposome-based vaccines

The use of liposomes as vaccine adjuvants was first described by Allison and Gregoriadis in 1974 (Allison & Gregoriadis, 1974). Since that time, numerous studies were performed and proved that liposomes can be used to enhance the immune response towards a large variety of peptide and protein antigens derived from various microbial pathogens as well as tumours.

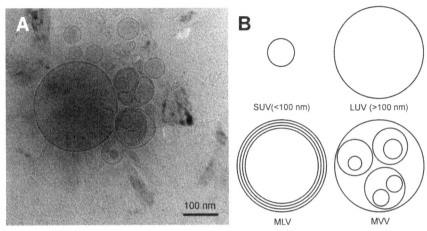

A) Photograph of various liposomal structures by cryoelectron microscopy. B) Schema of types of liposomes: SUV – small unilamellar vesicle, LUV - large unilamellar vesicle,, MLV – multilamellar vesicle, MVV – multivesicular vesicle

Fig. 4. Schematic representation of various morphological classes of liposomes and their real image obtained by cryoelectron microscopy.

The potential for the participation of liposome-based recombinant vaccines on the human and veterinary vaccine market is very promising (Adu-Bobie et al., 2003). Liposomal vaccines have been around for about 30 years and plenty of liposome variants have been developed; some of them with evident immune-stimulating properties and an attractive safety profile which resulted in registered products on the market or preparations in advanced stages of clinical testing. Liposomal hepatitis A vaccine is the first formulation of liposomes to become licensed for clinical use in humans (Gluck et al., 1992) (Hepatitis A - HepA, Epaxal http://www.crucell.com/Products-Epaxal). Epaxal liposomes contain influenza hemagglutinin protein which facilitates their binding and endocytosis by specific receptor on antigen presenting cells. Such forms of liposomes are called virosomes. Liposomes represent almost ideal carrier system for the preparation of synthetic vaccines due to their biodegradability and versatility as regards the incorporation of quite a number of various molecules having different physico-chemical properties (the size of the molecule, hydrophilicity or hydrophobicity, the electric charge).

The molecules and antigens can be either sterically entrapped into the liposomes (the internal aqueous space), or embedded into the lipid membrane (e.g., membrane-associated proteins/antigens) by hydrophobic interactions. Further, they can be attached to either the external or the internal membrane by electrostatic, covalent or metallo-chelating interactions. It is possible to encapsulate simultaneously various compounds into the liposomes: hydrophilised/lipophilised adjuvants (e.g., MPL A, CpG oligonucleotides, MDP and its analogues), soluble or membrane protein antigens, and ligands for the targeting to specific receptors on the antigen-presenting cells. Further, liposomes can be coated with mucoadhesive biopolymers, or undergo surface-charge modifications (e.g., by cationic lipids) (Altin & Parish, 2006).

As a great advantage, liposomes can be used for the preparation of self-assembling hybrid supramolecular nanosystems such as proteoliposomes, which can combine liposomal nanoparticles with suitable immunopotentiating/adjuvant molecules (e.g., MPL A, CpG oligonucleotides, MDP, and its analogues).

3.3 Liposomes as antigen carriers

Structural diversity of liposomes permits tailoring of liposome-based vaccines to obtain an optimal adjuvant effect for a particular antigen. Their safety profiles and ability to induce an immune response makes them likely to be included in vaccine formulations. Liposomal formulations offer several major advantages. They can (1) prevent degradation of the delivered antigens and adjuvants; (2) allow membrane proteins to reconstitute and preserve their antigen structure; (3) increase the antigenic effect of weak immunogens; (4) target antigen-presenting cells (APCs) and direct the antigen to MHCI or MHCII presentation; (5) contain the antigen and adjuvant molecules in one particle, thus functioning simultaneously as a delivery system and a vaccine adjuvant; and (6) reduce the antigen and adjuvant doses required for an immune response and controlled release. Therefore, liposomes provide a safe and effective platform for construction of subunit vaccines.

With respect to the physico-chemical nature of liposomes and phospholipid bilayers, the liposomes represent one of the most versatile structures for the preparation of drug delivery systems. Both hydrophobic and hydrophilic protein or peptide antigens can be associated with liposomes. Generally, antigens can be associated with liposomes in two ways and it is known that the encapsulated liposomal antigens induce a different immune response than the surface-linked antigens in both humoral (Shahum & Therien, 1988) and cell-mediated immunity (Fortin et al., 1996). If entrapped into the internal aqueous space of a liposome, the protein or peptide antigen is protected against proteolytic degradation and the antigen clearance is decreased. On the other hand, the liposomal membrane represents a barrier restricting the interaction of the antigen with and its recognition by B-cells. Especially, the stable multilamellar liposomes were found to be low immunogenic (Shek & Heath, 1983) and the antibody response reached is low or absent when the liposomes are made of lipids with a high transition temperature; in other words, when they are composed of saturated phospholipids (Gregoriadis et al., 1987). These liposomes are very stable in body fluids as well as in digestive tract and prevent a release of the entrapped antigen. Also the interaction of the encapsulated antigens with B-cells is limited. The fluidity of liposomes was found to be an important parameter also for the immune response towards a surface-linked antigen. Again, more fluidic liposomes composed of unsaturated phospholipids are more efficiently phagocytosed by APC and induce a one order of magnitude higher immune response than rigid liposomes composed of saturated phospholipids (Uchida & Taneichi, 2008).

Liposomes are potentially very useful for the construction of vaccination systems given their facile biodegradability and versatility as carriers for varieties of molecules having different physico-chemical properties (such as size, hydrophilicity, hydrophobicity, or net electrical charge). Liposomes also offer the possibility to associate or entrap simultaneously more than one type of molecules. Of particular interest to us has been the co-association of hydrophilic or lipophilic adjuvants (e.g., monophosphoryl lipid A [MPL A], CpG oligonucleotides, muramyl dipeptide (MDP), and/or MDP lipophilic analogues) with soluble or membrane protein antigens or ligands for the targeting of specific receptors on antigen-presenting cells. The molecules and antigens can be either sterically entrapped into the liposomes (the internal aqueous space), or embedded into the lipid membrane (e.g., membrane-associated proteins/antigens) by hydrophobic interactions (Fig. 5). The ligands for the targeting to specific receptors on the antigen-presenting cells can significantly enhance the intensity of the immune response (Altin & Parish, 2006). For the mucosal application, the liposomes can be coated with mucoadhesive biopolymers or modified with surface-charge modifiers (e.g., cationic lipids). In this way, liposomes become a versatile platform that represents a real multifunctional vaccination carrier.

The importance of liposomes for the effective co-administration of adjuvants could be demonstrated using MDP as an example. MDP has a weak immunoadjuvant activity in aqueous solution due to its rapid excretion into urine. Appropriate formulations of hydrophilic MDP in "water in oil" emulsions (Parant et al., 1979) or liposomes were used to harness its full adjuvant potenital (Tsujimoto et al., 1986). Some lipophilic derivatives of MDP like B30-MDP and MDP-Lys (L18) were synthesised and tested as adjuvants for recombinant hepatitis B surface antigen (Tamura et al., 1995) or influenza surface antigens hemagglutinin and neuraminidase (Nerome et al., 1990). We used new synthetic nonpyrogenic lipophilic analogues of norAbu-MDP modified at a peptide part by hydrophobic ligands (Fig. 2) and these well defined synthetic molecules were used for the first time in combination with metallochelating liposomes to construct an experimental recombinant vaccine. Surprisingly, we have found that at certain surface density of lipophilic analogues of norAbu-MDP (about 5 mol % of total liposomal lipid), the liposomes are promptly recognised and phagocytosed by human dendritic cells. The phagocytosis is about one order of magnitude higher than that of proteoliposomes or liposomes lacking norAbu-MDP adjuvant. This finding implicates an existence of receptors on dendritic cells, which can recognise some molecular pattern formed by the hydrophilic part of norAbu-MDP exposed on the liposomal surface (illustration of this phenomenon is in Fig. 12B).

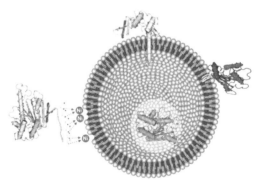

a) physical entrapment inside the liposome (blue protein); b) reconstitution of membrane protein in lipid bilayer via hydrophobic transmembrane domain (yellow protein); c) anchoring of lipidised protein onto liposomal surface or attachment of recombinant protein onto the liposomal surface by covalent bond using activated lipids (red protein); d) attachment of recombinant protein onto the liposomal surface by non-covalent bond using metallochelating lipids

Fig. 5. Association of protein antigen with liposome.

3.4 Methods of liposome preparation

The laboratory and industrial procedures for the liposome preparation have been established and liposomes have been approved by FDA for biomedical applications. A variety of procedures for the preparation of various types of liposomes has been developed and reported in several reviews and monographs (Gregoriadis, 1992; Woodle &

Papahadjopoulos, 1988). To classify these methods, they were arranged in three categories: 1) mechanical dispersion methods such as hand shaking or vortexing, sonication, and high pressure homogenisation; 2) detergent-solubilizing dispersion methods including solubilized lecithin dispersion with sodium cholate or octylglucoside; and 3) solvent dispersion methods such as ethanol injection, ether infusion, and reverse-phase evaporation. These primary processes can be linked with secondary processes such as high-pressure homogenisation or extrusion through polycarbonate filters of various pore size, which are easy ways to prepare liposomes of a desired size and morphology (Barnadas-Rodriguez & Sabes, 2001; Berger et al., 2001; Cullis, 1987; Hope et al., 1985; Perrett et al., 1991; Turanek, 1994; Woodle & Papahadjopoulos, 1988; Schneider et al., 1995).

The stability of proteins is limited and not all the methods are useful for the preparation of proteoliposomes, especially if the protein is to be entrapped inside the liposome. The detergent dilution method is characterized by very mild conditions during the process and is suitable for the reconstitution of membrane proteins. Because many recombinant proteins tend to precipitate, this method is also useful to work with these protein antigens. Next paragraph describes in detail a modified detergent dilution method for the preparation of proteoliposomes.

3.5 Preparation and characterization of metallochelating liposomes

Because the preparation of metallochelating liposomes represents a post-forming modification of liposomes, it avoids a denaturation of proteins owing to the process used for the liposome production. Therefore, nearly all the methods mentioned above could be used to prepare more or less monodisperse liposomes. Here we describe in brief a modification of the detergent removal method, which is suitable for the preparation of very monodisperse unilamellar liposomes that are useful for structural studies by various techniques (e.g., TEM, dynamic light scattering, and gel permeation chromatography).

When essentially unilamellar monodisperse liposomes of spherical shape are needed (which is a prerequisite for a precise monitoring of the proteoliposome formation by dynamic light scattering), the detergent removal method is preferred to the other methods. The method is based on the transformation of phospholipid micelles stabilized by detergent with high critical micellar concentration (CMC) (e.g., cholate) to desk micelles and finally to vesicles during the process of detergent removal (Zumbuehl & Weder, 1981) (Fig. 6). The mild conditions provided by this method are advantageous for the preparation of proteoliposomes, especially for the reconstruction of membrane proteins (Rigaud & Levy, 2003) like viral or bacterial antigens or recombinant his-tagged proteins that are often prone to precipitation.

There are many variants of the detergent removal method, e.g., the dilution of the solution of mixed micelles, gel permeation chromatography, a simple dialysis or the controlled one in a special apparatus, cross-flow filtration, adsorption on beads, etc. (Schubert, 2003). The application of the flow-through ultrafiltration cell represents a new approach to the detergent removal method (Masek et al., 2011a). The linkage of the cell with systems like FPLC facilitates automation of the whole procedure and manipulation with the sample. The full control over the dialysis rate and the removal of the undesired residua (e.g., detergent, organic solvents, protein solubilizers) is ensured and various steps like an addition of required components through an injection valve during various stages of the liposome formation are easy to perform without breaching the sterile conditions. In this case, the sterile filter inserted in front of the cell inlet ensures that the sterility is kept during the whole process (Fig. 7). The low dead volume of the cell is of great importance for the preparation of liposomes and proteoliposomes in small laboratory scale. However, this arrangement enables also very easy up-scaling of the whole technology. A precise control over the rate of the detergent removal yields a final liposomal preparation of high monodispersity (PDI within the range of 0.05 - 0.06), which is shown to be reached routinely (Fig. 8). This monodispersity is better than those obtained by the dialysis method performed in the dialysis bags or slides (produced by Pierce) (PDI ≈ 0.08-0.12).

The size of the mixed micelles (≈5-6 nm; see Fig. 8) used by us for the preparation of liposomes is in good correlation with the Small's mixed micellar model proposing the structure of a small phospholipid bilayer disc stabilised at its hydrophobic edges by the molecules of cholate (Small, 1971; Schubert, 2003).

The process of the formation of the monodisperse liposomes is in good accordance with the proposed kinetic model of the micelle-vesicle transition based on a rapid formation of disk-like intermediate micelles followed by a growth of these micelles up to their critical size and their subsequent closure to form vesicles. The final size of the liposomal preparation could be controlled by ionic strength of the buffer used for the preparation of the micelles (Fig. 9). An increase of the NaCl concentration reduces CMC of cholate and shields the negative charge of the mixed micelles. These two factors are responsible for the formation and stabilisation of the large discoid bilayer micelles that are transformed into the larger liposomes (Schubert, 2003).

Various additives like bilayer stabilising sugars (e.g., sucrose) or recombinant protein solubilizers (e.g., urea, guanidine) are compatible with this method and can shift the size of the liposomes into the required range (Walter et al., 2000). Some recombinant proteins (e.g., circovirus envelope protein), which tend to precipitate in the absence of stabilizing buffers (imidazole and urea stabilizing buffer) were successfully linked onto metallochelating liposomes by one-step procedure based on the addition of the protein into the mixed micelle solution prepared in protein stabilizing buffer and transforming into proteoliposomes during the ultrafiltration procedure (Turánek, unpublished results).

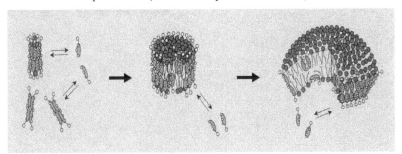

Adapted: R. Schubert, Methods Enzymol. 367 (2003) 46–70.

Fig. 6. Principle of detergent removal method and formation of liposomes from mixed bile salt-phospholipid micelles.

Small mixed micelles are fused in disc phospholipid micelles stabilized at edges by detergent. Further removal of detergent induces formation of large disc micelles which spontaneously vesiculate after reaching a critical size.

The transformation of micelles into liposomes during the ultrafiltration removal of cholate is a critical step affecting the final quality of liposomes. This process is easy to be monitored by DLS. The removal of cholate induced a formation of disc micelles, which was reflected by an increase of the micelle size and eventually by a formation of liposomes (Fig. 10). The process of liposome formation had been completed before the CMC of cholate was reached, as shown by the dashed vertical line. This line divides the flow-through volume axis into the left part, where micelles do predominantly exist and are transformed into liposomes, and the right part, where liposomes represent the main lipid form, while the residual detergent and other low molecular weight contaminants (e.g., traces of ethanol or tetrahydrofuran used to solubilize the lipids) are continuously removed by the process of ultrafiltration.

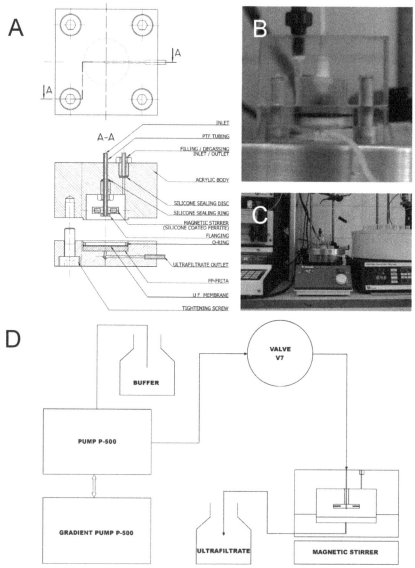

(A) Schematic illustration of the ultrafiltration cell. (B) Photograph of the ultrafiltration cell in detail (pink: LR-PE-labelled liposomes inside the cell). (C) Schematic illustration of the linkage of the ultrafiltration cell with the FPLC system. (D) Photograph of the system

Fig. 7. System for preparation of liposomes by removal of detergent using ultrafiltration.

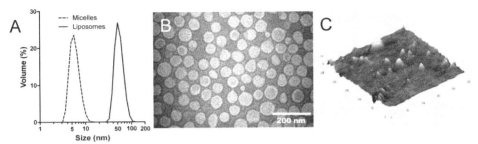

A) Size distribution of micelles and liposomes. The hydrodynamic diameters of the micelles and liposomes were determined by dynamic light scattering instrument NanoSizer NS (Malvern, UK) at 25 °C. Silica cuvette of 45-µl volume (Hellma, Germany) was used. (B) TEM micrograph of monodisperse liposomal preparation. (C) AFM micrograph of monodisperse liposomal preparation

Fig. 8. Size distribution of micelles and metallochelating liposomes analysed by dynamic light scattering and visualization of liposomes by TEM and AFM.

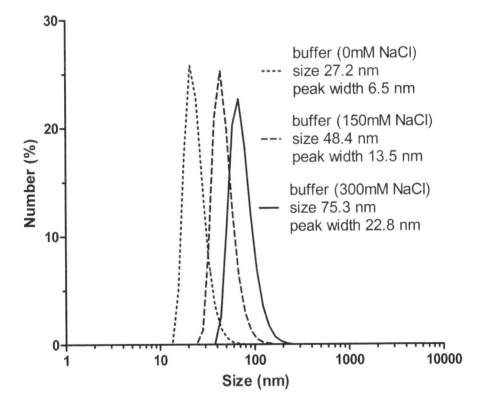

Fig. 9. Effect of ionic strength on the size distribution of liposomes prepared by detergent removal method.

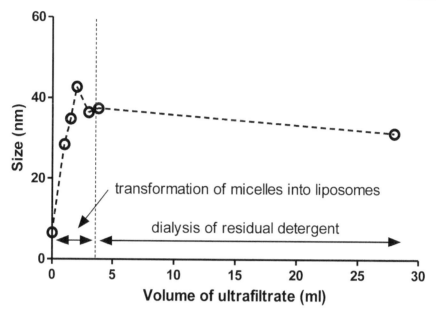

Fig. 10. Transformation of micelles into liposomes during ultrafiltration monitored by DLS. The dashed vertical line indicates the ultrafiltrate volume, when the CMC of sodium cholate was reached. This line divides the flow-through volume axis into the leftpart, where micelles do predominantly exist and are transformed into liposomes, and the right part, where liposomes represent the main lipid form, while residual detergent and ethanol/THF are continuously removed by the process of ultrafiltration.

4. Metallochelating bond and its application for construction of proteoliposomes

With respect to a potential application for the construction of vaccines, the question of *in vitro* and especially *in vivo* stability is of great importance. This problem could be divided into two fields. First, the stability of the liposomes themselves and second, the effect of the components presented in biological fluids (e.g., proteins and ions) on the stability of the metallochelating bond. It is beyond the scope of this chapter to address thoroughly this particular question. However, the GPC data indicate a good *in vitro* stability of the proteoliposomes containing recombinant His-tagged Outer surface protein C from *Borrelia burgdorferi* (rOspC) designated rOspC–HisTag during the chromatographic process, within which they experience a shear stress and dilution.

Also, the data on the incubation of rOspC-HisTag proteoliposomes in serum at 37 °C demonstrated the stability of the metallochelating bond linking the protein to the liposomal surface. In fact, *in vivo* fate of liposomes after the intradermal administration is different than that following an intravenous injection. First, dilution of proteoliposomes is not so rapid and second, the ratio of tissue fluid proteins to proteoliposomes is more favourable to proteoliposomes owing to their relatively high concentration at the site of application. Moreover, the flow rate of the tissue fluid within intradermal extracellular matrix is considerably lower than that of the muscle tissue or blood vessels. This fact is often overlooked. The stability of metallochelating bond probably depends also on the character of a particular protein. The study by Ruger shows that the single-chain Fv Ni-NTA-DOGS liposomes are unstable in human plasma and the majority of single-chain Fv fragments (anti CD 105) are released from the liposomal surface, which results in a lost of the specific targeting performance to the cells expressing a surface protein endoglin (CD 105) (Ruger et al., 2006). On the other hand, Ni-NTA3 –DTDA liposomes with single-chain Fv fragments (anti CD11c) bound onto the liposomal surface were able to target dendritic cells *in vitro* as well as *in vivo*. The application of the three-functional chelating lipid Ni-NTA3 –DTDA probably endows the metallochelating bond with a higher *in vivo* stability (van Broekhoven et al., 2004). Application of Ni-NTA-DOGS liposomes for the construction of experimental vaccine against sytemic *Candida* infection based on *Candida* Heat shock protein 90 kDa (rHSP90-HisTag) showed good stability in serum as well as strong immune response against recombinant rHSP90-HisTag antigen in mice (Masek et al., 2011b).

In vivo activity (immunogenicity) was also demonstrated for antigens associated with ISCOM particles via metallochelating lipid dipalmitoyliminodiacetic acid (Malliaros et al., 2004) and a peptide antigen associated with liposomes via Ni-NTA-DOGS (Chikh et al., 2002). Generally, metal ions, physico-chemical character of the metallochelating lipids and their surface density on the particles belong to the factors that could be optimized to get a required *in vivo* stability and, therefore, a strong immune response. The design and synthesis of new metallochelating lipids might accelerate a development and application of metallochelating liposomes for the construction of drug delivery systems and vaccines. Besides Ni^{2+}, other divalent ions such as Zn^{2+}, Co^{2+}, Fe^{2+}, and Cu^{2+}, have to be considered and experimentally tested as well.

A   B   C   D

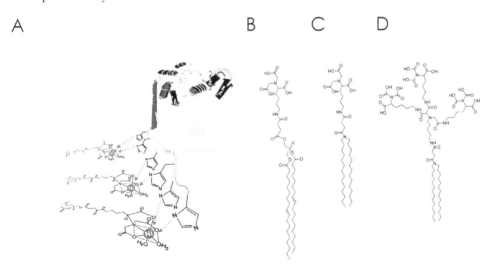

Fig. 11. Schematic illustration of recombinant His-Tagged protein bound onto the surface of metallochelating liposome (A) and formulae of metallochelating lipids NTA-DOGS (B), NTA–DTDA (C), trivalent NTA3 –DTDA (D).

5. Metallochelation liposomes for construction of experimental recombinant vaccines

Only few papers report the implementation of the metallochelating lipids in the attachment of the recombinant proteins or synthetic peptides with His-Tag anchor (short peptide consisting of 4 to 6 molecules of histidine). Both reversible character and high affinity of the metallochelating bonds are very useful for the preparation of various self-assembling supra-molecular structures useful for the construction of experimental vaccines (Chikh et al., 2002; Malliaros et al., 2004; Masek et al., 2011a; Masek et al., 2011b). As an example of synthetic liposome-based recombinant vaccine we can use metallochelating liposomes and recombinant antigen rOspc-6HisTag derived from the pathogen Borrelia burgdorferi.

5.1 rOspC antigen *Borrelia* as an example for construction of metallochelation liposome-based vaccines

Lyme disease or Lyme borreliosis is an infectious disease caused by spirochetes of the *Borrelia burgdorferi* sensu lato complex vectored by ticks of the genus *Ixodes*. At least three species are pathogenic for humans, *B. burgdorferi* sensu stricto, *B. afzelii*, and *B. garinii*. The initial stage of Lyme disease is commonly associated with skin rash occurring within few weeks after the tick bites. Later the infection can spread to bloodstream and insult of joints, heart, and nervous system. Although not common, some patients experience late stage symptoms like arthritis, nervous system complications, or acrodermatitis chronica atrophicans. Here, slower response to antibiotics therapy, sometimes taking weeks or months to recover, or eventually, incomplete resolution is observed. In a few cases, antibiotic-refractory complications persist for months to years after antibiotic therapy, most likely due to infection-induced autoimmunity. Therefore, alternative approaches such as preventive immunisation are needed, mainly in the endemic areas (Krupka M, 2007; Tilly et al., 2008).

Protective immune response to *Borrelia* involves non-specific activity of complement, phagocytic cells and *Borrelia*-specific Th1-dependent response leading to production of complement-activating antibodies, in mouse presented mostly by IgG2a (IgG2b). During natural infection, nevertheless, *Borrelia* and tick saliva modulate the immune response toward non-protective Th2 type response, associated with production of neutralizing, poorly opsonizing *Borrelia*-specific antibodies (Vesely et al., 2009). *Borrelia* outer surface proteins OspA and OspC are among the most promising antigens for elicitation of opsonizing antibodies. The applicability of OspA antigen is limited because *Borrelia* expresses it mainly in the tick and the antibodies thus should act outside of the vaccinee's organism (Pal et al., 2000). Therefore continuously high level of OspA-specific antibodies is required to prevent *Borrelia* transfer. In contrast, OspC is expressed during the transfer and the initial stage of infection. In this case the vaccine-induced immune memory has enough time to initiate the production of opsonizing antibodies preventing *Borrelia* spreading (Tilly et al., 2006).

OspC antigen can be used here as an example of reverse vaccinology approach. Full length recombinant OspC is difficult to prepare in high yield and purity. Production of Osp-s for vaccination purposes is hindered by low yield of fully processed lipidized Osp antigens or low immunogenicity of their non-lipidized versions. In our experiments, removing of N' terminal lipidation signal was associated with an increase of the recombinant protein yield and purity but, as demonstrated also for other *Borrelia* lipoproteins, a decrease in immunogenicity (Erdile & Guy, 1997; Gilmore et al., 2003; Lovrich et al., 2005; Weis et al., 1994). Induction of OspC-specific opsonizing antibodies to non – lipidised OspC could be enhanced by appropriate adjuvants and carriers like such as various modification of liposomes. It was reported that immunisation of mice with non-lipidated OspC in strong adjuvants (Complete Freund's Adjuvant, TiterMax, or Alum) could induce intense OspC-specific antibody responses (Earnhart et al., 2007; Earnhart and Marconi, 2007; Gilmore et al., 1996; Gilmore and Mbow, 1999; Ikushima et al., 2000). Here we demonstrated that similarly strong response could be elicited by immunisation of experimental mice with metallochelating nanoliposomes with entrapped lipophilic derivatives of norAbu-MDP as a potent adjuvant molecule.

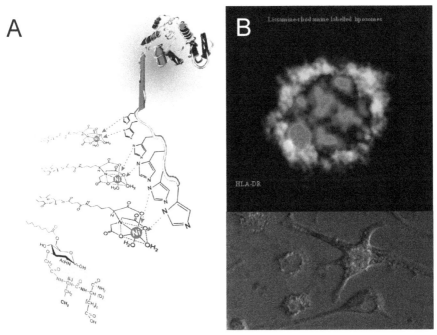

A) Scheme of metallochelating liposome with His-Tag recombinant protein antigen bound to the surface. Lipophilic nor-AbuMDP analogue as adjuvant is exposed on the surface of proteoliposome; B) Upper photograph (confocal microscop) shows human dendritic cell with phagocytosed prototypical vaccination nanoparticles cumulated in giant lysozomes (green – HLA-DR marked with antiHLA-DR antibodies; red – Lyssamine-rhodamine labelled vaccination particles). Lower photograph (Hofman modulation contrast and fluorescence microscopy) shows dendritic cells with accumulated vaccination particles inside (red fluorescence)

Fig. 12. Schematic drawing of prototypical recombinant protein vaccination nanoparticle and interaction with human dendritic cells.

5.2 Structure of OspC metallochelating proteoliposome

Protein-His-Tag/metallochelating lipid complex is anchored in phospholipid bilayer by its lipid moiety. Egg yolk or soya phosphatidyl choline contain large portion of unsaturated fatty acids, therefore DOGS-NTA lipid is freely miscible with these lipids and phase separation do not occur in liposomal bilayer. In other words, contribution of lipids to the formation of various protein structures on the liposomal surface is negligible.

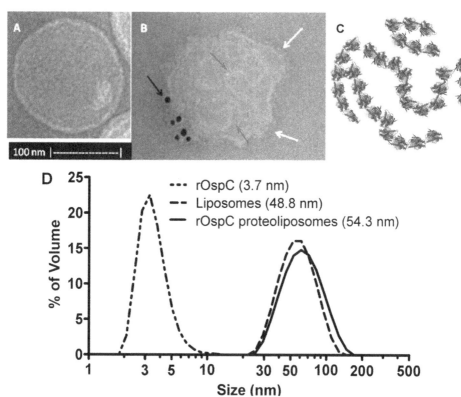

A) Structure of plain metallochelating liposome (TEM) B) Structure of metallochelating liposome with OspC bound onto the surface (white arrows – OspC); black arrow – OspC marked by 10nm immunogold particles; small red arrow – bead chain of OspC molecules; C) Schematic presentation of bead chain model of proteoliposome D) Size distribution and hydrodynamic diameter of OspC, plain metallochelating liposomes and OspC proteoliposomes analysed by DLS. The size distribution of parent monodispersed metallochelating liposomes (dashed line) was compared with that of OspC proteoliposomes (full line). As a reference, size distribution of OspC (dotted line) is shown. Numbers in brackets represent mean hydrodynamic diameters of particles

Fig. 13. Characterisation of OspC preoteoliposomes.

This is advantageous for study of protein-protein interaction of proteins anchored on the surface of liposomes. Moreover, anchoring of proteins via metallochelating bond produces highly oriented binding of proteins because the proteins are attached to liposomes exclusively by His-tagged end of the polypeptide chain. If some interaction between liposomal surface bound proteins exists, we can observe formation of various structures which are conditioned by the character and number of protein-protein interactions. Another important feature of metallochelating proteoliposomes is relatively high surface concentration of proteins. This concentration could be set by changing the DOGS-NTA/phosphatidyl choline ratio in the lipid mixture used for preparation of liposomes.

The ultrastructure of the rOspC-His-tag proteoliposomes was revealed by TEM (Fig. 7). rOspC-His tag as an example of the preoteoliposomal structure "Bead chain" model. Binding of individual molecules of OspC protein onto the liposomal surface is clearly visible on the rim of the liposome and immunogold staining confirmed the identity of OspC molecules as well as preservation of the epitopes recognised by polyclonal antibodies (Fig. 7b). In the case of recombinant OspC, TEM micrograph (Fig. 7b) showed that the individual molecules of OspC antigen were bound onto the surface of the liposome and some organisation in beads-like structures was revealed. This observation testifies against the simplification of the proteoliposomal structures and hence against accepting the simple schematic concept based on the random distribution on the liposomal surface.

Binding of OspC onto the surface of metallochelating liposomes was proved by an increase of hydrodynamic diameter as assayed by DLS. The increase of the size of proteoliposomes is well distinguished from plain liposomes, even if the increase of the size is only 5.5 nm (Fig.. 13D). This precise measurement was allowed by a preparation of parent monodispersed metallochelating liposomes and pointed to the importance of using monodispersive liposomal preparation for such a study (Fig. 8). In the case of homogenous coating of liposomes by OspC, the increase in the size should be of about 7.4 nm, theoretically. A lower increase of the size (5.5 nm) indicates only partial coating of the liposomal surface by the protein and this is in a good accordance with the structure revealed by TEM (Fig. 7b).

Binding of OspC onto metallochelating liposomes was confirmed also by GPC used as an independent method (Fig. 9) The liposomal fraction was separated from free protein and OspC was assayed by SDS PAGE followed by immunoblot (Fig. 9C). The vast majority of OspC was shown to be bound onto liposomes and was only slightly ripped from their surface by shearing forces taking place during penetrating through GPC column. The tailing character of the OspC elution profile is supportive to this explanation. Stability of the metallochelating bond in model biological fluid was studied by incubation in undiluted human serum. In spite of the presence of serum, it was estimated that more than 60% of OspC was still associated with liposomes. Based on this data, the half life of OspC proteoliposomes in serum was estimated to be at least 1 hour.

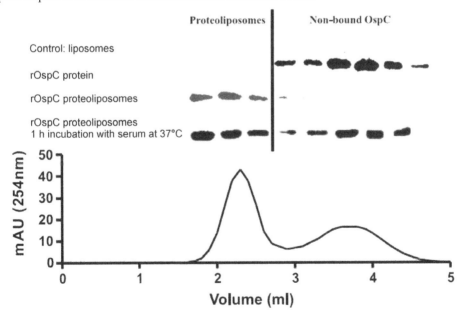

Fig. 14. Stability analysis of OspC proteoliposomes by GPC.

Non-bound OspC was separated from the proteoliposomes by gel permeation chromatography using Superose 6 column. OspC was detected in various fractions from GPC by immunoblot. OspC GPC elution profile is correlated with immunoblot assay.

5.3 Immunization experiments

New synthetic nonpyrogenic lipophilic analogues of NorAbu-MDP modified at a peptide part by two different hydrophobic ligands (Fig. 15) and for the first time these defined synthetic molecules were used in combination with metallochelating liposomes to construct an experimental recombinant vaccine. The important finding was that both MT06 and MT05 adjuvants exerted a high adjuvant effect comparable or better than MDP but proved itself as nonpyrogenic (rabbit pyrogenicity test) and safe. While alum induced stronger antibody response in IgG1 subtype, both MT06 derivatives and liposomal-MDP induced stronger immune response in both IgG2 subtypes. Interestingly, in comparison with MT06, the analogue MT05 induced stronger response in IgG1, IgG3 and IgM isotypes, respectively. This interesting finding pointed on the effect of lipophilic residues, which could not be supposed as the only accessory part of the molecule, but can significantly affect the quality of immune response. The position (peptide or sugar part) and the character (hydrophobicity and bulkiness) of the lipophilic function can affect the interaction with appropriate receptors as well as the metabolic degradation of the molecule. This aspect has not been described in the literature yet and is of interest for our understanding of the mechanism of action.

OspC itself did not elicit detectable OspC-specific antibodies of IgG, IgM, and IgA isotypes (Ig*). Similarly, when OspC bound onto the surface of liposomes was used for immunisation, only negligible increase of OspC-specific Ig* was detected after the second immunisation. In contrast, immunisation with OspC plus adjuvants (FCA, AlOH, MDP, MT05, and MT06) elicited strong OspC-specific antibody responses with ELISA titres of the same magnitude. The contributions of particular IgG isotypes for the immune response showed differences among various adjuvants (Fig. 15) (FCA was no tested for the IgG isotype response).

Mice (5 per group) were immunised by i.d. application of various liposome - adjuvant formulations of rOspC. Pooled sera from each group were used for ELISA analysis of specific antibodies titers. Naive sera were obtained before immunisation. ELISA plates were coated with 100 µl of rOspC (1 µg/ml), incubated with anti-mouse IgG1 (A), anti-mouse IgG2a (B), anti-mouse IgG2b (C), anti-mouse IgG3 (D), or anti-mouse IgM (E) and after addition of OPD plus H2O2, the absorbance was read at 490 nm on ELISA reader. The results are expressed as the end point titers +/- SD. Mean values are expressed in the table under each graph. Formulae of tested compounds in this experiment (F).

Although AlOH adjuvant induced strong OspC-specific antibody responses in total immunoglobulin level (Ig*) and IgG1 isotype (Fig. 15A), the response in complement-activating IgG isotypes (IgG2a and IgG2b; Fig. 15B,C) was only modest. In comparison with AlOH, the synthetic adjuvant MT06 when combined with liposomes-bounded OspC induced strong OspC-specific response in isotypes IgG2a and at lesser extent in IgG2b (Fig. 15B). Application of another synthetic adjuvant MT05 was associated with dominancy of IgG3 and IgM (Fig. 15D,E). Furthermore, we compare responses to synthetic norAbu-MDP adjuvants with the response to liposomes-bounded OspC plus MDP, which elicits strongest OspC-specific antibodies in IgG2b and in lesser extent in IgG2a isotype (Fig. 15B,C).

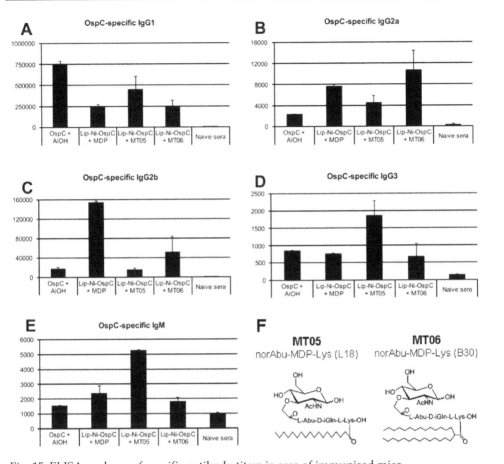

Fig. 15. ELISA analyses of specific antibody titers in sera of immunised mice.

6. Conclusion

Liposomes offer an interesting alternative to aluminium hydroxide and other adjuvant systems, especially in those cases, when the induction of cell immunity is of importance. Various methods for preparation of proteoliposomal vaccination particles are developed and metallochelating liposomes represent the newest and versatile approach to this problem. New nonpyrogenic lipophilic derivatives of norAbu-MDP have been shown to be potent adjuvants for week antigens like recombinant nonlipidized OspC and are suitable adjuvant components, together with other synthetic adjuvants like MPLA and CpG oligos, for construction of vaccines based on various liposomal platforms. Adjuvant potency of these new MDP analogues is comparable to MDP but they lack side effects related to MDP, like strong pyrogenicity and flu like syndromes. Also other synthetic adjuvans like (MPL-A and CpG oligonucleotides) or could be entrapped in metallochelating liposomes alone or in their combinations to precisely tailor the immune response towards particular antigen, especially if the Th1 response is of interest.

Moreover, liposomes are also applicable for non-invasive routes of application as mucosal and transdermal ones, and if rationally designed and applied with new synthetic adjuvants derived from PAMP, they are able to steer the immune system towards desired effective response. The most important observation was that in all vaccinated animals liposomal based vaccines did not induce any side effects. Application of modern physico-chemical and microscopic methods for study of the structure and stability of proteoliposomal vaccination particles is indispensable part of successful development of modern safe and effective vaccines.

7. Acknowledgment

This work was supported by grants: GAČR P304/10/1951; MSM 6198959223; MZE 0002716202; KAN 200520703 AVČR and KAN 200100801.

8. References

Adu-Bobie, J., Capecchi, B., Serruto, D., Rappuoli, R. & Pizza, M. (2003). Two years into reverse vaccinology. Vaccine, Vol.21, No.7-8, pp. 605-610, ISSN 0264-410X

Agostini, L., Martinon, F., Burns, K., McDermott, M.F., Hawkins, P.N. & Tschopp, J. (2004). NALP3 forms an IL-1 beta-Processing inflammasome with increased activity in Muckle-Wells autoinflammatory disorder. Immunity, Vol.20, No.3, pp. 319-325, ISSN 1074-7613

Allen, T.M., (1997). Liposomes - Opportunities in drug delivery. Drugs, Vol.54, No.4, pp. 8-14. ISSN 0012-6667

Allison, A.C. & Gregoriadis, G. (1974). Liposomes As Immunological Adjuvants. Nature, Vol.252, No.5480, pp. 252, ISSN 0028-0836

Altin, J.G. & Parish, C.R. (2006). Liposomal vaccines - targeting the delivery of antigen. Methods, Vol.40, No.1, pp. 39-52, ISSN 1046-2023

Anderson, P.M., (2006). Liposomal muramyl tripeptide phosphatidyl ethanolamine: ifosfamide-containind chemotherapy in osteosarcoma. Future Oncol, Vol.2, No.3, pp. 333-343

Anderson, P.M., Tomaras, M. & McConnell, K. (2010). Mifamurtide in Osteosarcoma-A Practical Review. Drugs of Today, Vol.46, No.5, pp. 327-337, ISSN 1699-3993

Azuma, I. & Seya, T. (2001). Development of immunoadjuvants for immunotherapy of cancer. International Immunopharmacology, Vol.1, No.7, pp. 1249-1259, ISSN 1567-5769

Bangham, A.D., Standish, M.M. & Watkins, J.C. (1965). Diffusion of Univalent Ions Across Lamellae of Swollen Phospholipids. Journal of Molecular Biology, Vol.13, No.1, pp. 238-252, ISSN 0022-2836

Barnadas-Rodriguez, R. & Sabes, M. (2001). Factors involved in the production of liposomes with a high-pressure homogenizer. International Journal of Pharmaceutics, Vol.213, No.1-2, pp. 175-186, ISSN 0378-5173

Berger, N., Sachse, A., Bender, J., Schubert, R. & Brandl, M. (2001). Filter extrusion of liposomes using different devices: comparison of liposome size, encapsulation efficiency, and process characteristics. International Journal of Pharmaceutics, Vol.223, No.1-2, pp. 55-68, ISSN 0378-5173

Chikh, G.G., Li, W.M., Schutze-Redelmeier, M.P., Meunier, J.C. & Bally, M.B. (2002). Attaching histidine-tagged peptides and proteins to lipid-based carriers through use of metal-ion-chelating lipids. *Biochimica et Biophysica Acta-Biomembranes*, Vol.1567, No.1-2, pp. 204-212, ISSN 0005-2736

Creaven, P.J., Cowens, J.W., Brenner, D.E., Dadey, B.M., Han, T., Huben, R., Karakousis, C., Frost, H., Lesher, D., Hanagan, J., Andrejcio, K. & Cushman, M.K. (1990). Initial Clinical-Trial of the Macrophage Activator Muramyl Tripeptide Phosphatidylethanolamine Encapsulated in Liposomes in Patients with Advanced Cancer. *Journal of Biological Response Modifiers*, Vol.9, No.5, pp. 492-498, ISSN 0732-6580

Cullis, P.R., Hope, M.J., Bally, M.B., Madden, T.P. & Mayer, L.D. (1987). Liposomes as pharmaceuticals, In: *Liposomes: From Biophysics to Therapeutics*, Ostro, M.J., (Ed.), 39-69, Dekker, ISBN 0-8247-7762-X, New York/Basel, USA/Switzerland

Darcissac, E.C.A., Vidal, V., Guillaume, M., Thebault, J.J. & Bahr, G.M. (2001). Clinical tolerance and profile of cytokine induction in healthy volunteers following the simultaneous administration of IFN-alpha and the synthetic immunomodulator murabutide. *Journal of Interferon and Cytokine Research*, Vol.21, No.9, pp. 655-661, ISSN 1079-9907

Earnhart, C.G., Buckles, E.L. & Marconi, R.T. (2007). Development of an OspC-based tetravalent, recombinant, chimeric vaccinogen that elicits bactericidal antibody against diverse Lyme disease spirochete strains. *Vaccine*, Vol.25, No.3, pp. 466-480, ISSN 0264-410X

Earnhart, C.G. & Marconi, R.T. (2007). An octavalent Lyme disease vaccine induces antibodies that recognize all incorporated OspC type-specific sequences. *Human Vaccines*, Vol.3, No.6, pp. 281-289, ISSN 1554-8619

Erdile, L.F. & Guy, B. (1997). OspA lipoprotein of Borrelia burgdorferi is a mucosal immunogen and adjuvant. *Vaccine*, Vol.15, No.9, pp. 988-995, ISSN 0264-410X

Fortin, A., Shahum, E., Krzystyniak, K. & Therien, H.M. (1996). Differential activation of cell-mediated immune functions by encapsulated and surface-linked liposomal antigens. *Cellular Immunology*, Vol.169, No.2, pp. 208-217, ISSN 0008-8749

Gilmore, R.D., Bacon, R.M., Carpio, A.M., Piesman, J., Dolan, M.C. & Mbow, M.L. (2003). Inability of outer-surface protein C (OspC)-primed mice to elicit a protective anamnestic immune response to a tick-transmitted challenge of Borrelia burgdorferi. *Journal of Medical Microbiology*, Vol.52, No.7, pp. 551-556, ISSN 0022-2615

Gilmore, R.D., Kappel, K.J., Dolan, M.C., Burkot, T.R. & Johnson, B.J.B. (1996). Outer surface protein C (OspC), but not P39, is a protective immunogen against a tick-transmitted Borrelia burgdorferi challenge: Evidence for a conformational protective epitope in OspC. *Infection and Immunity*, Vol.64, No.6, pp. 2234-2239, ISSN 0019-9567

Gilmore, R.D. & Mbow, M.L. (1999). Conformational nature of the Borrelia burgdorferi B31 outer surface protein C protective epitope. *Infection and Immunity*, Vol.67, No.10, pp. 5463-5469, ISSN 0019-9567

Girardin, S.E. & Philpott, D.J. (2004). The role of peptidoglycan recognition in innate immunity. *European Journal of Immunology*, Vol.34, No.7, pp. 1777-1782, ISSN 0014-2980

Gluck, R., Althaus, B., Berger, R., Just, M. & Cryz, S.J. (1992). Development, safety and immunogeincity of new inactivated hepatitis A vaccines - Effect of adjuvants, In: *Travel Medicine 2*, Lobel, H.O., Steffen, R., Kozarsky, P.E., (Eds.), 135-136, International Society of Travel Medicine, ISBN 0-323-03453-5, Atlanta, USA

Gregoriadis, G. (Ed.). (1992). *Liposome Technology*, CRC Press, Inc., ISBN 9780849367090, Boca Raton, USA

Gregoriadis, G. (1995). Engineering liposomes for drug delivery: Progress and problems. *Trends in Biotechnology*, Vol.13, No.12, pp. 527-537, ISSN 0167-7799

Gregoriadis, G., Davis, D. & Davies, A. (1987). Liposomes As Immunological Adjuvants - Antigen Incorporation Studies. *Vaccine*, Vol.5, No.2, pp. 145-151, ISSN 0264-410X

Hope, M.J., Bally, M.B., Webb, G. & Cullis, P.R. (1985). Production of Large Unilamellar Vesicles by A Rapid Extrusion Procedure - Characterization of Size Distribution, Trapped Volume and Ability to Maintain A Membrane-Potential. *Biochimica et Biophysica Acta*, Vol.812, No.1, pp. 55-65, ISSN 0006-3002

Ichihara, N., Kanazawa, R., Sasaki, S., Ono, K., Otani, T., Yamaguchi, F. & Une, T. (1988). Phase I study and clinical pharmacological study of muroctasin. *Arzneimittel-forschung/Drug Research*, Vol.38-2, No.7A, pp. 1043-1069, ISSN 0004-4172

Ikushima, M., Matsui, K., Yamada, F., Kawahashi, S. & Nishikawa, A. (2000). Specific immune response to a synthetic peptide derived from outer surface protein C of Borrelia burgdorferi predicts protective borreliacidal antibodies. *Fems Immunology and Medical Microbiology*, Vol.29, No.1, pp. 15-21, ISSN 0928-8244

Keefer, M.C., Graham, B.S., McElrath, M.J., Matthews, T.J., Stablein, D.M., Corey, L., Wright, P.F., Lawrence, D., Fast, P.E., Weinhold, K., Hsieh, R.H., Chernoff, D., Dekker, C. & Dolin, R. (1996). Safety and immunogenicity of Env 2-3, a human immunodeficiency virus type 1 candidate vaccine, in combination with a novel adjuvant, MTP-PE/MF59. *Aids Research and Human Retroviruses*, Vol.12, No.8, pp. 683-693, ISSN 0889-2229

Keitel, W., Couch, R., Bond, N., Adair, S., Vannest, G. & Dekker, C. (1993). Pilot Evaluation of Influenza-Virus Vaccine (Ivv) Combined with Adjuvant. *Vaccine*, Vol.11, No.9, pp. 909-913, ISSN 0264-410X

Krupka, M., Raska, M., Belakova, J., Horynova, M., Novotny, R. & Weigl, E. (2007). Biological aspects of Lyme disease spirochetes: unique bacteria of the Borrelia burgdorferi species group. *Biomed Pap Med Fac Univ Palacky Olomouc Czech Repub*, Vol.151, No.2, pp. 175-186,

Lovrich, S.D., Jobe, D.A., Schell, R.F. & Callister, S.M. (2005). Borreliacidal OspC antibodies specific for a highly conserved epitope are immunodominant in human lyme disease and do not occur in mice or hamsters. *Clinical and Diagnostic Laboratory Immunology*, Vol.12, No.6, pp. 746-751, ISSN 1071-412X

Malliaros, J., Quinn, C., Arnold, F.H., Pearse, M.J., Drane, D.P., Stewart, T.J. & Macfarlan, R.I. (2004). Association of antigens to ISCOMATRIX (TM) adjuvant using metal chelation leads to improved CTL responses. *Vaccine*, Vol.22, No.29-30, pp. 3968-3975, ISSN 0264-410X

Masek, J., Bartheldyova, E., Korvasova, Z., Skrabalova, M., Koudelka, S., Kulich, P., Kratochvilova, I., Miller, A.D., Ledvina, M., Raska, M. & Turanek, J. (2011). Immobilization of histidine-tagged proteins on monodisperse metallochelation liposomes: Preparation and study of their structure. *Analytical Biochemistry*, Vol.408, No.1, pp. 95-104, ISSN 0003-2697

Masek, J., Bartheldyova, E., Turanek-Knotigova, P., Skrabalova, M., Korvasova, Z., Plockova, J., Koudelka, S., Skodova, P., Kulich, P., Krupka, M., Zachova, K., Czernekova, L., Horynova, M., Kratochvilova, I., Miller, A.D., Zyka, D., Michalek, J., Vrbkova, J., Sebela, M., Ledvina, M., Raska, M. & Turanek, J. (2011). Metallochelating liposomes with associated lipophilised norAbuMDP as biocompatible platform for construction of vaccines with recombinant His-tagged antigens: Preparation, structural study and immune response towards rHsp90. *J Control Release*, Epub Ahead of print

McDonald, C., Inohara, N. & Nunez, G. (2005). Peptidoglycan signaling in innate immunity and inflammatory disease. *Journal of Biological Chemistry*, Vol.280, No.21, pp. 20177-20180, ISSN 0021-9258

Nerome, K., Yoshioka, Y., Ishida, M., Okuma, K., Oka, T., Kataoka, T., Inoue, A. & Oya, A. (1990). Development of A New Type of Influenza Subunit Vaccine Made by Muramyldipeptide Liposome - Enhancement of Humoral and Cellular Immune-Responses. *Vaccine*, Vol.8, No.5, pp. 503-509, ISSN 0264-410X

Pal, U., de Silva, A.M., Montgomery, R.R., Fish, D., Anguita, J., Anderson, J.F., Lobet, Y. & Fikrig, E. (2000). Attachment of Borrelia burgdorferi within Ixodes scapularis mediated by outer surface protein A. *Journal of Clinical Investigation*, Vol.106, No.4, pp. 561-569, ISSN 0021-9738

Parant, M., Parant, F., Chedid, L., Yapo, A., Petit, J.F. & Lederer, E. (1979). Fate of the Synthetic Immunoadjuvant, Muramyl Dipeptide (C-14-Labeled) in the Mouse. *International Journal of Immunopharmacology*, Vol.1, No.1, pp. 35-41, ISSN 0192-0561

Perrett, S., Golding, M. & Williams, W.P. (1991). A Simple Method for the Preparation of Liposomes for Pharmaceutical Applications - Characterization of the Liposomes. *Journal of Pharmacy and Pharmacology*, Vol.43, No.3, pp. 154-161, ISSN 0022-3573

Rappuoli, R. (2000). Reverse vaccinology. *Current Opinion in Microbiology*, Vol.3, No.5, pp. 445-450, ISSN 1369-5274

Rigaud, J.L. & Levy, D. (2003). Reconstitution of membrane proteins into liposomes. *Liposomes, PT A*, Vol. 372, pp. 65-86, ISSN 0076-6879

Ruger, R., Muller, D., Fahr, A. & Kontermann, R.E. (2006). In vitro characterization of binding and stability of single-chain Fv Ni-NTA-liposomes. *Journal of Drug Targeting*, Vol.14, No.8, pp. 576-582, ISSN 1061-186X

Schneider, T., Sachse, A., Rossling, G. & Brandl, M. (1995). Generation of contrast-carrying liposomes of defined size with a new continous high-pressure extrusion method. International Journal of Pharmaceutics, Vol.117, No.1, pp. 1-12, ISSN 0378-5173

Schubert, R. (2003). Liposome preparation by detergent removal. Liposomes, PT A, Vol.367, pp. 46-70, ISSN 0076-6879

Shahum, E. & Therien, H.M. (1988). Immunopotentiation of the Humoral Response by Liposomes - Encapsulation Versus Covalent Linkage. *Immunology*, Vol.65, No.2, pp. 315-317, ISSN 0019-2805

Shek, P.N. & Heath, T.D. (1983). Immune-Response Mediated by Liposome-Associated Protein Antigens .3. Immunogenicity of Bovine Serum-Albumin Covalently Coupled to Vesicle Surface. *Immunology*, Vol.50, No.1, pp. 101-106, ISSN 0019-2805

Small, D.M. (1971). Chemistry, In: *The Bile Acids: Chemistry, Physiology, and Metabolism*, Nair, P.P., Kritchevsky, D., (Ed.), 249-356, Plenum Press, New York, USA

Stadler, K., Masignani, V., Eickmann, M., Becker, S., Abrignani, S., Klenk, H.D. & Rappuoli, R. (2003). SARS - Beginning to understand a new virus. *Nature Reviews Microbiology*, Vol.1, No.3, pp. 209-218, ISSN 1740-1526

Tamura, M., Yoo, Y.C., Yoshimatsu, K., Yoshida, R., Oka, T., Ohkuma, K., Arikawa, J. & Azuma, I. (1995). Effects of Muramyl Dipeptide Derivatives As Adjuvants on the Induction of Antibody-Response to Recombinant Hepatitis-B Surface-Antigen. *Vaccine*, Vol.13, No.1, pp. 77-82, ISSN 0264-410X

Tilly, K., Krum, J.G., Bestor, A., Jewett, M.W., Grimm, D., Bueschel, D., Byram, R., Dorward, D., VanRaden, M.J., Stewart, P. & Rosa, P. (2006). Borrelia burgdorferi OspC protein required exclusively in a crucial early stage of mammalian infection. *Infection and Immunity*, Vol.74, No.6, pp. 3554-3564, ISSN 0019-9567

Tilly, K., Rosa, P.A. & Stewart, P.E.(2008). Biology of infection with Borrelia burgdorferi. *Infectious Disease Clinics of North America*, Vol.22, No.2, 217-+, ISSN 0891-5520

Traub, S., von Aulock, S., Hartung, T. & Hermann, C. (2006). MDP and other muropeptides - direct and synergistic effects on the immune system. *Journal of Endotoxin Research*, Vol.12, No.2, pp. 69-85, ISSN 0968-0519

Tsubura, E., Nomura, T., Niitani, H., Osamura, S., Okawa, T., Tanaka, M., Ota, K., Nishikawa, H., Masaoka, T., Fukuoka, M., Horiuchi, A., Furuse, K., Ito, M., Nagai, K., Ogura, T., Kozuru, M., Hara, N., Hara, K., Ichimaru, M. & Takatsuki, K. (1988). Restorative Activity of Muroctasin on Leukopenia Associated with Anticancer Treatment. *Arzneimittel-Forschung/Drug Research*, Vol.38-2, No.7A, pp. 1070-1074, ISSN 0004-4172

Tsujimoto, M., Kotani, S., Kinoshita, F., Kanoh, S., Shiba, T. & Kusumoto, S. (1986). Adjuvant Activity of 6-O-Acyl-Muramyldipeptides to Enhance Primary Cellular and Humoral Immune-Responses in Guinea-Pigs - Adaptability to Various Vehicles and Pyrogenicity. *Infection and Immunity*, Vol.53, No.3, pp. 511-516, ISSN 0019-9567

Turanek, J. (1994). Fast-Protein Liquid-Chromatography System As A Tool for Liposome Preparation by the Extrusion Procedure. *Analytical Biochemistry*, Vol.218, No.2, pp. 352-357, ISSN 0003-2697

Turanek, J., Kasna, A., Koudela, B., Ledvina, M. & Miller, A.D. (2005). Stimulation of innate immunity in newborn kids against Cryptosporidium parvum infection-challenge by intranasal/per-oral administration of liposomal formulation of N-L 18-norAbu-GMDP adjuvant. *Parasitology*, Vol.131, No.5, pp. 601-608, ISSN 0031-1820

Uchida, T. & Taneichi, M. (2008). Clinical application of surface-linked liposomal antigens. *Mini-Reviews in Medicinal Chemistry*, Vol.8, No.2, pp. 184-192, ISSN 1389-5575

van Broekhoven, C.L., Parish, C.R., Demangel, C., Britton, W.J. & Altin, J.G. (2004). Targeting dendritic cells with antigen-containing liposomes: A highly effective procedure for induction of antitumor immunity and for tumor immunotherapy. *Cancer Research*, Vol.64, No.12, pp. 4357-4365, ISSN 0008-5472

Vesely, D.L., Fish, D., Shlomchik, M.J., Kaplan, D.H. & Bockenstedt, L.K. (2009). Langerhans Cell Deficiency Impairs Ixodes scapularis Suppression of Th1 Responses in Mice. *Infection and Immunity*, Vol.77, No.5, pp. 1881-1887, ISSN 0019-9567

Wack, A. & Rappuoli, R. (2005). Vaccinology at the beginning of the 21st century. *Current Opinion in Immunology*, Vol.17, No.4, pp. 411-418, ISSN 0952-791

Walter, A., Kuehl, G., Barnes, K. & VanderWaerdt, G. (2000). The vesicle-to-micelle transition of phosphatidylcholine vesicles induced by nonionic detergents: effects of sodium chloride, sucrose and urea. *Biochimica et Biophysica Acta-Biomembranes*, Vol.1508, No.1-2, pp. 20-33, ISSN 0005-2736

Weis, J.J., Ma, Y. & Erdile, L.F. (1994). Biological activities of native and recombinant Borrelia burgdorferi outer surface protein A: dependence on lipid modification. *Infection and Immunity*, Vol.62, No.10, pp. 4632-4636, ISSN 0019-9567

Woodle, M.C. & Papahadjopoulos, D. (Eds.). (1988). Methods in Enyzmology, Academic Press, San Diego, USA

Zumbuehl, O. & Weder, H.G. (1981). Liposomes of Controllable Size in the Range of 40 to 180 Nm by Defined Dialysis of Lipid-Detergent Mixed Micelles. Biochimica et Biophysica Acta, Vol.640, No.1, pp. 252-262, ISSN 0006-3002

5

Genetic Modification of Domestic Animals for Agriculture and Biomedical Applications

Cai-Xia Yang and Jason W. Ross
Department of Animal Science, Iowa State University,
USA

1. Introduction

Transgenic technology has been applied mainly in the study of gene structure and function in model organisms and gene therapy for human diseases. Transgenic technology has potential for rapidly improving quantity and quality of agricultural products, compared to traditional selection and breeding methods in domestic animals that are time consuming when attempting to alter the desired allele frequency for specific traits. Additionally, transgenic animals can be used as biomedical research models or directly for human health, by producing recombinant pharmaceutical proteins and/or organs for xenotransplantation. Due to the advantage of bypassing the need of embryonic stem (ES) cells that are difficult to isolate in domestic animal species, cell-based method of transgenesis followed by somatic cell nuclear transfer (SCNT) is currently widely applied. However, due to the limitations in making genetic modifications and SCNT, producing genetically modified animals is still inefficient. Fortunately, the current advancement of new techniques and methods in both gene targeting (Urnov et al., 2010) and abilities to produce pluripotent stem cells (Voigt and Serikawa, 2009) holds great promises for this field.

In this chapter, we will review the recent progress and technical route of the cell-based method of transgenesis by SCNT and discuss the newly emerging methods to enrich the gene targeting frequency of somatic cells. We will also discuss factors to improve the efficiency of SCNT and our future perspectives on the promises of this field.

2. Recent progress and applications of transgenic domestic animals

2.1 Methods of creating genetically modified animals

2.1.1 Pronuclear microinjection /viral-mediated /sperm-mediated /ICSI (Intractyoplasmic sperm injection)- mediated gene transfer

Numerous methods have been successfully used to introduce genetic modifications and produce transgenic animals, including pronuclear microinjection of foreign DNA into zygotes (Hammer et al., 1985; Pursel and Rexroad, 1993), viral-mediated gene transfer (Chan et al., 1998; Cabot et al., 2001; Whitelaw et al., 2008), sperm-mediated gene transfer (Castro et al., 1991; Chang et al., 2002; Lavitrano et al., 2002 and 2006) and intracytoplasmic injection (ICSI) of a sperm head carrying foreign DNA (Perry et al., 1999; Osada et al., 2005; Moisvadi et al., 2009; García-Vázquez et al., 2010). Despite the proven successful application of these techniques, some problems, such as inefficiency and mosaicism (transgene not going into the germline) (Table 1) remain to be solved and limit the practical application of these methods.

Method	Advantages	Disadvantages	Reference
Pronuclear microinjection	The first method successfully used for different animal species.	Low embryo survival, low and random integration, multiple copies, high cost in domestic animals.	Hammer et al., 1985; Pursel and Rexroad, 1993
Viral-mediated DNA transfer	Infect both dividing and non-dividing cells, less damage by co-culture with zona-free zygotes or injection into the perivitelline space compared with pronuclear microinjection, high integration.	Limited DNA capacity, random integration.	Chan et al, 1998; Cabot et al., 2001; Whitelaw et al., 2008
Sperm-mediated DNA transfer	Relatively high efficiency as compared to pronuclear injection, low cost, ease of use.	No control of integration site.	Castro et al, 1991; Chang et al., 2002; Lavitrano et al., 2002; Lavitrano et al., 2006
Intracytoplasmic sperm injection -mediated DNA transfer	Allow introduction of very large DNA transgenes, relatively high efficiency as compared to pronuclear injection.	No control of integration site.	Perry et al., 1999; Osada et al., 2005; Moisvadi et al., 2009; Garcia-Vázquez et al., 2010

Table 1. The Advantages and disadvantages of different transgenic methods.

Gene targeting by homologous recombination often offers more precise and site-specific integration, sometimes at single nucleotide level. This is particularly important since single nucleotide changes can be a common culprit for some of the human diseases, which require more precise manipulation to build biomedical models using transgenic animals.

2.1.2 Cell-based transgenesis via SCNT

The first live animal by SCNT was produced in 1997, "Dolly" (Wilmut et al., 1997), demonstrating the ability of a differentiated somatic cell to produce live offspring following nuclear remodeling and reprogramming by an oocyte. Then, the birth of "Polly" in the same year (Schnieke et al., 1997), the first transgenic sheep produced by transfer of nuclei from transfected fetal fibroblasts, demonstrated a route to create transgenic cloned animals. This cell-based method of transgenesis by SCNT can bypass the absence of ES cells, offers the reliability of germline transmission by avoiding mosaic transgene integration, and provides the only currently used strategy to knock out a gene in domestic animals (reviewed by Ross et al., 2009a). Currently, SCNT using transgenic cells cultured *in vitro* as a source of donor nuclei is becoming the most utilized technique to produce the transgenic domestic animals. However, while the advantages and success of this strategy are well documented (Table 2), the procedure is still labor-intensive and inefficient. Recently, some new techniques have been reported that may have the potential to improve the production efficiency of cloned transgenic domestic animals by increasing the efficiency of gene targeting or nuclear remodeling and reprogramming following SCNT.

2.2 Utilization of transgenic models

The potential for transgenic domestic animals to benefit humans is not only in agricultural production by providing more and better agricultural products for human consumption but also in biomedicine, such as for producing recombinant pharmaceutical proteins, making organs suitable for xenotransplantation and establishing human disease models. An overview list of the transgenic domestic animals produced via SCNT is given in Table 2 to demonstrate their applications.

2.2.1 Improved animal agriculture production

Increased utilization of domestic animals and their products requires breeding and selection strategies for specific traits. However, classical breeding and genetic selection have some disadvantages, such as the inability to control gene frequency of desired genotypes coupled with long generation intervals. The application of transgenic technology offers a powerful tool to rapidly improve agriculture production by developing domestic animals that express desired traits via genome manipulation strategies. Previous studies have demonstrated the practical application of transgenesis to improve numerous agricultural traits of domestic animals, including increased growth rate (Pursel et al., 1999), increased meat quality (Saeki et al., 2004), enhanced disease resistance (Lo et al., 1991; Clements et al., 1994), and better milk production and composition (Wheeler et al., 2001; Reh et al., 2004). Nevertheless, utilization of the cell-based method of transgenesis via SCNT has also been successfully used to alter characteristics of pork quality (Lai et al., 2006), enhance disease resistance (Denning et al., 2001a; Wall et al., 2005; Richt et al., 2007) and improve milk composition (Brophy et al., 2003).

Species	Application	Donor cells	Key Molecule	Construct	DNA delivery method	Reference
Sheep	Bioreactor	serum starved fetal fibroblasts	human clotting factor IX (FIX)	pMIX1	Lipofection	Schnieke et al., 1997
	Disease model	serum starved fetal fibroblasts	alpha(I) procollagen (COL1A1)	COL1-1/COL1-2	Lipofection	McCreath et al., 2000
	Xenotransplantation; disease resistance	serum starved fetal fibroblasts	alpha(1,3)galactosyl transferase (GGTA1), prion protein (PrP)	GGTA1; PrP	Electroporation	Denning et al., 2001a
Cattle	Marker gene	actively dividing fetal fibroblasts	β-galactosidase–neomycin, cytomegalovirus (CMV)	pCMV/ βGEO	Not described	Cibelli et al., 1998
	Bioreactor	serum starved fetal/ear fibroblasts	bovine prochymosin coding gene, αS1-casein promoter	TFF1/TFF2	Lipofection	Zakhartchenko et al., 2001
	GFP-reporter	dividing fetal fibroblast	enhanced, humanized version of the GFP-reporter	CEEGFP	Lipofection	Bordignon et al., 2003
	Bioreactor	serum starved fetal fibroblasts and G1 fibroblasts	β-casein and Κκ-casein	$CSN2^{A3}$ / $CSN2/3^{B}$	Lipofection	Brophy et al., 2003
	Bioreactor	fetal fibroblasts	bispecific single-chain variable fragment (biscFV) molecule with anti-human CD28 × anti-human melanoma specificity	Bi-scFV r28M	Lipofection	Grosse-Hovest et al., 2004
	Bioreactor	fetal fibroblasts	growth hormone (GH)	hGH vector	Not described	Salamone et al., 2006
	Bioreactor	fetal oviduct epithelial cells	human α-lactalbumin	ptLa4-EGFP-NEO	Electroporation	Wang et al., 2008
	Bioreactor	fetal fibroblasts	human lactoferrin	hLF BAC	Microinjection/ Electroporation/ Lipofection	Yang et al., 2008
Goat	eGFP	fetal fibroblasts	green fluorescent protein reporter gene	CEeGFP	Lipofection	Keefer et al., 2001
Pig	Xenotransplantation	fetal fibroblasts	alpha-1,3-galactosyltransferase locus	pGalGT	Electroporation	Lai et al., 2002a
	Xenotransplantation	fetal fibroblasts	alpha-1,3-galactosyltransferase locus	pPL654 and pPL657	Electroporation	Dai et al., 2002
	Disease model	fetal fibroblasts	cystic fibrosis (CF); CF transmembrane conductance receptor (CFTR)	CFTR-null and CFTR-ΔF508	Adeno-associated virus (AAV)	Rogers et al., 2008
	Disease model	ear fibroblasts	Alzheimer's disease; neuronal variant of the human amyloid precursor protein gene with the Swedish mutation	pPDGFbEGFP	Lipofection	Kragh et al., 2009
	Xenotransplantation	fetal fibroblasts	human A20 gene	pCAGGSEhA20-IRESNEO	Electroporation	Oropeza et al., 2009
	Disease model	fetal fibroblasts	Huntington's disease	pCAG-HTT-2A-ECFP	Electroporation	Yang et al., 2010
Rabbit	eGFP	serum starved adult fibroblasts	green fluorescent protein reporter gene	pEGFP-C1	Lipofection	Li et al., 2009

Table 2. Overview on successful transgenic domestic animals via SCNT.

2.2.2 Xenotransplantation

Due to the potential application of human organ transplantation and the growing gap between the demand and availability of organs for transplantation, the pig has long been considered as an alternative source to provide organs for humans. In contrast to organs from other animal species, domestic pig organs share many similarities to those of humans including size, anatomy and physiology. However, despite these similarities, significant immunological barriers exist impeding the success of pig to human xenotransplantation (Cooper et al., 2008), in addition to concerns regarding the transmission of pig specific viruses to the human genome (Magre et al., 2003). The immunological obstacles of xenotransplantation include rapid hyperacute rejection (HAR), delayed acute vascular rejection (AVR), and the cellular immune response that occurs within weeks (Auchincloss and Sachs, 1998). Two transgenic strategies have been successfully applied to overcome HAR. One is to express human proteins that inhibit the complement cascade in transgenic pigs (Fodor et al., 1994; Cozzi and White, 1995; Diamond et al., 2001).   For example, transgenic expression of human complement inhibitor CD59, CD46 and DAF (decay-accelerating factor, also named as CD55) in pigs prolongs survival rates from minutes to days and months following heart and kidney transplantation into baboons or monkeys by blocking the damage from HAR (Diamond et al., 1996 and 2001; Byrne et al., 1997; Zaidi et al., 1998; Bhatti et al., 1998; Chen et al., 1999). The second strategy to avoid HAR is to knockout the genes that induce the production of antigenic structures (α-gal-epitopes) on the surface of pig organs (Lai et al., 2002a; Dai et al., 2002; Phelps et al., 2003; Yamada et al., 2005). α-gal-epitopes on endothelial cells of porcine transplanted organs can be recognized by human xenoreactive natural antibodies (XNA) and activate the HAR cascade (Galili, 1993). Genetically engineering of pigs to lower or inhibit the expression level of XNA targets is thought to be a promise way to eliminate the HAR. Following the successful production of α-1,3-galactosyltransferase knockout pigs (Lai et al., 2002a; Dai et al., 2002; Phelps et al., 2003; Yamada et al., 2005), van Poll et al. (2010) recently showed that exposure of isolated xenogeneic pig liver sinusoidal endothelial cells (LSECs) from α-1,3-galactosyltransferase-deficient pigs to human and baboon serum reduces IgM binding and complement activation levels as compared to wild-type pig LSECs. However, Diswall et al. (2010) found a different reactivity pattern of baboon and human serum to pig glycolipid antigens isolated from α-1,3-galactosyltransferase knockout and wild-type pig hearts and kidneys, suggesting that non-human primates may not be an ideal model for modeling pig to human xenotransplantation. If HAR is controlled, the next obstacle to xenotransplantation is AVR which is due to the loss of porcine thrombomodulin in xenograft rejection or the inability of porcine thrombomodulin to activate human protein C. One of genetic engineering strategies to overcome AVR is to express human thrombomodulin in pigs. Petersen et al. (2006) showed the production of transgenic cloned pigs using CD59/DAF and human thrombomodulin triple transgenic adult donor cells.

With regard to risks associated with xenotransplantation, previous studies have shown that the risk of cross-transmission of pig endogenous retrovirus (PERV) to human patients or nonhuman primate recipients is low (Paradis et al., 1999; Switzer et al., 2001), although it has been found PERV can infect human cells in culture (Patience et al., 1997). Never-the-less, some investigators have been working to further reduce the possibility by creating pigs with suppressed expression of endogenous retroviruses (Ramsoondar et al., 2009). In addition to PERV, herpesvirus is another concern regarding biosafety in xenotransplantation (Mueller et al., 2011). Overall, dramatically increased knowledge will facilitate the clinical application of transgenic strategies of pig to human xenotransplantation.

2.2.3 Production of recombinant proteins

The mammary gland and blood of transgenic domestic animals, including sheep, goats, cows, pigs and rabbits, have been successfully used as bioreactor to produce numerous recombinant proteins, such as antibodies (Grosse-Hovest et al., 2004), growth factors (Schnieke et al., 1997) and pharmaceuticals (reviewed by Melo et al., 2007). Using various mammary gland-specific or blood-specific promoters to drive the expression of specific protein-coding genes, transgenic domestic animals can continuously produce the recombinant proteins in large quantities in their milk or blood. Recombinant proteins, including human von Willebrand factor (Lee et al., 2009), human erythropoietin (Park et al., 2006), human insulin-like growth factor-I (Monaco et al., 2005), human factor VIII (Paleyanda et al., 1997) and bovine alpha-lactalbumin (Bleck et al., 1998) have been produced in the milk of transgenic pigs. Transgenic goats, capable of synthesizing human butyrylcholinesterase (Huang et al., 2007) and human longer acting tissue plasminogen activator (Ebert et al., 1991) in their milk have also been created. Human salmon calcitonin in milk of transgenic rabbits (McKee et al., 1998); human factor IX (Schnieke et al., 1997) and alpha-1-antitrypsin (Wright et al., 1991) in milk of transgenic sheep; and human lactoferrin (van Berkel et al., 2002; Yang et al., 2008), human growth hormone (Salamone et al., 2006) and human α-lactalbumin (Wang et al., 2008) in milk of transgenic cows are all additional examples of using transgenic domestic animals and the mammary gland as a bioreactor for production of recombinant proteins. Table 2 summarizes some recombinant proteins expressed in milk or blood of cloned transgenic domestic animals. One of the major advantages of using domestic animals for this purpose is that the produced protein is thought to undergo more accurate posttranslational processing to ensure their biological activity. While this application of transgenic technology to produce recombinant protein products is rapidly developing, research efforts exploring the efficacy of these products are still needed.

2.2.4 Biomedical models of human diseases

Another important application of genetically modified domestic animals is to create better and novel biomedical models of human diseases. Pig models of different human diseases, including retinitis pigmentosa, cardiovascular disease, diabetes, Huntington's disease, cystic fibrosis and Alzheimer's disease have been well discussed by Prather et al. (2008). Many of these biomedical models created by SCNT are listed in Table 2.

3. Technical aspects of cell-based transgenesis by SCNT

The general procedure of cell-based transgenesis via SCNT is to construct a DNA vector, deliver the vector into cultured somatic cells, select transgenic cell lines, utilize SCNT and transfer cloned embryo into surrogates (Figure 1).

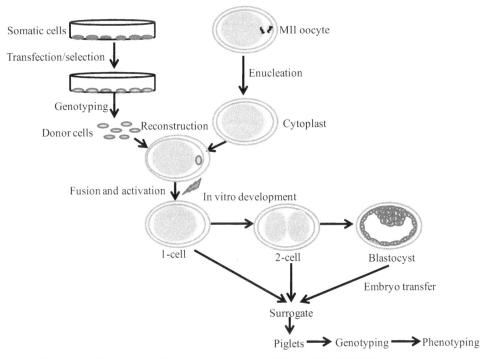

Fig. 1. Technical diagram of cell-based transgenesis followed by SCNT in pigs.

3.1 Vector construction

Currently, the whole genome sequence of several domestic animals, including cattle, poultry, cats, dogs, horses, pigs and rabbits is available via the ensembl database (www.ensembl.org), providing researchers useful sequence information for a large number of genes useful in designing DNA constructs for transgenic genome modification. Precise design and construction of DNA constructs is critical efficiently creating transgenic domestic animals. Transgenesis involves adding a gene to a host genome (transgenic), physically deleting a specific region of the host genome making a non-functional gene (knock-out), replacing an active gene by another active gene (knock-in), or introducing a point mutation (point mutation knock-in). Depending on the objective of the transgenic modification, different strategies of vector design need to be carefully considered to ensure success.

3.1.1 Transgenic vs. Knock-out vs. Knock-in

The design strategy for the transgenic vector, which is based on random integration, essentially includes a gene ORF (open reading frame), a promoter element and the appropriate RNA processing component(s). The promoter is a major transcriptional regulatory element that normally includes regulatory elements and the transcriptional start site typically located in 5' sequence of the gene. Utilization of specific promoter with a transgene enables tissue or cell specific expression and can significantly impact the expression efficiency of the transgene. Carefully choosing a well-characterized promoter will enable the precise control of transgene expression. The gene ORF is usually derived from the cDNA for the protein of interest, which includes translational start (ATG) and stop codons. Examination of extra sequence, if any, existing between the transcriptional start site and the translational start codon should be performed to ensure the absent of potential regulatory elements. Furthermore, consideration should be given to remove non-coding sequence of genes to avoid the introduction of the regulatory elements although this may also have consequences impacting mRNA stability. Reliable transgene expression is not only regulated by the sequence in the expression vector, but also by intrinsic factors in the host genome following transgene integration. Several additional points should be considered in transgenic vector construction. (i) How GC rich regulatory sequence are, especially CpG islands in the promoters can have significant implications for the expression of the transgene as methylation of CpG islands can inactivate the promoter and silence the transgene expression. (ii) Inclusion of specific elements that favor mRNA maturation and transfer to the cytoplasm, as it has been demonstrated that the inclusion of an intron in the transgenic vector can increase the transgene expression level (Choi et al., 1991; Duncker et al., 1997). These genomic regions may have critical sequence motifs affecting mRNA splicing and accumulation. (iii) Removal of unnecessary plasmid DNA sequence used in recombinant DNA cloning. In mice, it was found that plasmid sequence existing in the transgenic vector can decrease transgene expression (Kjer-Nielsen et al., 1992). Furthermore, the local chromatin status of the transgenic locus can also affect expression meaning that the integration of the same transgene in different genomic locations can have profound effects on the expression level of the transgene.

The primary strategy for targeted genome modifications, including knock-out and knock-in applications, is to use homologous recombination to introduce precise, site-specific genome alterations. In knock-out targeting vector systems, the primary approach is to delete DNA fragments (entire gene or partial deletion) important for the gene function by homologous recombination. Compared with the knock-out targeting vector, in addition to the targeting arms and positive selection cassette, the knock-in targeting vector includes the extra replacement cassette that will replace the target gene with a new gene (a set of genes or a point mutation). Targeting vectors relying on homologous recombination contain 5' and 3' homologous arms flanking a positive selection cassette (Rogers et al., 2008; Sun et al., 2008). Several principles should be considered when designing a successful targeting vector. (i) Avoid excessive repetitive DNA. In mouse ES cells it has been demonstrated that the excessive repetitive DNA within the targeting vector can significantly reduce targeting frequency (Wu et al., 2008). (ii) Use isogenic DNA as the source for producing exogenous homologous arms. While gene targeting in domestic species by using non-isogenic DNA as a source for targeting arms is possible (McCreath et al., 2000; Denning et al., 2001a and b; Kuroiwa et al., 2004; Marques et al., 2006; Richt et al., 2007), the use of isogenic DNA can largely improve the efficiency of gene targeting (te Riele et al., 1992). (iii) Increase the length of continuous exogenous homologous arms. The efficiency of gene targeting was generally found to be increased with the length of targeting arms in mouse ES cells (Hasty et al., 1991a). (iv) Use multiple cell lines; when targeting the CFTR (cystic fibrosis transmembrane conductance receptor) gene in pig fetal fibroblast cell lines, Rogers et al. (2008) demonstrated drastic differences in the targeting efficiencies between cell lines derived from littermate pig fetuses.

3.1.2 Tissue specificity and inducible promoters

As mentioned, the promoter used in a transgenic DNA construct determines when, where and to what extent the transgene is expressed. Hundreds of promoters can be isolated for the expression of transgenes, and are generally classified as constitutive promoters, tissue/cell-specific or Developmental stage-specific promoters and inducible promoters. Constitutive promoters, such as commonly used SV40 (simian virus 40 promoter), CMV (cytomegalovirus immediate-early promoter), PGK (mouse phosphoglycerate kinase 1 promoter) and CAGG (chicken β-Actin promoter coupled with CMV early enhancer) promoters for mammalian systems (Qin et al., 2010), can continuously drive transgene expression in all tissues and species. Tissue/cell-specific or developmental stage-specific promoters can restrict transgene expression to specific tissue(s) or only during certain developmental stages. For example, the 5.5-Kb osteopontin (OPN) promoter has been used to drive GFP expression in transgenic mice in the same cell-specific and developmental stage-specific manner as endogenous OPN expression (Higashibata et al., 2004). Inducible promoters, as their name suggests, may be activated by the presence of endogenous or exogenous factors. Exogenous factors include chemical compounds such as antibiotics or physical factors such as heat and light. For example, TRE promoter (tetracycline-responsive element promoter) can be activated by the rtTA (reverse tetracycline-controlled transcriptional activator) in a doxycycline-inducible manner (Qin et al., 2010). Antibiotic-induced promoters are the most commonly used in animal genetic modification because of easy manipulation. Inducible promoters provide a very useful tool in animal genetic engineering to turn on or off transgene expression in a particular tissue or at certain developmental stages.

3.1.3 Positive/negative selection strategies

Considering the rarity of a homologous recombination event relative to random integration of the targeting vector, an efficient targeting vector design should incorporate a good selection strategy, providing a powerful tool to improve the frequency of targeted colonies and reduce screening cell lines that result from random integration. Promoter-less gene targeting vector, also referred to the promoter-trapping method, has been used to enrich the gene targeting events in somatic cells in pigs and sheep (McCreath et al., 2000; Denning et al., 2001a and b). In a promoter-trapping vector, the selectable gene lacks its own promoter but it becomes activated from the target gene promoter after correctly integrating into the genome. In a fibroblast cell line, promoter-less vectors can enrich targeting frequency 5,000- to 10,000-fold (Hanson and Sedivy, 1995). Despite this potential improvement, the major limitation of the promoter-trapping method is that it requires active transcription of the targeted gene to drive expression of the selectable marker used in the targeting vector. Thus, if the target gene is only active in cell types that are difficult to culture, it is nearly impossible to target the gene locus using this method. Compared with promoter-trapping method, a more widely used strategy includes utilization of both positive and negative selection (PNS) (Jin et al., 2003; Kuroiwa et al., 2004; Richt et al., 2007). A positive selection gene (antibiotic-resistant gene such as neo) in the targeting vector is needed to select cells with an integrated construct a negative selection gene (cytotoxic genes such as the thymidine kinase (TK) gene or the diphtheria toxin A-chain (DT-A) gene) to further select against random integration event. The negative selection marker in the targeting vector usually is placed downstream of homologous arms and recombined away during the process of homologous recombination. The enrichment of the PNS selection is the ratio of clones recovered with the positive selection only (PS) versus the positive and negative dual selections. The PNS strategy can be used to target both active and inactive gene loci. It has been shown that the targeting efficiency at the COL1A1 locus in fibroblasts by the promoter-trapping strategy is 15.7-fold higher than by PS only (Marques et al., 2006).

3.2 Delivery of DNA vector into cultured somatic cells

Following vector construction and preparation of the DNA construct for delivery into a somatic cell, it is important to identify an efficient method of DNA delivery into somatic cells as efficiency between delivery methods varies greatly. There are numerous methods to introduce an exogenous DNA construct into somatic cells, which can be categorized as liposome-mediated DNA transfer (Hyun et al., 2003a; Lee et al., 2005), electroporation (Dai et al., 2002; Ramsoondar et al., 2003; Watanabe et al., 2005) and viral-mediated delivery (Lai et al., 2002b; Rogers et al., 2008).

3.2.1 Liposome-mediated DNA delivery

Liposome-mediated DNA transfer can easily transfect a large number of somatic cells without the need of specialized equipment and expertise, compared with other methods. Lee et al. (2005) demonstrated the efficiency of gene transfection with a plasmid containing the enhanced green fluorescence protein gene into fetal-derived bovine fibroblast cells by lipids was significantly higher than that obtained by electroporation. They also validated that transfection efficiency in fetal-derived bovine fibroblast cells, regardless of the delivery methods, was significantly higher than delivering DNA into cumulus-derived fibroblast cells and adult ear skin-derived fibroblast cells, establishing that both delivery method and cell line origin affect the efficiency of gene transfection. Using liposome-mediated DNA delivery followed by SCNT, genetically modified pigs (Hyun et al., 2003a) and sheep (McCreath et al., 2000) have successfully been created.

3.2.2 Electroporation

In contrast to liposome-mediated DNA delivery, electroporation has been widely used for the delivery of exogenous DNA into the cytoplasm of somatic cells to generate genetically modified cell lines for nuclear transfer. Electroporation has been utilized to successfully provide genetically modified donor cells for SCNT to create transgenic cloned domestic animals, including cattle (Kuroiwa et al., 2004), goat (Yu et al., 2006; Zhu et al., 2009), pig (Dai et al., 2002; Lai et al., 2002a; Ramsoondar et al., 2003; Watanabe et al., 2005) and sheep (Denning et al., 2001a). Ross et al. (2010a) indentified optimal electroporation conditions (three 1 ms pulses of 300 V to 200 µL of 1×10^6 cells/mL in the presence of 12.5µg DNA/mL), which can consistently deliver DNA vector into the 65-80% surviving porcine fetal fibroblasts and have been used to produce healthy, viable transgenic piglets (Ross et al., 2009b). In adult rhesus macaque fibroblasts, it has been demonstrated that electroporation can generate more transfected cells than liposome-mediated methods (Meehan et al., 2008), which is consistent with other similar comparisons (Yáñez and Porter, 1999). Of the numerous delivery methods, electroporation was demonstrated to have the greatest efficiency in generating targeted cell lines via homologous recombination (Vasquez et al., 2001). Targeting the hypoxanthine guanine phosphoribosyl transferase (HPRT) locus, Mir and Piedrahita (2004) demonstrated that the electroporation of a DNA construct with a nuclear localization signal into s-phase synchronized cells can increase targeting efficiency sevenfold and decrease random integration events 54-fold in primary fetal bovine fibroblasts. Later, Meehan et al. (2008) confirmed this method by successful gene targeting of the HPRT locus in adult rhesus macaque fibroblasts achieved by electroporation of S-phase synchronized cells with a construct containing a SV40 enhancer.

3.2.3 Viral-mediated delivery

Because viruses have the natural ability to stably transfect somatic cells with high efficiency, utilizing viral particles to delivery exogenous DNA into somatic cells has been widely successful. In contrast to liposome-mediated delivery and electroporation, which deliver linear, double-strand DNA into the cytoplasm of somatic cells, viral delivery of exogenous DNA delivers a high number of the linear, intact single-strand DNA molecules into the nucleus of cells. One report has demonstrated the efficient targeting of the PRNP gene encoding the prion protein PrP in bovine fetal fibroblasts by adeno-associated virus (AAV) vectors (Hirata et al., 2004). Also, the same transfection method followed by SCNT was used to successfully produce the CFTR-null and CFTR-DeltaF508 heterozygous pigs and CFTR-deficient ferrets (Rogers et al., 2008; Sun et al., 2008). Despite high transfection efficiency and production of transgenic cloned animals using viral-mediated delivery method, the need to produce high concentrations of virus particles in addition to limitations of the size of DNA capable of being delivered via virus limits the application of this method.

3.3 Selection and characterization of transgenic cell lines
3.3.1 Selection by marker

Following DNA construct delivery into somatic cells, transfected cells are cultured 24-48h in the absence of selection, followed by selection. Selection agents are chosen according to the DNA construct and added in cell-type specific concentrations into the cell culture medium. Typically, G418 (Geneticin) is used when the neomycin resistance gene is the positive selection marker and gangciclovir when using the TK gene for negative selection. Due to the limited lifespan of many of the somatic cell lines that are typically used for nuclear transfer and the significant amount of time to produce clonal colonies, it is important to perform the genetic screening by PCR or southern blotting as early as possible, prior to cell senescence.

3.3.2 Genotyping by PCR or southern blotting

Screening selected colonies by PCR for a gene targeting event typically involves using one PCR primer specific to the host genome sequence and the other primer specific to the selectable marker between the arms of homology in the targeting vector. This approach provides a simple, rapid and highly sensitive method to identify the gene targeting event in the transfected cells (Gómez-Rodríguez et al., 2008). Furthermore, PCR can be performed using the lysate from the small amount of cells, and also can be facilitated by pooled analysis of multiple cell lines. These types of PCR analysis sometimes require optimization as the amplicon is typically several thousand bp depending on the length of the targeting arms. Following the initial screening via PCR, the targeted colonies can be further expanded for analysis by southern blotting. A strategy for identifying targeted cells by southern blotting should be incorporated in the vector design. Two DNA probes on each side of the wild type gene, but outside the targeting vector, are designed to detect a change in fragment size resulting from either introduction or elimination. It is necessary to verify that the probes for southern blotting work well on wild-type genomic DNA from the somatic cells before the targeting experiment. Also, it is important to perform southern blotting analysis using both probes to confirm the double-crossover event since homologous recombination can occur at only one end of the targeting vector (Hasty et al., 1991b; Moens et al., 1992). Of the methods of screening for gene targeting in somatic cells, southern blotting is considered to be the golden approach to identify correctly targeted colonies despite being time-consuming and relatively expensive compared to PCR approaches.

3.3.3 Fluorescence In situ hybridization

Fluorescence in situ hybridization (FISH) is most commonly used to identify the number of integration sites following random integration of a transgene. FISH has been successfully used to detect the number of bacterial artificial chromosome (BAC) sequences integrated and chromosomal location(s) in mouse ES cells (Yang and Seed, 2003). Although the chromosomal location can be shown directly by FISH, Gómez-Rodríguez et al. (2008) found that screening of BAC-based constructs by FISH can be prone to false positives because of small pieces of integrated DNA that are below the limits of detection by fluorescent hybridization.

3.4 SCNT using genetically modified somatic cells

Once the cell line containing the appropriate genetic modification has been identified, those cells can then be used for SCNT as described in Figure 1. SCNT is a process in which the nucleus of a somatic cell is transferred into the cytoplasm of an enucleated oocyte. The hereby reconstructed SCNT embryos can be activated to initiate development and then are transferred into a synchronized surrogate mother immediately or after short-term *in vitro* culture. In addition to the technical factors involved in the SCNT process, the status of the donor cells and the quality of unfertilized oocytes are considered to largely affect the overall efficiency of the SCNT. This point will be discussed more below.

4. Improvement of the efficiency of cell-based transgenesis via SCNT

Since cell-based transgenesis via SCNT relies extensively on both transgenic and SCNT techniques, the efficiency could be improved by increasing SCNT efficiency and/or gene transfer efficiency.

4.1 Factors affecting SCNT efficiency

SCNT is a technique that requires precise skills in micromanipulation. In addition to technical skills, numerous other factors also impact SCNT efficiency, including donor cell types, quality of the recipient oocytes, cell cycle synchronization of both donor cell and recipient cytoplasm, the epigenetic status of the donor cell, method of reconstructed zygote activation, epigenetic reprogramming following activation, in vitro culture conditions of reconstructed embryos and embryo transfer into surrogate mothers. Here, we will focus on the cell cycle synchronization and methods being used to improve nuclear remodeling and reprogramming during SCNT.

4.1.1 Type of cell and synchronization of cell and recipient oocyte

Different types of donor cells and the different origins of recipient oocyte have been successfully used to produce the cloned animals in domestic species (Table 2). Generally, nuclei from less differentiated donor cells demonstrate greater SCNT efficiency than those from differentiated donor cells (Hill et al., 2000; Lee et al., 2003). SCNT embryos created from *in vivo*-derived MII oocytes have better developmental potential than *in vitro*-maturated MII oocytes; and *in vitro*-maturated oocytes from sexually mature animals have been demonstrated to more efficiently produce cloned embryos than oocytes *in vitro*-matured from prepubertal animals (Lai et al., 2002a; Lee et al., 2003; Hyun et al., 2003b). Since the first SCNT animal was produced by transferring the nuclei from quiescent cells (G0), synchronization of the donor nuclei into G0/G1 is thought to be crucial to the success of SCNT (Wilmut et al., 1997). The G0 cells, with lower transcription activity and different chromatin configuration in contrast to the cells at other stages of the cell cycle, may be more responsive to factors inside the recipient oocyte cytoplasm that impact the nuclear remodeling and reprogramming process following SCNT. The cell cycle of cultured somatic cells can be synchronized by serum starvation, contact inhibition and chemical treatments (Cho et al., 2005; Gibbons et al., 2002). Serum starvation to induce the donor cells into G0 is widely used for the production of cloned transgenic animals (Table 2). However, recent studies have demonstrated that synchronizing the cell cycle of donor cells by serum starvation can cause apoptosis (Dalman et al., 2010) and reduce blastocyst production in cattle (Miranda Mdos et al., 2009). The production of the cloned transgenic calves from non-quiescent fetal fibroblasts demonstrated that the synchronization of the donor cells is not an absolute requirement for SCNT success (Cibelli et al., 1998). However, the relative cell cycle combination of both donor cells and recipient oocyte, not just donor cells, is thought to be important for the maintenance of correct ploidy and the subsequent development of reconstructed embryos (reviewed by Campbell, 1999).

4.1.2 Methods to improve nuclear remodeling and reprogramming

Following transfer of a donor nucleus into a recipient oocyte cytoplasm and subsequent activation, nuclear remodeling events of the chromatin structure, such as changes of DNA methylation patterns and histone modifications, result in the reprogramming of gene expression to recapitulate developmental patterns observed in a normal fertilized embryo (Whitworth and Prather, 2010). In contrast to less differentiated donor nuclei, the relatively high level of DNA methylation and low histone acetylation exist in the chromatin of the highly differentiated nuclei. These epigenetic modifications are used to maintain the temporal and spatial patterns of gene expression specific to the cell type or developmental stage. When differentiated nuclei are transferred into the enucleated oocyte cytoplasm, correctly establishing normal patterns of zygotic gene expression is crucial to the full term development of SCNT animals. However, numerous studies have demonstrated improper reprogramming of genes in embryos and tissues of domestic animals following SCNT (Wrenzycki et al., 2001; Pfister-Genskow et al., 2005; Aston et al., 2010; Ross et al., 2010b). Thus, various strategies are under development to facilitate and promote appropriate nuclear reprogramming of the transferred nucleus following nuclear transfer or to pre-program the genome of the donor nucleus prior to SCNT. Of these strategies, one widely used is to treat reconstructed SCNT embryos, but not the donor cells, with histone deacetylase inhibitors (HDACi), such as trichostatin A (TSA), 6-(1,3-dioxo-1H, 3H-benzo [de] isoquinolin-2-yl)-hexanoic acid hydroxyamide (Scriptaid), sodium butyrate and valproic acid. It has been demonstrated that histone deacetylase inhibitor treatment after SCNT can improve both *in vitro* development of SCNT embryos to the blastocyst stage and *in vivo* development to term following embryo transfer (Li et al., 2008; Cervera et al., 2009; Zhao et al., 2009a, 2010; Das et al., 2010; Himaki et al., 2010a; Miyoshi et al., 2010). However, these treatments may be with some level of toxicity, as one group has reported that offspring from TSA-treated rabbit embryos did not survive to adulthood (Meng et al., 2009). The exact mechanism by which the HDACi treatment significantly improves the cloning efficiency remains largely unknown, although it has been shown to increase levels of global histone acetylation after HDACi treatment which may subsequently change the structure of chromatin and improve nuclear reprogramming (Shi et al., 2008; Iager et al., 2008; Zhao et al., 2010; Das et al., 2010).

Accruing investigations have demonstrated that abnormal DNA methylation patterns contribute to the lower developmental competency of SCNT derived embryos (Kang et al., 2001, 2002; Bourc'his et al., 2001; Santos et al., 2003; Wrenzycki et al., 2006), suggesting the inability of oocyte to fully restore the DNA methylation pattern of differentiated donor nuclei to that of normal totipotent 1-cell stage embryos. Thus, the second method used to assist epigenetic reprogramming is to reduce the DNA methylation level in donor cells or reconstructed embryos by treating them with the DNA methyl-transferase inhibitor such as 5-aza-2'-deoxycytidine (5-aza-dC). Unfortunately, previous studies have demonstrated that 5-aza-dC treatment of donor cells or cloned embryos does not improve the *in vitro* and *in vivo* development of SCNT derived embryos (Enright et al., 2005; Tsuji et al., 2009). However, combining the treatment of donor cells and embryos with both TSA and 5-aza-dC resulted in improved blastocyst development (Ding et al., 2008). Furthermore, an additional study has demonstrated enhanced gene targeting frequency in ES cells with low genomic methylation levels, suggesting the epigenetic status of targeted loci may influence the efficiency of gene targeting by affecting the accessibility for the homologous recombination machinery (Domínguez-Bendala and McWhir, 2004).

An additional strategy to promote developmental reprogramming of cloned embryos is to treat them with latrunculin A (LatA), an actin polymerization inhibitor. One group has reported that post-activation treatment with LatA is effective to improve *in vitro* developmental capacity of gene-modified cloned miniature pig embryos and embryos treated with LatA have the ability to develop into fetuses (Himaki et al., 2010b). Pre-reprogramming donor nuclei prior to SCNT has also been attempted. Rathbone et al (2010) reported that pretreatment of permeabilized ovine fetal fibroblasts with a cytoplasmic extract produced from germinal vesicle (GV) stage *Xenopus laevis* oocytes improves the live birth rate, but not development to blastocyst stage or pregnancy rate following embryo transfer.

4.2 Methods to improve gene transfer efficiency

Production of transgenic domestic animals has been widely accomplished, however, several limitations remain. Typically, plasmid based DNA constructs are limiting in the size of exogenous DNA to transfer and creating animals with targeted genetic modifications has been significantly more challenging. Thus, it is important to continue development of new strategies that broaden application and increase the efficiency of creating targeted genetic modifications in domestic animals.

4.2.1 Artificial chromosomes as DNA transfer vector

In contrast to plasmid based vector, artificial chromosomes have the capacity to carry Megabase-sized pieces of DNA that are maintained as autonomous, replicating chromosomes. Artificial chromosome vectors include a centromere, two telomeres and origins of replication (Robl et al., 2007). A 10 Mb human artificial chromosome (HAC) vector containing the entire unarranged sequences of the human immunoglobulin heavy and light chain loci (1-1.5 Megabase for each locus) have been transferred into bovine fibroblasts using a microcell-mediated chromosome transfer approach. Following SCNT using selected cells, trans-chromosomal cloned bovine offspring were produced that expressed human immunoglobulin proteins in the blood. HAC was retained as an independent chromosome with the proportion of cells ranged from 78 to 100% in most animals and HAC retention rate has not changed over several years. This system provides a useful tool to produce human therapeutic polyclonal antibodies using trans-chromosomal cloned domestic animals (Kuroiwa et al., 2002; Robl et al., 2007). In domestic animals, the swine artificial chromosome (SAC) (about 310 kb) containing pig centromeric DNA and the neomycin resistance gene was constructed and introduced into pig cell lines, and one positive clone was characterized, showing the possibility for producing transgenic pigs for xenotransplantation and other purposes (Poggiali et al., 2002).

4.2.2 Zinc finger nucleases (ZFNs) and transcription activator-like effector nucleases (TALNs)

The low frequency of homologous recombination hampers rapid progress and wide application of gene targeting in domestic animals. The generation of a site-specific double-stranded DNA break (DSB) within the desired locus can facilitate gene targeting by resulting in additions and deletions causing inactivation of gene function. Naturally occurring DNA-binding proteins, including zinc finger proteins (ZFPs) and meganucleases, have been engineered to bind site-specific DNA sequence. Zinc finger nucleases (ZFNs) combine the specific DNA-binding domain (ZFP) with the non-specific cleavage domain of the restriction endonuclease *FokI* and offer powerful tools to create a site-specific DSB to facilitate local homologous recombination. The ZFN induced DSB can lead to incorporating exogenous DNA in a site specific manner by utilization of a homologous recombination targeting vector that overlaps with the DSB region. Additionally, DSBs repaired by non-homologous end-joining (NHEJ) can result in loss of a single nucleotide, multiple nucleotides, or small regions; all capable of rendering a targeted gene dysfunctional. For increasing the specificity of DNA binding, multiple zinc fingers, each recognizing and binding to a 3-bp sequence of DNA nucleotides can be linked in tandem to recognize a unique genomic locus. In human somatic cells, custom-designed ZFNs yielded more than 18% targeting efficiency at the X-linked interleukin-2 receptor gamma gene locus and about 7% of the cells possessed a bi-allelic gene modification (Urnov et al., 2005). ZFNs can also promote the addition of novel DNA sequence into a targeted endogenous locus of human cells at a frequency ranging from 5% to 15% depending on the size of extra-chromosomal DNA (Moehle et al., 2007). Thus, the efficiency of gene targeting using ZFNs offers the possibility of gene therapy for human genetic disease and an approach for improvement of genetic engineering in domestic animals. Recently, targeted gene disruption of exogenous EGFP gene was achieved in porcine somatic cells using ZFNs (Watanabe et al., 2010) and transgenic EGFP knockout pigs were produced using ZFNs followed by SCNT (Whyte et al., 2011).

The appropriate design of a site-specific ZFN is critical to successfully introduce ZFN-mediated genetic modifications. Of the available zinc-finger engineering methods, modular assembly is the most easily performed method, but is also associated with high failure rates to yield a functional three–zinc finger array for the majority of potentially targetable sites (Ramirez et al, 2008). The second method is to combine selection-based methods with ZFNs made using modular assembly. While effective, this method is labor intensive and requires additional expertise. The third method, referred to as OPEN (Oligomerized Pool ENgineering) is based on bacterial 2-hybrid (B2H) selection and has been proven to be a rapid platform for plant and human cells with high targeting efficiency (ranging from 1 to 50% at different loci) and less toxicity compared to modular assembly system. However, utilization of OPEN requires an archive of pre-selected zinc-finger pools and *E. coli* selection (Maeder et al., 2008). Furthermore, due to the challenge of engineering the endonucleases, orthophenanthroline (OP, a DNA cleaving molecule) was conjugated with triplex-forming oligonucleotides (TFOs, sequence-specific binding capacity) to induce targeted DSBs and stimulate mutations at the target site in approximately 10% of treated human cells (Cannata et al., 2008). TFO conjugating to OP or other DNA cleaving molecules may provide a useful tool to induce targeted gene modification because triplex-forming sequences are frequent in mammalian genes. While ZFN-driven gene targeting can be much more efficient than homologous recombination-based methods, the design and development of highly specific ZFNs remain difficult because of the lack of a simple correspondence between amino acid sequence and DNA recognition sequence.

Recently, several groups have shown that transcription activator–like effectors (TALEs) from the bacterial genus *Xanthomonas* contain a central domain of tandem repeats that can be readily engineered to bind virtually any DNA sequence (Boch et al., 2009; Christian et al., 2010; Morbitzer et al., 2010). The structure of the central protein domain, highly conserved in all the known TALEs, includes 17.5 tandem repeats with 34 amino acids per repeat. In each repeat monomer of a TALE, only amino acid positions 12 and 13 are hypervariable (repeat variable diresidues) (Boch et al., 2011), which can specifically recognize a single nucleotide in the target site (Boch et al., 2009; Moscou et al., 2009). Thus, the correspondence between each repeat variable diresidues and the binding nucleotide in DNA sequence opens the possibility to create novel sequence-specific DNA binding proteins by rearrangement of TALE repeats. The engineered hybrid TALE nucleases (TALNs), produced by fusion of the *FokI* endonuclease domain with the high-specificity DNA-binding domains of TALEs, can bind and create targeted DSBs in tobacco and yeast (Mahfouz et al., 2011; Li et al., 2011), showing the feasibility of engineering TALE-based hybrid nucleases capable of generating site-specific genome modification. Recently, Miller et al. (2011) reported the generation of discrete edits or small deletions within endogenous human NTF3 and CCR5 genes and the insertion of 46-bp sequence at CCR5 locus into the genome of human K562 cells using designed TALNs, demonstrating the effective application of TALNs to modify endogenous genes. While the simple DNA-binding code of TALEs enables easier design strategies as compared to ZFPs, the repetitive nature of TALE DNA-binding domains results in difficulty to efficiently synthesize new TALEs by currently used vector construction methods. To overcome this problem, Zhang et al. (2011) recently developed a new strategy to construct repeat domains of TALEs by hierarchical ligation.

4.3 Induced pluripotent stem (iPS) cells

Owing to the lack of ES cells in domestic animals, it is difficult to replicate strategies routinely used to create genetically modified mice. As an alternative, cell-based transgenesis via SCNT is currently used to produce genetically modified domestic animals. Recent advancements in the ability to generate induced pluripotent stem (iPS) cells may open another potential strategy to improve the efficiency of SCNT in domestic animals. Induced pluripotent stem cells in mice and human were successfully generated by reprogramming somatic cells with viral delivery of a combination of four defined transcription factors including Sox2, Oct4, Klf4, and c-Myc or Sox2, Oct4, Nanog, and Lin28 (Takahashi and Yamanaka, 2006; Takahashi et al., 2007; Yu et al., 2007). The iPS cells are similar to ES cells in morphology, biochemistry, gene expression and the ability to differentiate into many cell types and self renew (Wernig et al., 2007; Lowry et al., 2008). Furthermore, subsequent studies have optimized existing procedures and discovered novel reprogramming protocols to generate the iPS cells, including non-integrating viruses (Stadtfeld et al., 2008), non-viral vectors (Okita et al., 2008), non-integration episomal vectors (Yu et al., 2009) and RNA-induced reprogramming (Warren et al., 2010), which greatly decrease the biosafety concerns associated with the application of iPS cells. The use of iPS cells can benefit animal transgenesis in several aspects. (i) Stable genetic modification in iPS cells may be more efficient compared to somatic cells as is the case with ES cells. (ii) Genetically modified iPS cells can be used to produce chimeric animals since it has been reported that iPS cells produce viable, live-born and fertile mice offspring through tetraploid complementation (Zhao et al., 2009b; Boland et al., 2009). (iii) The use of genetically modified iPS cells as donors may increase the efficiency of cell-based transgenesis via SCNT owing to the pluripotent status of iPS cells. (iv) In contrast to the limited lifespan of somatic cells, true iPS cells are immortalized. These advantages, coupled to successful derivation of iPS cells from domestic animals (Esteban et al., 2009; Ezashi et al., 2009; Wu et al., 2009) presents a new opportunity to produce transgenic animals using iPS cells.

5. Summary

While the potential opportunities of transgenic domestic animals in biomedicine and agriculture are significant, current procedures, including cell-based transgenesis via SCNT, to produce genetically modified domestic animals are not without limitations. The combination of new technologies, including ZFNs/TALNs to enhance targeted genome modification and iPS cells and other strategies to improve epigenetic remodeling of SCNT embryos, represent pathways for improving the success rates of current genome manipulation strategies resulting in transgenic domestic animals. These modern approaches may have limitations of their own, such as the difficulty and high cost to design, produce and validate the target-specific ZFPs, constructing custom-designed TALEs, and maintenance of iPS cells. However, despite these limitations, we expect to see these strategies become widely utilized as a result of the potential opportunities that utilization of these strategies offers to the field of targeted genome manipulation in domestic animals.

6. References

Aston, K.I., Li, G.P., Hicks, B.A., Sessions, B.R., Davis, A.P., Rickords, L.F., Stevens, J.R., White, K.L. 2010. Abnormal levels of transcript abundance of developmentally important genes in various stages of preimplantation bovine somatic cell nuclear transfer embryos. Cell Reprogram. 12(1): 23-32.

Auchincloss, H. Jr., Sachs, D.H. 1998. Xenogeneic transplantation. Annu Rev Immunol 16: 433-470.

Bhatti, F.N., Zaidi, A., Schmoeckel, M., Cozzi, E., Chavez, G., Wallwork, J., White, D.J., Friend, P.J. 1998. Survival of life-supporting HDAF transgenic kidneys in primates is enhanced by splenectomy. Transplant Proc 30(5): 2467.

Bleck, G.T., White, B.R., Miller, D.J., Wheeler, M.B. 1998. Production of bovine alpha-lactalbumin in the milk of transgenic pigs. J Anim Sci 76(12):3072-3078.

Boch, J., Scholze, H., Schornack, S., Landgraf, A., Hahn, S., Kay, S., Lahaye, T., Nickstadt, A., Bonas, U. 2009. Breaking the code of DNA binding specificity of TAL-type III effectors. Science 326(5959): 1509-1512.

Boch, J. 2011. TALEs of genome targeting. Nat Biotechnol 29(2): 135-136.

Boland, M. J., Hazen, J. L., Nazor, K. L., Rodriguez, A.R., Gifford, W., Martin, G., Kupriyanov, S., Baldwin, K.K. 2009. Adult mice generated from induced pluripotent stem cells. Nature 461 (7260): 91-94.

Bordignon, V., Keyston, R., Lazaris, A., Bilodeau, A.S., Pontes, J.H., Arnold, D., Fecteau, G., Keefer, C., Smith, L.C. 2003. Transgene expression of green fluorescent protein and germ line transmission in cloned calves derived from in vitro-transfected somatic cells. Biol Reprod 68: 2013-2023.

Bourc'his, D., Le Bourhis, D., Patin, D., Niveleau, A., Comizzoli, P., Renard, J.P., Viegas-Pequignot, E. 2001. Delayed and incomplete reprogramming of chromosome methylation patterns in bovine cloned embryos. Curr Biol 11: 1542-1546.

Brophy, B., Smolenski, G., Wheeler, T., Wells, D., L'Huillier, P., Laible, G. 2003. Cloned transgenic cattle produce milk with higher levels of beta-casein and kappa-casein. Nat Biotechnol 21(2):157-162.

Byrne, G.W., McCurry, K.R., Martin, M.J., McClellan, S.M., Platt, J.L., Logan, J.S. 1997. Transgenic pigs expressing human CD59 and decay-accelerating factor produce an intrinsic barrier to complement mediated damage. Transplantation 63: 149.

Cabot, R.A., Kühholzer, B., Chan, A.W., Lai, L., Park, K.W., Chong, K.Y., Schatten, G., Murphy, C.N., Abeydeera, L.R., Day, B.N., Prather, R.S. 2001. Transgenic pigs produced using in vitro matured oocytes infected with a retroviral vector. Anim Biotechnol 12(2):205-214.

Campbell, K.H. 1999. Nuclear transfer in farm animal species. Semin Cell Dev Biol. 10(3):245-252.

Cannata, F., Brunet, E., Perrouault, L., Roig, V., Ait-Si-Ali, S., Asseline, U., Concordet, J.P., Giovannangeli, C. 2008. Triplex-forming oligonucleotide-orthophenanthroline conjugates for efficient targeted genome modification. Proc Natl Acad Sci USA 105(28): 9576-9581.

Castro, F.O., Hernandez, O., Uliver, C., Solano, R., Milanes, C., Aguilar, A., Perez, R., De Armas, R., Herrera, N., Fuente, J.D.L. 1991. Introduction of foreign DNA into the spermatozoa of farm animals. Theriogenology 34: 1099-1110.

Cervera, R.P., Martí-Gutiérrez, N., Escorihuela, E., Moreno, R., Stojkovic, M. 2009. Trichostatin A affects histone acetylation and gene expression in porcine somatic cell nucleus transfer embryos. Theriogenology 72(8): 1097-1110.

Chan, A.W., Homan, E.J., Ballou, L.U., Burns, J.C., Bremel, R.D. 1998. Transgenic cattle produced by reverse-transcribed gene transfer in oocytes. Proc Natl Acad Sci USA 95(24): 14028-14033.

Chang, K., Qian, J., Jiang, M., Liu, Y.H., Wu, M.C., Chen, C.D., Lai, C.K., Lo, H.L., Hsiao, C.T., Brown, L., Bolen, J. Jr., Huang, H.I., Ho, P.Y., Shih, P.Y., Yao, C.W., Lin, W.J., Chen, C.H., Wu, F.Y., Lin, Y.J., Xu, J., Wang, K. 2002. Effective generation of transgenic pigs and mice by linker based sperm-mediated gene transfer. BMC Biotechnol 2: 5.

Chen, R.H., Naficy, S., Logan, J.S., Diamond, L.E., Adams, D.H. 1999. Hearts from transgenic pigs constructed with CD59/DAF genomic clones demonstrate improved survival in primates. Xenotransplantation 6: 194.

Cho, S.R., Ock, S.A., Yoo, J.G., Mohana kumar, B., Choe, S.Y., Rho, G.J. 2005. Effect of confluent, Roscovitine treatment and serum starvation on the cell-cycle synchronization of bovine foetal fibroblasts. Reprod Domest Anim 40: 171-176.

Choi, T., Huang, M., Gorman, C., Jaenisch, R. 1991. A generic intron increases gene expression in transgenic mice. Molecular Cellular Biology 11: 3070-3074.

Christian, M., Cermak, T., Doyle, E.L., Schmidt, C., Zhang, F., Hummel, A., Bogdanove, A.J., Voytas, D.F. 2010. Targeting DNA double-strand breaks with TAL effector nucleases. Genetics 186(2): 757-761.

Cibelli, J.B., Stice, S.L., Golueke, P.J., Kane, J.J., Jerry, J., Blackwell, C., Ponce de Leon, F. A., Robl. J.M. 1998. Cloned transgenic calves produced from nonquiescent fetal fibroblasts. Science 280: 1256-1258.

Clements, J.E., Wall, R.J., Narayan, O., Hauer, D., Schoborg, R., Sheffer, D., Powell, A., Carruth, L.M., Zink, M.C., Rexroad, C.E. 1994. Development of transgenic sheep that express the visna virus envelope gene. Virology 200: 370-380.

Cooper, D.K., Ezzelarab, M., Hara, H., Ayares, D. 2008. Recent advances in pig-to-human organ and cell transplantation. Expert Opin Biol Ther 8(1): 1-4.

Cozzi, E., White, D.J.G. 1995. The generation of transgenic pigs as potential organ donors for humans. Nat Med 1: 964-966.

Dai, Y., Vaught, T.D., Boone, J., Chen, S.H., Phelps, C.J., Ball, S., Monahan, J.A., Jobst, P.M., McCreath, K.J., Lamborn, A.E., Cowell-Lucero, J.L., Wells, K.D., Colman, A., Polejaeva, I.A., Ayares, D.L. 2002. Targeted disruption of the alpha1,3-galactosyltransferase gene in cloned pigs. Nat Biotechnol 20: 251-255.

Dalman, A., Eftekhari-Yazdi, P., Valojerdi, M., Shahverdi, A., Gourabi, H., Janzamin, E., Fakheri, R., Sadeghian, F., Hasani, F. 2010. Synchronizing Cell Cycle of Goat Fibroblasts by Serum Starvation Causes Apoptosis. Reprod Domest Anim. 45: e46–e53.

Das, Z.C., Gupta, M.K., Uhm, S.J., Lee, H.T. 2010. Increasing histone acetylation of cloned embryos, but not donor cells, by sodium butyrate improves their in vitro development in pigs. Cell Reprogram 12(1): 95-104.

Denning, C., Burl, S., Ainslie, A., Bracken, J., Dinnyes, A., Fletcher, J., King, T., Ritchie, M., Ritchie, W.A., Rollo, M., de Sousa, P., Travers, A., Wilmut, I., Clark, A.J. 2001a. Deletion of the lpha(1,3)galactosyl transferase (GGTA1) gene and the prion protein (PrP) gene in sheep. Nat Biotechnol 19. 559-562.

Denning, C., Dickinson, P., Burl, S., Wylie, D., Fletcher, J. Clark, A.J. 2001b. Gene targeting in primary fetal fibroblasts from sheep and pig. Cloning Stem Cells 3(4): 221-231.

Diamond LE, McCurry KR, Martin MJ, McClellan SB, Oldham ER, Platt JL, Logan JS. 1996. Characterization of transgenic pigs expressing functionally active human CD59 on cardiac endothelium. Transplantation 61(8): 1241-1249.

Diamond, L.E., Quinn, C.M., Martin, M.J., Lawson, J., Platt, J.L., Logan, J.S. 2001. A human CD46 transgenic pig model system for the study of discordant xenotransplantation. Transplantation 71(1): 132-142.

Ding X, Wang Y, Zhang D, Wang Y, Guo Z, Zhang Y. 2008. Increased pre-implantation development of cloned bovine embryos treated with 5-aza-2'-deoxycytidine and trichostatin A. Theriogenology 70(4): 622-630.

Diswall, M., Angström, J., Karlsson, H., Phelps, C.J., Ayares, D., Teneberg, S., Breimer, M.E. 2010. Structural characterization of alpha1,3-galactosyltransferase knockout pig heart and kidney glycolipids and their reactivity with human and baboon antibodies. Xenotransplantation 17(1): 48-60.

Domínguez-Bendala, J., McWhir, J. 2004. Enhanced gene targeting frequency in ES cells with low genomic methylation levels. Transgenic Res 13(1): 69-74.

Duncker, B.P., Davies, P.L., Walker, V.K. 1997. Introns boost transgene expression in Drosophila melanogaster. Mol Gen Genet 254: 291-296.

Ebert, K.M., Selgrath, J.P., DiTullio, P., Denman, J., Smith, T.E., Memon, M.A., Schindler, J.E., Monastersky, G.M., Vitale, J.A., Gordon, K. 1991. Transgenic production of a variant of human tissue-type plasminogen activator in goat milk: generation of transgenic goats and analysis of expression. Biotechnology (N Y) 9(9): 835-838.

Enright, B.P., Sung, L.Y., Chang, C.C., Yang, X., Tian, X.C. 2005. Methylation and acetylation characteristics of cloned bovine embryos from donor cells treated with 5-aza-2'-deoxycytidine. Biol Reprod 72(4): 944-948.

Esteban, M. A., Xu, J., Yang, J., Peng, M., Qin, D., Li, W., Jiang, Z., Chen, J., Deng, K., Zhong, M., Cai, J., Lai, L., Pei, D. 2009. Generation of induced pluripotent stem cell lines from Tibetan miniature pig. J Biol Chem 284(26): 17634-17640.

Ezashi, T., Telugu, B. P., Alexenko, A. P., Sachdev, S., Sinha, S., Roberts, R. M. 2009. Derivation of induced pluripotent stem cells from pig somatic cells. Proc Natl Acad Sci USA 106(27): 10993-10998.

Fodor, W.L., Williams, B.L., Matis, L.A., Madri, J.A., Rollins, S.A., Knight, J.W., Velander, W., Squinto, S.P. 1994. Expression of a functional human complement inhibitor in a transgenic pig as a model for the prevention of xenogeneic hyperacute organ rejection. Proc Natl Acad Sci USA 91(23): 11153-11157.

Galili, U. 1993. Interaction of the natural anti-Gal antibody with alpha-galactosyl epitopes: a major obstacle for xenotransplantation in humans (see comments). Immunol Today 14: 480-482.

García-Vázquez, F.A., Ruiz, S., Matás, C., Izquierdo-Rico, M.J., Grullón, L.A., De Ondiz, A., Vieira, L., Avilés-López, K., Gutiérrez-Adán, A., Gadea, J. 2010. Production of transgenic piglets using ICSI-sperm-mediated gene transfer in combination with recombinase RecA. Reproduction 140(2): 259-272.

Gibbons, J., Arat, S., Rzucidlo, J., Miyoshi, K., Waltenburg, R., Respess, D., Venable, A., Stice, S. 2002. Enhanced survivability of cloned calves derived from roscovitine-treated adult somatic cells. Biol Reprod 66: 895-900.

Gómez-Rodríguez, J., Washington, V., Cheng, J., Dutra, A., Pak, E., Liu, P., McVicar, D.W., Schwartzberg, P.L. 2008. Advantages of q-PCR as a method of screening for gene targeting in mammalian cells using conventional and whole BAC-based constructs. Nucleic Acids Res. 36(18): e117.

Grosse-Hovest, L., Müller, S., Minoia, R., Wolf, E., Zakhartchenko, V., Wenigerkind, H., Lassnig, C., Besenfelder, U., Müller, M., Lytton, S.D., Jung, G., Brem, G. 2004. Cloned transgenic farm animals produce a bispecific antibody for T cell-mediated tumor cell killing. Proc Natl Acad Sci USA 101(18): 6858-6863.

Hammer, R.E., Pursel, V.G., Rexroad, C.E. Jr, Wall, R.J., Bolt, D.J., Ebert, K.M., Palmiter, R.D., Brinster, R.L. 1985. Production of transgenic rabbits, sheep and pigs by microinjection. Nature 315(6021): 680-683.

Hanson, K.D., Sedivy, J.M. 1995. Analysis of biological selections for high-efficiency gene targeting. Mol Cell Biol. 15(1): 45-51.

Hasty, P., Rivera-Pérez, J., Bradley, A. 1991a. The length of homology required for gene targeting in embryonic stem cells. Mol Cell Biol. 11(11): 5586-5591.

Hasty, P., Rivera-Perez, J., Chang, C., Bradley, A. 1991b. Target frequency and integration pattern for insertion and replacement vectors in embryonic stem cells. Mol Cell Biol 11: 4509-4517.

Higashibata, Y., Sakuma, T., Kawahata, H., Fujihara, S., Moriyama, K., Okada, A., Yasui, T., Kohri, K., Kitamura, Y., Nomura, S. 2004. Identification of promoter regions involved in cell- and developmental stage-specific osteopontin expression in bone, kidney, placenta, and mammary gland: an analysis of transgenic mice. J Bone Miner Res 19(1): 78-88.

Hill, J.R., Winger, Q.A., Long, C.R., Looney, C.R., Thompson, J.A., Westhusin, M.E. 2000. Development rates of male bovine nuclear transfer embryos derived from adult and fetal cells. Biol Reprod 62(5): 1135-1140.

Himaki, T., Yokomine, T.A., Sato, M., Takao, S., Miyoshi, K., Yoshida, M. 2010a. Effects of trichostatin A on in vitro development and transgene function in somatic cell nuclear transfer embryos derived from transgenic Clawn miniature pig cells. Anim Sci J 81(5): 558-563.

Himaki, T., Mori, H., Mizobe, Y., Miyoshi, K., Sato, M., Takao, S., Yoshida, M. 2010b. Latrunculin A dramatically improves the developmental capacity of nuclear transfer embryos derived from gene-modified clawn miniature pig cells. Cell Reprogram 12(2): 127-131.

Hirata, R.K., Xu, C., Dong, R., Miller, D.G., Ferguson, S., Russell, D.W. 2004. Efficient PRNP gene targeting in bovine fibroblasts by adeno-associated virus vectors. Cloning Stem Cells 6(1): 31-36.

Huang, Y.J., Huang, Y., Baldassarre, H., Wang, B., Lazaris, A., Leduc, M., Bilodeau, A.S., Bellemare, A., Côté, M., Herskovits, P., Touati, M., Turcotte, C., Valeanu, L., Lemée, N., Wilgus, H., Bégin, I., Bhatia, B., Rao, K., Neveu, N., Brochu, E., Pierson, J., Hockley, D.K., Cerasoli, D.M., Lenz, D.E., Karatzas, C.N., Langermann, S. 2007. Recombinant human butyrylcholinesterase from milk of transgenic animals to protect against organophosphate poisoning. Proc Natl Acad Sci USA 104(34): 13603-13608.

Hyun, S., Lee, G., Kim, D., Kim, H., Lee, S., Nam, D., Jeong, Y., Kim, S., Yeom, S., Kang, S., Han, J., Lee, B., Hwang, W. 2003a. Production of nuclear transfer-derived piglets using porcine fetal fibroblasts transfected with the enhanced green fluorescent protein. Biol Reprod 69(3): 1060–1068.

Hyun, S.H., Lee, G.S., Kim, D.Y., Kim, H.S., Lee, S.H., Kim, S., Lee, E.S., Lim, J.M., Kang, S.K., Lee, B.C., Hwang, W.S. 2003b. Effect of maturation media and oocytes derived from sows or gilts on the development of cloned pig embryos. Theriogenology 59(7): 1641-1649.

Iager, A.E., Ragina, N.P., Ross, P.J., Beyhan, Z., Cunniff, K., Rodriguez, R.M., Cibelli, J.B. 2008. Trichostatin A improves histone acetylation in bovine somatic cell nuclear transfer early embryos. Cloning Stem Cells 10(3): 371-379.

Jin, D.I., Lee, S.H., Choi, J.H., Lee, J.S., Lee, J.E., Park, K.W., Seo, J.S. 2003. Targeting efficiency of a-1,3-galactosyl transferase gene in pig fetal fibroblast cells. Exp Mol Med. 35(6): 572-577.

Kang, Y.K., Koo, D.B., Park, J.S., Choi, Y.H., Chung, A.S., Lee, K.K., Han, Y.M. 2001. Aberrant methylation of donor genome in cloned bovine embryos. Nat Genet 28: 173-177.

Kang, Y.K., Park, J.S., Koo, D.B., Choi, Y.H., Kim, S.U., Lee, K.K., Han, Y.M. 2002. Limited demethylation leaves mosaic-type methylation states in cloned bovine pre-implantation embryos. EMBO J 21: 1092-1100.

Keefer, C. L., Baldassarre, H., Keyston, R., Wang, B., Bhatia, B., Bilodeau, A.S., Zhou, J.F., Leduc, M., Downey, B.R., Lazaris, A., Karatzas, C.N. 2001. Generation of dwarf goat (Capra hircus) clones following nuclear transfer with transfected and nontransfected fetal fibroblasts and in vitro-matured oocytes. Biol Reprod 64: 849-856.

Kjer-Nielsen, L., Holmberg, K., Perera, J.D., McCluskey, J. 1992. Impaired expression of chimaeric major histocompatibility complex transgenes associated with plasmid sequences. Transgenic Research 1: 182-187.

Kragh, P.M., Nielsen, A.L., Li, J., Du, Y., Lin, L., Schmidt, M., Bogh, I.B., Holm, I.E., Jakobsen, J.E., Johansen, M.G., Purup, S., Bolund, L., Vajta, G., Jorgensen, A.L. 2009. Hemizygous minipigs produced by random gene insertion and handmade cloning express the Alzheimer's disease-causing dominant mutation APPsw. Transgenic Res 18: 545-558.

Kuroiwa, Y., Kasinathan, P., Choi, Y.J., Naeem, R., Tomizuka, K., Sullivan, E.J., Knott, J.G., Duteau, A., Goldsby, R.A., Osborne, B.A., Ishida, I., Robl, J.M. 2002. Cloned transchromosomic calves producing human immunoglobulin. Nat Biotechnol 20: 889-894.

Kuroiwa, Y., Kasinathan, P., Matsushita, H., Sathiyaselan, J., Sullivan, E.J., Kakitani, M., Tomizuka, K., Ishida, I., Robl, J.M. 2004. Sequential targeting of the genes encoding immunoglobulin-mu and prion protein in cattle. Nat Genet 36(7): 775-780.

Lai, L., Kolber-Simonds, D., Park, K.W., Cheong, H.T., Greenstein, J.L., Im, G.S., Samuel, M., Bonk, A., Rieke, A., Day, B.N., Murphy, C.N., Carter, D.B., Hawley, R.J., Prather, R.S. 2002a. Production of a-1, 3-galactosyltransferase knockout pigs by nuclear transfer cloning. Science 295: 1089-1092.

Lai, L., Park, K.W., Cheong, H.T., Kuhholzer, B., Samuel, M., Bonk, A., Im, G.S., Rieke, A., Day, B.N., Murphy, C.N., Carter, D.B., Prather, R.S. 2002b. Transgenic pig expressing the enhanced green fluorescent protein produced by nuclear transfer using colchicine-treated fibroblasts as donor cells. Mol Reprod Dev 62(3): 300-306.

Lai, L., Kang, J.X., Li, R., Wang, J., Witt, W.T., Yong, H.Y., Hao, Y., Wax, D.M., Murphy, C.N., Rieke, A., Samuel, M., Linville, M.L., Korte, S.W., Evans, R.W., Starzl, T.E., Prather, R.S., Dai, Y. 2006. Generation of cloned transgenic pigs rich in omega-3 fatty acids. Nat. Biotechnol 24: 435-436.

Lavitrano, M., Bacci, M.L., Forni, M., Lazzereschi, D., Di Stefano, C., Fioretti, D., Giancotti, P., Marfé, G., Pucci, L., Renzi, L., Wang, H., Stoppacciaro, A., Stassi, G.,

Sargiacomo, M., Sinibaldi, P., Turchi, V., Giovannoni, R., Della Casa, G., Seren, E., Rossi, G. 2002. Efficient production by sperm-mediated gene transfer of human decay accelerating factor (hDAF) transgenic pigs for xenotransplantation. Proc Natl Acad Sci USA 99(22): 14230-14235.

Lavitrano, M., Busnelli, M., Cerrito, M.G., Giovannoni, R., Manzini, S., Vargiolu, A. 2006. Sperm-mediated gene transfer. Reprod Fertil Dev 18(1-2): 19-23.

Lee, G.S., Hyun, S.H., Kim, H.S., Kim, D.Y., Lee, S.H., Lim, J.M., Lee, E.S., Kang, S.K., Lee, B.C, Hwang, W.S. 2003. Improvement of a porcine somatic cell nuclear transfer technique by optimizing donor cell and recipient oocyte preparations. Theriogenology 59(9): 1949-1957.

Lee, H.G., Lee, H.C., Kim, S.W., Lee, P., Chung, H.J., Lee, Y.K., Han, J.H., Hwang, I.S., Yoo, J.I., Kim, Y.K., Kim, H.T., Lee, H.T., Chang, W.K., Park, J.K. 2009. Production of recombinant human von Willebrand factor in the milk of transgenic pigs. J Reprod Dev 55(5): 484-490.

Lee, S.L., Ock, S.A., Yoo, J.G., Kumar, B.M., Choe, S.Y., Rho, G.J. 2005. Efficiency of gene transfection into donor cells for nuclear transfer of bovine embryos. Mol Reprod Dev 72(2): 191-200.

Li, J., Svarcova, O., Villemoes, K., Kragh, P.M., Schmidt, M., Bøgh, I.B., Zhang, Y., Du, Y., Lin, L., Purup, S., Xue, Q., Bolund, L., Yang, H., Maddox-Hyttel, P., Vajta, G. 2008. High in vitro development after somatic cell nuclear transfer and trichostatin A treatment of reconstructed porcine embryos. Theriogenology 70(5): 800-808.

Li ,S., Guo, Y., Shi, J., Yin, C., Xing, F., Xu, L., Zhang, C., Liu, T., Li, Y., Li, H., Du, L., Chen, X. 2009. Transgene expression of enhanced green fluorescent protein in cloned rabbits generated from in vitro-transfected adult fibroblasts. Transgenic Res 18: 227-235.

Li. T., Huang, S., Jiang, W.Z., Wright, D., Spalding, M.H., Weeks, D.P., Yang, B. 2011. TAL nucleases (TALNs): hybrid proteins composed of TAL effectors and FokI DNA-cleavage domain. Nucleic Acids Res 39(1): 359-372.

Lo, D., Pursel, V., Linton, P.J., Sandgren, E., Behringer, R., Rexroad, C., Palmiter, R.D., Brinster, R.L. 1991. Expression of mouse IgA by transgenic mice, pigs and sheep. Eur. J. Immunol 21: 1001–1006.

Lowry, W.E., Richter, L., Yachechko, R., Pyle, A.D., Tchieu, J., Sridharan, R., Clark, A.T., Plath, K. 2008. Generation of human induced pluripotent stem cells from dermal fibroblasts. Proc Natl Acad Sci USA 105(8): 2883-2888.

Maeder, M.L., Thibodeau-Beganny, S., Osiak, A., Wright, D.A., Anthony, R.M., Eichtinger, M., Jiang, T., Foley, J.E., Winfrey, R.J., Townsend, J.A., Unger-Wallace, E., Sander, J.D., Müller-Lerch, F., Fu, F., Pearlberg, J., Göbel, C., Dassie, J.P., Pruett-Miller, S.M., Porteus, M.H., Sgroi, D.C., Iafrate, A.J., Dobbs, D., McCray, P.B. Jr., Cathomen, T., Voytas, D.F., Joung, J.K. 2008. Rapid "open-source" engineering of customized zinc-finger nucleases for highly efficient gene modification. Mol Cell 31(2): 294-301.

Magre, S., Takeuchi, Y., Bartosch, B. 2003. Xenotransplantation and pig endogenous retroviruses. Rev Med Virol 13(5): 311-329.

Mahfouz, M.M., Li, L., Shamimuzzaman, M., Wibowo, A., Fang, X., Zhu, J.K. 2011. De novo-engineered transcription activator-like effector (TALE) hybrid nuclease with novel DNA binding specificity creates double-strand breaks. Proc Natl Acad Sci USA 108(6): 2623-2628.

Marques, M.M., Thomson, A.J., McCreath, K.J., McWhir, J. 2006. Conventional gene targeting protocols lead to loss of targeted cells when applied to a silent gene locus in primary fibroblasts. J Biotechnol 125(2): 185-193.

McCreath, K. J., Howcroft, J., Campbell, K. H., Colman, A., Schnieke, A. E., Kind, A. J. 2000. Production of gene-targeted sheep by nuclear transfer from cultured somatic cells. Nature 405: 1066-1069.

McKee, C., Gibson, A., Dalrymple, M., Emslie, L., Garner, I., Cottingham, I. 1998. Production of biologically active salmon calcitonin in the milk of transgenic rabbits. Nat Biotechnol16(7): 647-651.

Meehan, D.T., Zink, M.A., Mahlen, M., Nelson, M., Sanger, W.G., Mitalipov, S.M., Wolf, D.P., Ouellette, M.M., Norgren Jr, R.B. 2008. Gene targeting in adult rhesus macaque fibroblasts. BMC Biotechnol 8: 31.

Melo, E.O., Canavessi, A.M., Franco, M.M., Rumpf, R. 2007. Animal transgenesis: state of the art and applications. J Appl Genet 48(1): 47-61.

Meng, Q., Polgar, Z., Liu, J., Dinnyes, A. 2009. Live birth of somatic cell-cloned rabbits following trichostatin A treatment and cotransfer of parthenogenetic embryos. Cloning Stem Cells 11(1): 203-208.

Miller, J.C., Tan, S., Qiao, G., Barlow, K.A., Wang, J., Xia, D.F., Meng, X., Paschon, D.E., Leung, E., Hinkley, S.J., Dulay, G.P., Hua, K.L., Ankoudinova, I., Cost, G.J., Urnov, F.D., Zhang, H.S., Holmes, M.C., Zhang, L., Gregory, P.D., Rebar, E.J. 2011. A TALE nuclease architecture for efficient genome editing. Nat Biotechnol 29(2): 143-148.

Mir, B., Piedrahita, J.A. 2004. Nuclear localization signal and cell synchrony enhance gene targeting efficiency in primary fetal fibroblasts. Nucleic Acids Res 32(3): e25.

Miranda Mdos, S., Bressan, F.F., Zecchin, K.G., Vercesi, A.E., Mesquita, L.G., Merighe, G.K., King, W.A., Ohashi, O.M., Pimentel, J.R., Perecin, F., Meirelles, F.V. 2009. Serum-starved apoptotic fibroblasts reduce blastocyst production but enable development to term after SCNT in cattle. Cloning Stem Cells 11(4): 565-573.

Miyoshi, K., Mori, H., Mizobe, Y., Asasaka, E., Ozawa, A., Yoshida, M., Sato, M. 2010. Valproic acid enhances in vitro development and oct-3/4 expression of miniature pig somatic cell nuclear transfer embryos. Cell Reprog 12: 67-74.

Moehle, E.A., Rock, J.M., Lee, Y.L., Jouvenot, Y., DeKelver, R.C., Gregory, P.D., Urnov, F.D., Holmes, M.C. 2007. Targeted gene addition into a specified location in the human genome using designed zinc finger nucleases. Proc Natl Acad Sci USA 104(9): 3055-3060.

Moens, C. B., Auerbach, A. B., Conlon, R. A., Joyner, A. L., Rossant, J. 1992. A targeted mutation reveals a role for N-myc in branching morphogenesis in the embryonic mouse lung. Genes Dev 6: 691-704.

Moisvadi, S., Kaminski, J.M., Yanagimachi, R. 2009. Use of intracytoplasmic sperm injection (ICSI) to generate transgenic animals. Comp Immunol Microbiol Infect Dis 32(2): 47-60.

Monaco, M.H., Gronlund, D.E., Bleck, G.T., Hurley, W.L., Wheeler, M.B., Donovan, S.M. 2005. Mammary specific transgenic over-expression of insulin-like growth factor-I (IGF-I) increases pig milk IGF-I and IGF binding proteins, with no effect on milk composition or yield. Transgenic Res 14(5): 761-773.

Morbitzer, R., Römer, P., Boch, J., Lahaye, T. 2010. Regulation of selected genome loci using de novo-engineered transcription activator-like effector (TALE)-type transcription factors. Proc Natl Acad Sci USA 107(50): 21617-21622.

Moscou, M.J., Bogdanove, A.J. 2009. A simple cipher governs DNA recognition by TAL effectors. Science 326(5959): 1501.

Mueller, N.J., Takeuchi, Y., Mattiuzzo, G., Scobie, L. 2011. Microbial safety in xenotransplantation. Curr Opin Organ Transplant 16(2): 201-206.

Okita, K., Nakagawa, M., Hyenjong, H., Ichisaka, T., Yamanaka, S. 2008. Generation of mouse induced pluripotent stem cells without viral vectors. Science 322(5903): 949-953.

Oropeza, M., Petersen, B., Carnwath, J.W., Lucas-Hahn, A., Lemme, E., Hassel, P., Herrmann, D., Barg-Kues, B., Holler, S., Queisser, A-L., Schwinzer, R., Hinkel, R., Kupatt, C., Niemann, H. 2009. Transgenic expression of the human A20 gene in cloned pigs provides protection against apoptotic and inflammatory stimuli. Xenotransplantation 16: 522-534.

Osada, T., Toyoda, A., Moisyadi, S., Akutsu, H., Hattori, M., Sakaki, Y., Yanagimachi, R. 2005. Production of inbred and hybrid transgenic mice carrying large (>200 kb) foreign DNA fragments by intracytoplasmic sperm injection, Mol Reprod Dev 72: 329-335.

Paleyanda, R.K., Velander, W.H., Lee, T.K., Scandella, D.H., Gwazdauskas, F.C., Knight, J.W., Hoyer, L.W., Drohan, W.N., Lubon, H. 1997. Transgenic pigs produce functional human factor VIII in milk. Nat Biotechnol 15(10): 971-975.

Paradis, K., Langford, G., Long, Z., Heneine, W., Sandstrom, P., Switzer, W.M., Chapman, L.E., Lockey, C., Onions, D., Otto, E. 1999. Search for cross-species transmission of porcine endogenous retrovirus in patients treated with living pig tissue. Science 285(5431): 1236-1241.

Park, J.K., Lee, Y.K., Lee, P., Chung, H.J., Kim, S., Lee, H.G., Seo, M.K., Han, J.H., Park, C.G., Kim, H.T., Kim, Y.K., Min, K.S., Kim, J.H., Lee, H.T., Chang, W.K. 2006. Recombinant human erythropoietin produced in milk of transgenic pigs. J Biotechnol 122(3): 362-371.

Patience, C., Takeuchi, Y., Weiss, R.A. 1997. Infection of human cells by an endogenous retrovirus of pigs. Nat Med 3(3): 282-286.

Perry, A.C.F., Wakayama, T., Kishikawa, H., Kasai, T., Okabe, M., Toyoda, Y., Yanagimachi, R. 1999. Mammalian transgenesis by intracytoplasmic sperm injection. Science 284:1180-1183.

Petersen, B., Kues, W., Lucas-Hahn, A., Queisser, A.L., Lemme, E., Hoelker, M., Carnwath, J.W., Niemann, H. 2006. Generation of pigs transgenic for hCD59/DAF and human thrombomodulin by somatic nuclear transfer. Reprod Fertil Dev 18: 142.

Pfister-Genskow, M., Myers, C., Childs, L.A., Lacson, J.C., Patterson, T., Betthauser, J.M., Goueleke, P.J., Koppang, R.W., Lange, G., Fisher, P., Watt, S.R., Forsberg, E.J., Zheng, Y., Leno, G.H., Schultz, R.M., Liu, B., Chetia, C., Yang, X., Hoeschele, I., Eilertsen, K.J. 2005. Identification of differentially expressed genes in individual bovine preimplantation embryos produced by nuclear transfer: improper reprogramming of genes required for development. Biol Reprod 72(3): 546-555.

Phelps, C.J., Koike, C., Vaught, T.D., Boone, J., Wells, K.D., Chen, S.H., Ball, S., Specht, S.M., Polejaeva, I.A., Monahan, J.A., Jobst, P.M., Sharma, S.B., Lamborn, A.E., Garst, A.S.,

Moore, M., Demetris, A.J., Rudert, W.A., Bottino, R., Bertera, S., Trucco, M., Starzl, T.E., Dai, Y., Ayares, D.L. 2003. Production of alpha 1,3-galactosyltransferase-deficient pigs. Science 299(5605): 411-414.

Poggiali, P., Scoarughi, G.L., Lavitrano, M., Donini, P., Cimmino, C. 2002. Construction of a swine artificial chromosome: a novel vector for transgenesis in the pig. Biochimie 84(11): 1143-1150.

Prather, R.S., Shen, M., Dai, Y. 2008. Genetically modified pigs for medicine and agriculture. Biotechnology and Genetic Engineering Reviews 25: 245-266.

Pursel, V.G., Rexroad, C.E. Jr. 1993. Recent progress in the transgenic modification of swine and sheep. Mol Reprod Dev 36(2): 251-254.

Pursel, V.G., Wall, R.J., Mitchell, A.D., Elsasser, T.H., Solomon, M.B., Coleman, M.E., Mayo, F., Schwartz, R. J. 1999. Expression of insulin-like growth factor-1 in skeletal muscle of transgenic pigs. In 'Transgenic Animals in Agriculture'. (Eds J. D. Murray, G. B. Anderson, A. M. Oberbauer and M. M. McGloughlin.) pp. 131-144. (CABI Publishing: New York.)

Qin, J.Y., Zhang, L., Clift, K.L., Hulur, I., Xiang, A.P., Ren, B.Z., Lahn, B.T. 2010. Systematic Comparison of Constitutive Promoters and the Doxycycline-Inducible Promoter. PLoS ONE 5(5): e10611.

Ramirez, C.L., Foley, J.E., Wright, D.A., Müller-Lerch, F., Rahman, S.H., Cornu, T.I., Winfrey, R.J., Sander, J.D., Fu, F., Townsend, J.A., Cathomen, T., Voytas, D.F., Joung, J.K. 2008. Unexpected failure rates for modular assembly of engineered zinc fingers. Nat Methods 5(5): 374-375.

Ramsoondar, J.J., Machaty, Z., Costa, C., Williams, B.L., Fodor, W.L., Bondioli, K.R. 2003. Production of alpha 1, 3-galactosyltransferase-knockout cloned pigs expressing human alpha 1, 2-fucosylosyltransferase. Biol Reprod 69(2): 437-445.

Ramsoondar, J., Vaught, T., Ball, S., Mendicino, M., Monahan, J., Jobst, P., Vance, A., Duncan, J., Wells, K., Ayares, D. 2009. Production of transgenic pigs that express porcine endogenous retrovirus small interfering RNAs. Xenotransplantation 16(3): 164-180.

Rathbone, A.J., Fisher, P.A., Lee, J.H., Craigon, J., Campbell, K.H. 2010. Reprogramming of ovine somatic cells with Xenopus laevis oocyte extract prior to SCNT improves live birth rate. Cell Reprogram. 12(5): 609-616.

Reh, W.A., Maga, E.A., Collette, N.M., Moyer, A., Conrad-Brink, J.S., Taylor, S.J., DePeters, E.J., Oppenheim, S., Rowe, J.D., BonDurant, R.H., Anderson, G.B., Murray, J.D. 2004. Hot topic: using a stearoyl-CoA desaturase transgene to alter milk fatty acid composition. J Dairy Sci 87: 3510-3514.

Richt, J.A., Kasinathan, P., Hamir, A.N., Castilla, J., Sathiyaseelan, T., Vargas, F., Sathiyaseelan, J., Wu, H., Matsushita, H., Koster, J., Kato, S., Ishida, I., Soto, C., Robl, J.M., Kuroiwa, Y. 2007. Production of cattle lacking prion protein. Nat Biotechnol 25(1): 132-138.

Robl, J.M., Wang, Z., Kasinathan, P., Kuroiwa, Y. 2007. Transgenic animal production and animal biotechnology. Theriogenology 67(1): 127-133.

Rogers, C.S., Hao, Y., Rokhlina, T., Samuel, M., Stoltz, D.A., Li, Y., Petroff, E., Vermeer, D.W., Kabel, A.C., Yan, Z., Spate, L., Wax, D., Murphy, C.N., Rieke, A., Whitworth, K., Linville, M.L., Korte, S.W., Schnieke, A.E., Kind, A.J., Ritchie, W.A., Mycock, K., Scott, A.R., Ritchie, M., Wilmut, I., Colman, A., Campbell, K.H. 1997. Human factor

IX transgenic sheep produced by transfer of nuclei from transfected fetal fibroblasts. Science 278: 2130-2133.

Rogers, C.S., Hao, Y., Roklina, T., Samuel, M., Stoltz, D.A., Li, Y., Petroff, E., Vermeer, D.W., Kabel, A.C., Yan, Z., Spate, L., Wax, D., Murphy, C.N., Rieke, A., Whitworth, K., Linville, M.L., Korte, S.W., Engelhardt, J.F., Welsh, M.J., Prather, R.S. 2008. Production of CFTR-null and CFTR-deltaF508 heterozygous pigs by adeno-associated virus-mediated gene targeting and somatic cell nuclear transfer. J Clin Invest 118(4): 1571-1577.

Ross, J.W., Prather, R.S., Whyte, J.J., Zhao, J. 2009a. Cloning and transgenics: progress and new approaches in domestic animals. CAB reviews: perspectives in agriculture, veterinary science. Nutr Nat Resour 4(38): 13.

Ross, J.W., Zhao, J., Walters, E.M., Samuel, M., Narfstrom, K., Jeong, M., DeMarco, M.A., McCall, M.A., Kaplan, H.J., Prather, R.S. 2009b. Somatic cell nuclear transfer to create a miniature swine model of retinitis pigmentosa, April 18-22, New Orleans, LA

Ross, J.W., Whyte, J.J., Zhao, J., Samuel, M., Wells, K.D., Prather, R.S. 2010a. Optimization of square-wave electroporation for transfection of porcine fetal fibroblasts. Transgenic Res 19(4): 611-620.

Ross, P.J., Wang, K., Kocabas, A., Cibelli, J.B. 2010b. Housekeeping gene transcript abundance in bovine fertilized and cloned embryos. Cell Reprogram 12(6): 709-717.

Saeki, K., Matsumoto, K., Kinoshita, M., Suzuki, I., Tasaka, Y., Kano, K., Tagachui, Y., Mikami, K., Hirabayashi, M., Kashiwazaki, N., Hosoi, Y., Murata, N., Iritani, A. 2004. Functional expression of a Delta12 fatty acid desaturase gene from spinach in transgenic pigs. Proc. Natl. Acad. Sci. USA 101: 6361-6366.

Salamone, D., Barañao, L., Santos, C., Bussmann, L., Artuso, J., Werning, C., Prync, A., Carbonetto, C., Dabsys, S., Munar, C., Salaberry, R., Berra, G., Berra, I., Fernández, N., Papouchado, M., Foti, M., Judewicz, N., Mujica, I., Muñoz, L., Alvarez, S.F., González, E., Zimmermann, J., Criscuolo, M., Melo, C. 2006. High level expression of bioactive recombinant human growth hormone in the milk of a cloned transgenic cow. J Biotechnol. 124(2): 469-472.

Santos, F., Zakhartchenko, V., Stojkovic, M., Peters, A., Jenuwein, T., Wolf, E., Reik, W., Dean, W. 2003. Epigenetic marking correlates with developmental potential in cloned bovine preimplantation embryos. Curr Biol 13: 1116-1121.

Schnieke, A.E., Kind, A.J., Ritchie, W.A., Mycock, K., Scott, A.R., Ritchie, M., Wilmut, I., Colman, A., Campbell, K.H. 1997. Human factor IX transgenic sheep produced by transfer of nuclei from transfected fetal fibroblasts. Science 278(5346). 2130-2133.

Shi, L.H., Ai, J.S., Ouyang, Y.C., Huang, J.C., Lei, Z.L., Wang, Q., Yin, S., Han, Z.M., Sun, Q.Y., Chen, D.Y. 2008. Trichostatin A and nuclear reprogramming of cloned rabbit embryos. J Anim Sci 86(5): 1106-1113.

Sun, X., Yan, Z., Yi, Y., Li, Z., Lei, D., Rogers, C.S., Chen, J., Zhang, Y., Welsh, M.J., Leno, G.H., Engelhardt, J.F. 2008. Adeno-associated virus-targeted disruption of the CFTR gene in cloned ferrets. J Clin Invest 118(4): 1578-1583.

Stadtfeld, M., Nagaya, M., Utikal, J., Weir, G., Hochedlinger, K. 2008. Induced pluripotent stem cells generated without viral integration. Science 322(5903): 945-949.

Switzer, W.M., Michler, R.E., Shangmugam, V., Matthews, A., Hussain, A.I., Wright A, Sandstrom, P., Chapman, L.E., Weber, C., Safley, S., Denny, R.R., Navarro, A.,

Evans, V., Norin, A.J., Kwiatkowski, P., Heneine, W. 2001. Lack of cross-species transmission of porcine endogenous retrovirus infection to nonhuman primate recipients of porcine cells, tissues and organs. Transplantation 71: 959-965.

Takahashi, K., Yamanaka, S. 2006. Induction of pluripotent stem cells from mouse embryonic and adult fibroblast cultures by defined factors. Cell 126 (4): 663-676.

Takahashi, K., Tanabe, K., Ohnuki, M., Narita, M., Ichisaka, T., Tomoda, K., Yamanaka, S. 2007. Induction of pluripotent stem cells from adult human fibroblasts by defined factors. Cell 131(5): 861-872.

te Riele, H., Maandag, E.R., Berns, A. 1992. Highly efficient gene targeting in embryonic stem cells through homologous recombination with isogenic DNA constructs. Proc Natl Acad Sci USA. 89(11): 5128-5132.

Tsuji, Y., Kato, Y., Tsunoda, Y. 2009. The developmental potential of mouse somatic cell nuclear-transferred oocytes treated with trichostatin A and 5-aza-2'-deoxycytidine. Zygote 17(2): 109-115.

Urnov, F.D., Miller, J.C., Lee, Y.L., Beausejour, C.M., Rock, J.M., Augustus, S., Jamieson, A.C., Porteus, M.H., Gregory, P.D., Holmes, M.C. 2005. Highly efficient endogenous human gene correction using designed zinc-finger nucleases. Nature 435(7042): 646-651.

Urnov, F.D., Rebar, E.J., Holmes, M.C., Zhang, H.S., Gregory, P.D. 2010. Genome editing with engineered zinc finger nucleases. Nat Rev Genet 11(9): 636-646.

van Berkel, P.H., Welling, M.M., Geerts, M., van Veen, H.A., Ravensbergen, B., Salaheddine, M., Pauwels, E.K., Pieper, F., Nuijens, J.H., Nibbering, P.H. 2002. Large scale production of recombinant human lactoferrin in the milk of transgenic cows. Nat Biotechnol 20(5): 484-487.

van Poll, D., Nahmias, Y., Soto-Gutierrez, A., Ghasemi, M., Yagi, H., Kobayashi, N., Yarmush, M.L., Hertl, M. 2010. Human immune reactivity against liver sinusoidal endothelial cells from GalTα(1,3)GalT-deficient pigs. Cell Transplant 19(6): 783-789.

Vasquez, K.M., Marburger, K., Intody, Z., Wilson, J.H. 2001. Manipulating the mammalian genome by homologous recombination. Proc Natl Acad Sci USA. 98(15): 8403-8410.

Voigt, B., Serikawa, T. 2009. Pluripotent stem cells and other technologies will eventually open the door for straightforward gene targeting in the rat. Dis Model Mech 2(7-8): 341-343.

Wall, R.J., Powell, A., Paape, M.J., Kerr, D.E., Bannermann, D.D., Pursel, V.G., Wells, K.D., Talbot, N., Hawk, H. 2005. Genetically enhanced cows resist intramammary *Staphylococcus aureus* infection. Nat Biotechnol 23: 445-451.

Wang, J., Yang, P., Tang, B., Sun, X., Zhang, R., Guo, C., Gong, G., Liu, Y., Li, R., Zhang, L., Dai, Y., Li, N. 2008. Expression and characterization of bioactive recombinant human α-lactalbumin in the milk of transgenic cloned cows. J Dairy Sci 91(12): 4466-4476.

Warren, L., Manos, P.D., Ahfeldt, T., Loh, Y.H., Li, H., Lau, F., Ebina, W., Mandal, P.K., Smith, Z.D., Meissner, A., Daley, G.Q., Brack, A.S., Collins, J.J., Cowan, C., Schlaeger, T.M., Rossi, D.J. 2010. Highly efficient reprogramming to pluripotency and directed differentiation of human cells with synthetic modified mRNA. Cell Stem Cell 7(5): 618-630.

Watanabe, S., Iwamoto, M., Suzuki, S., Fuchimoto, D., Honma, D., Nagai, T., Hashimoto, M., Yazaki, S., Sato, M., Onishi, A. 2005. A novel method for the production of transgenic cloned pigs: electroporation-mediated gene transfer to non-cultured cells and subsequent selection with puromycin. Biol Reprod 72(2): 309-315.

Watanabe, M., Umeyama, K, Matsunari, H., Takayanagi, S., Haruyama, E., Nakano, K., Fujiwara, T., Ikezawa, Y., Nakauchi, H., Nagashima, H. 2010. Knockout of exogenous EGFP gene in porcine somatic cells using zinc-finger nucleases. Bio chem Biophys Res Commun 402(1): 14-18.

Wernig, M., Meissner, A., Foreman, R., Brambrink, T., Ku, M., Hochedlinger, K., Bernstein, B.E., Jaenisch, R. 2007. In vitro reprogramming of fibroblasts into a pluripotent ES-cell-like state. Nature 448(7151): 318-324.

Wheeler, M.B., Bleck, G.T., Donovan, S.M. 2001. Transgenic alteration of sow milk to improve piglet growth and health. Reproduction 58 (Suppl.): 313-324.

Whitelaw, C.B., Lillico, S.G., King, T. 2008. Production of transgenic farm animals by viral vector-mediated gene transfer. Reprod Domest Anim 43 Suppl 2: 355-358.

Whitworth, K.M., Prather, R.S. 2010. Somatic cell nuclear transfer efficiency: how can it be improved through nuclear remodeling and reprogramming? Mol Reprod Dev 77(12): 1001-1015.

Whyte, J.J., Zhao, J., Wells, K.D., Samuel, M.S., Whitworth, K.M., Walters, E.M., Laughlin, M.H., Prather, R.S. 2011. Gene targeting with zinc finger nucleases to produce cloned eGFP knockout pigs. Mol Reprod Dev 78(1): 2.

Wilmut, I., Schnieke, A.E., McWhir, J., Kind, A.J., Campbell, K.H. 1997. Viable offspring derived from fetal and adult mammalian cells. Nature 385(6619): 810-813.

Wrenzycki, C., Wells, D., Herrmann, D., Miller, A., Oliver, J., Tervit, R., Niemann, H. 2001. Nuclear transfer protocol affects messenger RNA expression patterns in cloned bovine blastocysts. Biol Reprod 65(1): 309-317.

Wrenzycki, C., Herrmann, D., Gebert, C., Carnwath, J.W., Niemann, H. 2006. Gene expression and methylation patterns in cloned embryos. Methods Mol Biol 348: 285-304.

Wright, G., Carver, A., Cottom, D., Reeves, D., Scott, A., Simons, P., Wilmut, I., Garner, I., Colman, A. 1991. High level expression of active human alpha-1-antitrypsin in the milk of transgenic sheep. Biotechnology (N Y) 9(9): 830-834.

Wu, S., Ying, G.X., Wu, Q., Capecchi, M.R. 2008. A protocol for constructing gene targeting vectors: generating knockout mice for the cadherin family and beyond. Nature Protocols 3: 1056-1076.

Wu, Z., Chen, J., Ren, J., Bao, L., Liao, J., Cui, C., Rao, L., Li, H., Gu, Y., Dai, H., Zhu, H., Teng, X., Cheng, L., Xiao, L. 2009. Generation of pig induced pluripotent stem cells with a drug inducible system. J Mol Cell Biol 1(1): 46-54.

Yamada, K., Yazawa, K., Shimizu, A., Iwanaga, T., Hisashi, Y., Nuhn, M., O'Malley, P., Nobori, S., Vagefi, P.A., Patience, C., Fishman, J., Cooper, D.K., Hawley, R.J., Greenstein, J., Schuurman, H.J., Awwad, M., Sykes, M., Sachs, D.H. 2005. Marked prolongation of porcine renal xenograft survival in baboons through the use of α1,3-galactosyltransferase gene-knockout donors and the cotransplantation of vascularized thymic tissue. Nat Med 11: 32-34.

Yáñez, R.J., Porter, A.C. 1999. Influence of DNA delivery method on gene targeting frequencies in human cells. Somat Cell Mol Genet 25(1): 27-31.

Yang, Y., Seed, B. 2003. Site-specific gene targeting in mouse embryonic stem cells with intact bacterial artificial chromosomes. Nat Biotechnol 21(4): 447-451.

Yang, P., Wang, J., Gong, G., Sun, X., Zhang, R., Du, Z., Liu, Y., Li, R., Ding, F., Tang, B., Dai, Y., Li, N. 2008. Cattle mammary bioreactor generated by a novel procedure of transgenic cloning for large-scale production of functional human lactoferrin. PloS One 3(10): e3453.

Yang, D., Wang, C.E., Zhao, B., Li, W., Ouyang, Z., Liu, Z., Yang, H., Fan, P., O'Neill, A., Gu, W., Yi, H., Li, S., Lai, L., Li, X.J. 2010. Expression of Huntington's disease protein results in apoptotic neurons in the brains of cloned transgenic pigs. Hum Mol Genet 19(20): 3983-3994.

Yu, G., Chen, J., Yu, H., Liu, S., Chen, J., Xu, X., Sha, H., Zhang, X., Wu, G., Xu, S., Cheng, G. 2006. Functional disruption of the prion protein gene in cloned goats. J Gen Virol 87: 1019-1027.

Yu, J., Vodyanik, M. A., Smuga-Otto, K., Antosiewicz-Bourget, J., Frane, J.L., Tian, S., Nie, J., Jonsdottir, G.A., Ruotti, V., Stewart, R., Slukvin, I.I., Thomson, J.A. 2007. Induced pluripotent stem cell lines derived from human somatic cells. Science 318(5858): 1917-1920.

Yu, J., Hu, K., Smuga-Otto, K., Tian, S., Stewart, R., Slukvin, I.I., Thomson, J.A. 2009. Human induced pluripotent stem cells free of vector and transgene sequences. Science 324(5928): 797-801.

Zaidi, A., Schmoeckel, M., Bhatti, F., Waterworth, P., Tolan, M., Cozzi, E., Chavez, G., Langford, G., Thiru, S., Wallwork, J., White, D., Friend, P. 1998. Life-supporting pig-to primate renal xenotransplantation using genetically modified donors. Transplantation 65: 1584.

Zakhartchenko, V., Mueller, S., Alberio, R., Schernthaner, W., Stojkovic, M., Wenigerkind, H., Wanke, R., Lassnig, C., Mueller, M., Wolf, E., Brem, G. 2001. Nuclear transfer in cattle with non-transfected and transfected fetal or cloned transgenic fetal and postnatal fibroblasts. Mol Reprod Dev 60: 362-369.

Zhang, F., Cong, L., Lodato, S., Kosuri, S., Church, G.M., Arlotta, P. 2011. Efficient construction of sequence-specific TAL effectors for modulating mammalian transcription. Nat Biotechnol 29(2): 149-153.

Zhao, J., Ross, J.W., Hao, Y., Spate, L.D., Walters, E.M., Samuel, M.S., Rieke, A., Murphy, C.N., Prather, R.S. 2009a. Significant improvement in cloning efficiency of an inbred miniature pig by histone deacetylase inhibitor treatment after somatic cell nuclear transfer. Biol Reprod 81(3): 525-530.

Zhao, J., Hao, Y., Ross, J.W., Spate, L.D., Walters, E.M., Samuel, M.S., Rieke, A., Murphy, C.N., Prather, R.S. 2010. Histone deacetylase inhibitors improve in vitro and in vivo developmental competence of somatic cell nuclear transfer porcine embryos. Cell Reprogram 12(1): 75-83.

Zhao, X. Y., Li, W., Lv, Z., Liu, L., Tong, M., Hai, T., Hao, J., Guo, C.L., Ma, Q.W., Wang, L., Zeng, F., Zhou, Q. 2009b. iPS cells produce viable mice through tetraploid complementation. Nature 461(7260): 86-90.

Zhu, C., Li, B., Yu, G., Chen, J., Yu, H., Chen, J., Xu, X., Wu, Y., Zhang, A., Cheng, G. 2009. Production of Prnp -/- goats by gene targeting in adult fibroblasts. Transgenic Res 18(2):163-171.

iPS Cells: Born-Again Stem Cells for Biomedical Applications

Ambrose Jon Williams and Vimal Selvaraj
Cornell University,
USA

1. Introduction

1.1 Stemness

The fertilized egg, also known as the zygote, is a cell of total potential and plasticity and gives rise to the embryo and extra-embryonic tissues, and ultimately the whole adult organism. This property has since been termed *totipotency*, although the transition from fertilized egg to differentiated cells of the adult tissues (somatic cells) is not direct, progressing instead through lineages of successively more differentiated and committed intermediates towards the final cell type. Thus, the zygote gives rise to the trophoblast cells and inner cell mass (ICM) of the blastocyst stage embryo, the ICM gives rise to the primordial cells committed to the ectodermal, mesodermal and endodermal lineages. To give an example lineage, the ectoderm cells give rise to the neural crest stem cells, then neural stem cells, oligodendrocyte precursors, and finally oligodendrocytes which myelinate and form the white matter of the central nervous system. Each step is more specialized and less plastic than the base or 'stem' of the branch before it. Unlike the totipotent zygote, these 'stem cells' retain the ability to self-renew in addition to their plasticity. In addition to giving rise to somatic tissues during embryogenesis, the biological role of stem cells in an adult organism is to regenerate tissues lost to injury, disease or age.

1.2 History

The first stem cells discovered were the originator cells of teratomas, a rare tumor that comprised of multiple tissue types and was associated with embryonal carcinoma (EC). These EC stem cells have the then-unusual ability to self-renew indefinitely, as well as give rise to tissues from each of the three germinal layers (Kleinsmith and Pierce, 1964), a property termed *pluripotency*. Since then, many more stem cell types of varying potency have been discovered, including two more pluripotent cell types: the embryonic stem cell (ESC) is a non-cancerous analogue of the EC stem cell which is derived instead from the ICM of blastocyst stage embryos (Evans and Kaufman, 1981), and most recently induced pluripotent stem cells (iPSCs) which are produced from somatic cells that have been reprogrammed to a pluripotent state (Takahashi and Yamanaka, 2006). Pluripotent cells have since been demonstrated to have enormous potential for regenerative medicine, disease research and genetic engineering.

2. Applications of pluripotent stem cells

2.1 Animals from pluripotent cells

ESCs are the prototypical pluripotent stem cell and thus the most thoroughly characterized. They can self-renew indefinitely and are effectively immortal in cell culture. Although they lack the self-organizing capabilities of the fertilized egg, they can form any tissue in the adult organism as demonstrated by two key studies: injecting ESCs into blastocysts gives rise to chimaeric animals with tissues contributed by the injected ESCs as well as the original ICM (Moustafa and Brinster, 1972); injection of ESCs into blastocysts that have been rendered tetraploid (four genome copies, and therefore genomically incapable of forming a complete organism) produces animals wholly derived from the injected ESCs (Eggan et al., 2002). The latter technique is possible because tetraploid blastocysts retain the structural organization of a normal blastocyst, and although the tetraploid ICM will inevitably die out or senesce (and be replaced by the injected cells), the trophoblast component retains its function despite tetraploidy since trophoblasts eventually fuse and become polyploid anyway upon embryonic implantation into the maternal uterus. These properties are shared with all pluripotent cells, EC cells injected into blastocysts can also give rise to chimaeric animals (Mintz and Illmensee, 1975). Because of this potential, ESCs very quickly became a focus of applied research.

2.2 ESCs in genetic engineering and animal disease modeling

Modern reproduction techniques make it possible for a single ESC to give rise to a whole animal, greatly simplifying the process of genetically engineering animals. Previously, animals were bred extensively to isolate beneficial random mutations fertilized eggs were microinjected with DNA for random genomic integration (Gordon et al., 1980), or engineered animals were derived from nuclear-cloned somatic cells that had been engineered to the desired genotype; such a technique was used to generate cattle that lacked the prion protein and were thus made completely immune to bovine spongiform encephalopathy (BSE; mad cow disease, which transmits to humans as the variant Creutzfeldt-Jakob)(Richt et al., 2006). ESCs are easier to genetically engineer due to their infinite self-renewability, allowing a very small number of drug- or marker-selected cells to regenerate a whole culture or stable cell line. This technique has been used to generate a variety of mouse genetic models including sickle cell disease (Wu et al., 2006), thalassemia (Ciavatta et al., 1995), microcephaly (Pulvers et al., 2010), and T-cell lymphoma (Pechloff et al., 2010), as well as a p53-knockout rat for cancer research (Tong et al., 2010).

2.3 *In vitro* disease modeling using pluripotent cells

A major obstacle to disease research is the difficulty of acquiring diseased cells for study, usually because they are difficult to obtain from a living patient. For example, neurons are not easily obtained from a patient afflicted with Down syndrome, making detailed cell biology study of the neuronal basis for mental retardation impossible, and limiting our understanding of this disorder to more superficial behavioral neurological or postmortem pathological descriptions. However, a Down syndrome human ESC line as well as lines for other chromosomal trisomies have recently been derived (Biancotti et al., 2010), as has a human ESC line homozygous for Sickle Cell Disease (Pryzhkova et al., 2010). All were generated from embryos rejected by preimplantation genetic diagnosis (PGD) screens following *in vitro* fertilization (IVF). These lines allow cell culture study of diseased neurons, or any other cell type, by differentiating diseased ESCs into any cell type of interest; however researchers are still limited by the small number of diseased human ESC lines available.

Cloned embryos can be derived from adult cells using somatic cell nuclear transfer (SCNT), a technique made famous by the cloning of the sheep Megan, Morag and Dolly in the 1990s (Wilmut et al., 1997). It has been proposed that new diseased human ESC lines can be derived using this technique to make cloned embryos from diseased patients, and then harvesting them to create novel diseased ESC lines for disease study. At the time of this writing, SCNT for this application (Therapeutic Cloning) is currently legal in the United States and the European Union, but its legal status in these states as well as elsewhere across the world has been subject to numerous prior and continuing legal challenges. Although several large organizations continue to research this technology, it has been supplanted in recent years by alternate techniques for deriving patient-specific pluripotent stem cells.

2.4 Therapeutic potential of pluripotent stem cells

Pluripotent stem cells have been studied as, and shown great potential to be, a source of cell replacement therapies in a myriad of disease and injury models. Several human ESC lines have been differentiated into high-purity cardiomyocyte cultures that improve cardiac performance when transplanted into infracted rat hearts (Caspi et al., 2007). ESCs have also been differentiated into neural precursors and neurons including dopaminergic neurons which reverse the disease progression of Parkinsonian rats (Yang et al., 2008). In a model of spinal cord injury, ESC-derived oligodendrocytes transplanted into crushed rat spinal cords successfully restored locomotive function to the animals. Pancreatic beta cells, the insulin-secreting cells whose absence causes type I diabetes mellitus, have also been derived from ESCs and cure the diabetic phenotype of the mouse streptozotocin-induced model of diabetes upon transplantation (Kim et al., 2003). These are but a choice selection of the vast amount of scientific literature detailing the regenerative potential of ESCs.

At the time of this writing, two clinical trials are underway for ESC-based regenerative therapies in humans: an evaluation of human ESC-derived oligodendrocyte precursors to rescue neurologically complete spinal cord injury conducted by Geron Corporation, and ESC-derived retinal-pigmented epithelium for treatment of macular degeneration and Stargardt's macular dystrophy, which are major causes of blindness, conducted by Advanced Cell Technology Incorporated. A third proposed clinical trial is currently in the approval process between the Food and Drug Administration and applicant California Stem Cell Incorporated for ESC-derived motor neurons as a cure for type I spinal muscular atrophy, the leading genetic cause of infant mortality. These trials represent the first step in the direct evaluation of the therapeutic potential of pluripotent stem cells in human patients.

2.5 Pitfalls and obstacles to the use of ESCs

Transplants of ESC-derived tissues and biological devices are just as subject to immune rejection as conventional organ transplants, even more so due to the limited selection of human ESC lines. Although the engineering of non-immunogenic ESCs has been the subject of many academic initiatives and company startups, ongoing clinical and preclinical research for ESC-therapies is focused, for the mean time, on immune-privileged regions of the body: specifically the brain, eye and spinal cord. A second scientific concern is the purity of ESC-derived transplants because of the hazard posed by contaminating undifferentiated

ESCs that, if transplanted, can proliferate and form teratomas. The elimination of these leftover ESCs has been approached by several strategies: purification of differentiated cells by labeling and cell sorting (Pruszak et al., 2007), the engineering of special "suicide gene"-containing ESCs (Schuldiner et al., 2003), and the treatment of cells to be transplanted with chemotherapeutics (Bieberich et al., 2004). The concomitant destruction of stem cells by anti-cancer therapies reflects the generalized similarity between stem cells and cancer cells [reviewed in (Reya et al., 2001)].

As many as seven human embryos are sacrificed for each new human ESC line derived (Thomson et al., 1998); while the ethics of this are philosophically subjective they have nonetheless given rise to numerous high-profile legal challenges to continued ESC research and funding. In addition, the patent on derivation of human ESC lines is held by the Wisconsin Alumni Research Foundation. Until its expiration in 2016, commercial users wishing to use Wisconsin ESC ("WiCell") technologies might also be required to pay a royalty.

3. Induced pluripotency

3.1 Discovery

The laboratory of Shinya Yamanaka demonstrated in 2006 that somatic cells can be reprogrammed back to a primordial phenotype functionally identical to ESCs, and termed these reprogrammed cells induced pluripotent stem cells (iPSCs)(Takahashi and Yamanaka, 2006). These iPSCs have a morphology, growth and gene expression characteristics that are indistinguishable from ESCs. Like ESCs they also form teratomas consisting of tissues from all three germ layers when injected into immunodeficient animals (Takahashi and Yamanaka, 2006), and give rise to entire animals when injected into tetraploid blastocysts (Kang et al., 2009). iPSCs also have stable telomere lengths like ESCs (Marion et al., 2008) as well as an epigenetic state reflecting reversion back to pluripotency, although traces of the donor cell's epigenetic imprint are retained in early-passage iPSCs (Kim et al., 2010).

Pluripotency is typically induced by overexpressing in somatic cells the stem cell genes Oct3/4, Sox2, cMyc and Klf4 (Takahashi and Yamanaka, 2006) (collectively termed the Yamanaka factors) or by an alternate combination of Oct3/4, Sox2, Nanog and Lin28 (Yu et al., 2007) (the Thomson factors; this repertoire has not been extensively replicated in the literature). Retroviruses or lentiviruses are the standard vectors for inserting and over-expressing these transgenes for a period of 2-3 weeks. During which the formation of early ESC-like colonies are observed [Figure 1]. These colonies stain positively for alkaline phosphatase, a marker which distinguishes undifferentiated cells from fibroblasts, and when clonally selected and propagated they express the ESC markers SSEA-1 and Oct3/4 and assume a phenotype indistinguishable from ESCs (Takahashi and Yamanaka, 2006).

3.2 Molecular mechanisms of induced pluripotency

Induced pluripotency is a remarkably successful technique, although our understanding of the underlying mechanisms are limited. The Yamanaka combination of reprogramming factors wasn't arrived at by a serendipitous leap of understanding, but instead careful and methodical experimentation. When Yamanaka sought to reprogram skin cells into ESCs, he began with a list of 24 candidate genes identified by extensive review of ESC literature. Overexpression in fibroblasts for two weeks gave rise to ESC-like colonies expressing the pluripotency marker Fbxo15. After a yearlong process of elimination, his lab was able to replicate this result with just 4 genes: Oct3/4, Sox2, cMyc and Klf4 (Takahashi and Yamanaka, 2006).

It has been known for some time that Oct3/4, Sox2 and Nanog comprise the core of the pluripotency transcriptional network, as the deficiency in either one causes ES cells to lose pluripotency (Avilion et al., 2003; Mitsui et al., 2003; Nichols et al., 1998). It is interesting to note, however, that too much Oct3/4 or Sox2 can also disrupt pluripotency. As little as a two-fold excess in either causes ESCs to differentiate (Kopp et al., 2008; Niwa et al., 2000). Oct3/4, Sox2 and Nanog all occupy each others' promoters, and more than 90% of promoters bound to by Oct3/4 and Sox2 are also occupied by Nanog (Boyer et al., 2005). Although all three are required for pluripotency, Nanog overexpression is not required for induced pluripotency. Adding Nanog to the mix, however, increases reprogramming, as does combining the Yamanaka and Thomson reprogramming repertoires (Liao et al., 2008).

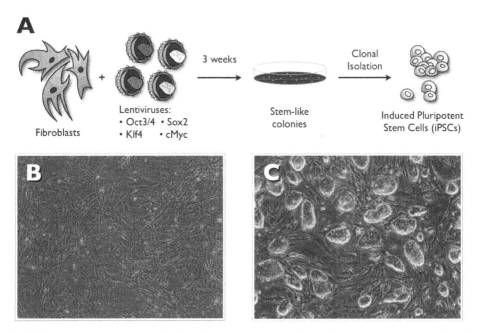

Fig. 1. Generation of iPSCs. A: Although there are a number of ways to generate iPS Cells, the model reprogramming experiment uses lentiviral vectors to integrate the Yamanaka factor transgenes into skin fibroblasts and over-express them for a period of three weeks. After this period of time, stem cell-like colonies become apparent in the reprogrammed culture which, following selection and characterization, will give rise to stable iPS Cell lines. B: Fibroblasts before reprogramming have typical morphology and grow in confluent cell monolayers. C: iPS Cells, however, grow in dense, elevated and round colonies with the characteristic "glass edge." They are microscopically indistinguishable from ESC cultures.

The use of the RNA-binding protein Lin28 as one of the Thomson factors suggested a role for microRNAs during the reprogramming process, and since then a number of pluripotency-regulating microRNAs have been identified (Zhong et al., 2010). Of particular interest is the miR-302 family of microRNAs, which induce stem cell-like plasticity when overexpressed in skin cancer cells (Lin et al., 2008).

Reprogramming to pluripotency is accompanied by the demethylation of promoter regions of known pluripotency genes, and this is observed in both iPSCs (Park et al., 2008b) and cloned embryos generated by SCNT (Lan et al., 2010). In partially or incompletely reprogrammed cells generated by either method, this demethylation is incomplete (Bourchis et al., 2001; Takahashi and Yamanaka, 2006). Likewise, chromatin alterations are also observed in reprogrammed cells as well as cloned embryos, and both of these processes are enhanced by histone deacetylase inhibitors (Han et al., 2010). Histone demethylation, particularly at the promoters of pluripotency genes, is also observed but it is not understood how this occurs during induced pluripotency.

3.3 The cancer generalization

Despite the success and reproducibility of this reprogramming technique, the permanent integration of additional copies of stem cell genes with high expression promoters poses a significant oncogenic hazard; in the earlier studies one in five chimaeric mice derived from iPSCs died from tumors resulting from spontaneous reactivation of reprogramming genes (Okita et al., 2007). Although the Yamanaka and Thomson factors are highly expressed in ESCs, they are either oncogenes themselves or associated with a poor clinical outcome when detected in cancers. cMyc specifically is one of the most well-characterized oncogenes, but Oct3/4 expression in animals also results in death due to extreme proliferation of undifferentiated progenitors (Hochedlinger et al., 2005). [Oct3/4 actually has no known role outside of pluripotent biology and when conditionally deleted in adult animals results in no detectable phenotype or defect in healing (Lengner et al., 2007)]. Oct3/4 (Gidekel et al., 2003) and Sox2 (Gangemi et al., 2009) are associated with cancer cell proliferation and tumor progression, while Klf4 has been linked to an invasive progression and metastasis in epithelial cancers (Pandya et al., 2004).

While the carcinogenic hazard introduced by genetic insertion of the Yamanaka factors led to a search for alternative reprogramming techniques, the generalization that ESCs and iPSCs biologically resemble cancer cells gave rise to a new line of thought: emulating oncogenesis to enhance reprogramming. Both SV40 Large T Antigen and TERT have been shown to enhance reprogramming when included in the Yamanaka factor repertoire (Park et al., 2008b), as has the knockout or knockdown of the tumor suppressor p53 [simultaneously discovered by 5 separate groups and reviewed in (Krizhanovsky and Lowe, 2009)]. The finding that adult stem cell populations, including hematopoietic (Eminli et al., 2009), keratinocyte (Aasen et al., 2008) and neural (Kim et al., 2008) stem cells, reprogram more easily than the more differentiated cells further down their lineages is also consistent with the understanding that adult stem cells are most prone to becoming cancerous. Although these studies contribute greatly to our understanding of induced pluripotency and stem cell biology, incorporating them into current techniques to enhance reprogramming has, until quite recently, been impossible, as doing so would greatly enhance the oncogenic hazards.

In recent years, a number of alternative techniques have emerged that induce pluripotency without genomic integration of the Yamanaka factors. Plasmid vectors have been used to induce pluripotency (Okita et al., 2008), however this technique has low reprogramming efficiency and half of the putative iPSCs generated contained some form of genomic integration. Adenoviruses also achieve reprogramming at a lowered efficiency; however a fraction of reprogrammed cells displayed karyotypic abnormalities (Stadtfeld et al., 2008).

Two other approaches favor transgene insertion followed by excision upon completion of reprogramming but also have their shortfalls: a retrotransposon vector which very rarely completely excises from the genome (Woltjen et al., 2009), and Cre-Lox recombination leaving behind residual sequences (Soldner et al., 2009). Most promising, however, are two DNA-free methods of reprogramming: direct delivery of recombinant transcription factors to the donor cells (Kim et al., 2009) and transfection with modified mRNAs encoding the Yamanaka factors (Warren et al., 2010). Although these last two methods achieve reprogramming with reduced efficiency, they circumvent the permanent oncogenic hazard presented by genomic integration of the Yamanaka factors (and any supplemental genes as well) and are most likely to be used for translational iPSC applications.

3.4 Advantages of induced pluripotency

Because they are derived from somatic cells and thus genetically autologous to the donor, iPSCs circumvent most of the obstacles, which have prevented clinical implementation of ESC technology. Moreover, being patient-specific they are not subject to immune rejection, and because induced pluripotency is an embryo-free method of deriving new pluripotent cell lines they are not subject to the funding restrictions or ethical controversies on ESC-derivation. The field of iPSC research is unlikely to see the sort of legal challenges that ESC research has, having drawn endorsements from social conservatives including Republicans in America and the Catholic Church.

iPSCs have several advantages in addition to overcoming ESC-specific obstacles. On a technique level, iPSCs are easier to derive than ESCs, and iPSC lines have already been derived from several species for which no ESC lines exist (Esteban et al., 2009; Li et al., 2009; Tomioka et al., 2010; Wu et al., 2009). This is because the optimal conditions for deriving ESCs vary across species [for example, human ESCs are maintained with basic fibroblast growth factor, while mouse ESCs are maintained with leukemia inhibitory factor], while induced pluripotency is conserved across mammals. Practically, this means induced pluripotency can facilitate easier genetic engineering of animals, as the most consistent and controlled techniques involve engineering of pluripotent stem cells. iPSC-based engineering of cattle and pigs is therefore becoming a new focus of the field [reviewed in (Telugu et al., 2010)].

Induced pluripotency has also become the key technique for deriving diseased pluripotent cells. Whereas several ESC lines modeling karyotypic abnormalities (Biancotti et al., 2010) and sickle cell disease (Pryzhkova et al., 2010) exist, derivation of new diseased ESC lines is limited to PGD-rejected embryos from *in vitro* fertilization. However, in the three years since induced pluripotency was first described in humans, iPSC lines representative of a large number of genetic diseases have been derived including amyotrophic lateral sclerosis, adenosine deaminase severe combined immunodeficiency, Shwachman-Bodian-Diamond syndrome, Gaucher disease type III, muscular dystrophies, Parkinson's disease, Huntington's disease, juvenile-onset type-1 diabetes, Down's syndrome, Lesch-Nyhan syndrome carrier, Fanconi anemia, spinal muscular atrophy, long-QT syndrome and familial dysautonomia and LEOPARD syndrome (Dimos et al., 2008; Ebert et al., 2009; Park et al., 2008a; Raya et al., 2009; Lee et al., 2009; Carvajal-Vergara et al., 2010; Moretti et al., 2010). The utility of these lines is their ability to give rise to diseased tissue *in vitro* for study as well as drug testing [Figure 2], whereas previously, research on these diseases and many others has been impeded.

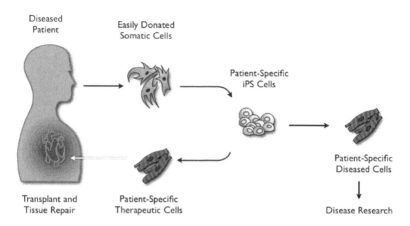

Fig. 2. Biomedical applications of iPSCs. Because iPSCs are patient-specific, neither they nor any cells derived from them are subject to transplant rejection. In this paradigm, easily obtained somatic cells such as skin fibroblasts or blood lymphocytes are reprogrammed to iPSCs, and therapeutic cells are derived from them to regenerate damaged or diseased tissue. In the case of the patient with a genetic disease, patient-specific iPSCs allow the derivation of genetically diseased tissues, which then could be subject to downstream research.

4. Reprogramming techniques

4.1 SCNT and induced pluripotency

The first experiments to demonstrate that a terminally differentiated phenotype could be reprogrammed to pluripotency took place in the mid 1990s with the generation of cloned animals, the most famous of which were Dolly the sheep and her counterparts. Megan and Morag (Wilmut et al., 1997). In these experiments, donor cell nuclei were transplanted into enucleated oocytes, a technique termed somatic cell nuclear transfer (SCNT). Upon receiving a diploid genome, the re-nucleated oocytes assumed a zygote phenotype, which formed embryos and eventually whole animals. Although these experiments were critical in showing somatic cells can be epigenetically reprogrammed to pluripotency, animals cloned by SCNT were argued to begin life not with the long telomeres typical of early-stage embryonic cells, but instead with the aged and shortened telomeres of their nuclear donors-in the case of Dolly, an adult mammary epithelial cell. However, this claim remains controversial and has been refuted by a study that used cloned cattle (Lanza et al., 2000), and a recent demonstration that mice could be continuously cloned through 15 generations without any appreciable age-related issues or cloning efficiency (Thuan et al., 2010).

Yamanaka's demonstration, ten years later, that somatic cells could be directly reprogrammed into a pluripotent state by the over-expression of exogenous Oct3/4, Sox2, cMyc and Klf4, did not suffer from telomere problems because the resulting iPSCs express telomerase at high levels, quickly extending the reprogrammed cells' telomeres to ESC-like lengths. However, introducing these transgenes using integrating retroviral vectors also brought about the risk of oncogenesis, and subsequent studies have attempted to circumvent this through one of three approaches: inserting the transgenes in a transient or easily-excisable manner, reprogramming through the use of fewer transgenes, or reprogramming without the use of genome-integrating vectors.

4.2 Excision of reprogramming transgenes

Variations of the Yamanaka technique induce transient expression using non-integrating plasmids (Okita et al., 2008) or adenoviral vectors (Stadtfeld et al., 2008) and have produced isogenic pluripotent stem cells after many treatments, but with poor yields (in both cases, < 0.0005%). Poor reprogramming efficiency makes necessary large donor sample sizes and high multiplicities of vector transfection. Additionally, half of iPSCs reprogrammed by plasmids had genomic integration. Although the incidence of this is reduced by the use of Adenoviral vectors, karyotypic abnormalities were observed in 23% of iPSC lines produced. Nonetheless, these experiments demonstrate reprogramming without genome integration. Retrotransposon elements (Woltjen et al., 2009) and cre-lox excision (Soldner et al., 2009) have also been used to insert and subsequently remove integrated transgenes, however complete removal of transgenes occurred at only 2% efficiency in the retrotransposon-reprogrammed cells and 18% of excisions still left trace sequences in the genomic DNA. The efficiency of cre-lox excision was not demonstrated, but was consistently noted to leave proviral trace sequences.

4.3 Use of fewer transgenes

Reprogramming without using c-Myc has been demonstrated (Nakagawa et al., 2008) at 1.6% the efficiency of the standard technique of when all four genes are used. c-Myc is the most hazardous of the reprogramming factors, and its omission in this experiment reduced tumor formation in mouse chimeras past the observation period. The histone deacytelase inhibitor valproate has been shown to enhance reprogramming efficiency >100-fold (Huangfu et al., 2008a) and can substitute for the Klf4 transgene in reprogramming (Huangfu et al., 2008b), at reduced efficiency. The histone methyltransferase inhibitor BIX01294 and L-channel Ca^{2+} agonist BayK8644, when used together, can replace the Sox2 transgene (Shi et al., 2008) at reduced efficiency. No approach has yet replaced Oct3/4, which appears to be absolutely essential for reprogramming and is also the most reliable pluripotency marker. Reprogramming of different somatic cell types with fewer transgenes has yielded varying results. Neuronal stem cells (NSCs) have been reprogrammed to pluripotency using just Oct3/4 (Kim, 2009) and at improved efficiencies using Oct3/4 and Sox2 (Kim et al., 2008).

4.4 Direct treatment with pluripotency factors

Despite concerns with telomere length, SCNT remains the gold standard of induced pluripotency demonstrating that induced pluripotency can be accomplished through an entirely non-genetic approach (albeit with very low efficiency). This is echoed by recent experiments: somatic cells, when fused with pluripotent cells (either ESCs or iPSCs) always produce binucleate cells with both nuclei in a pluripotent state (Sumer et al., 2009). Furthermore, induced pluripotency has been demonstrated at extremely low (0.006%) efficiencies through the use of recombinant pluripotent factors (Kim et al., 2009) with multiple treatments over an 8-week period.

Permeablization of somatic cells using bacterial pore-forming toxins, and then treating them with whole cell extracts from the desired cell phenotype has shown some promise in epigenetic reprogramming, but so far no such experiments have generated induced pluripotent stem cells. In one such study, human 293T epithelial cells treated with T cell extracts, assume a phenotype similar to T-cells, and a neuronal-like phenotype when treated with neural precursor extracts (Hakelien et al., 2002). Numerous attempts have been made to induce pluripotency by using extracts from ESCs, ECs, iPSCs and oocytes (Taranger et al., 2005; Zhu et al., 2010), however none of these have succeeded in producing stable iPSC lines.

5. Critical thinking on induced pluripotency

Although iPSCs are highly similar to ESCs in biology and function, an increasing body of literature describes defects and subtle differences between the two pluripotent cell types. Effective characterization of these phenomena is critical and will likely give rise to new and more stringent criteria by which reprogrammed cells can be evaluated for suitability.

One of the earliest characterizations of reprogrammed cells was the demonstration of DNA demethylation on the promoters of genes involved in pluripotency, such as Oct3/4 and Nanog (Okita et al., 2007; Takahashi and Yamanaka, 2006). Although these promoters exhibited near-total demethylation, indicating an activation of gene transcription, the demethylation was rarely as complete as the pattern observed in ESCs, and this was particularly true for early-passage (newly-created) iPSCs. Although these differences were small, genome-wide analysis of methylation patterns outside of these specific genes has revealed significant errors in epigenetic reprogramming (Lister et al., 2011). On a genomic scale, iPSC and ESC methylomes are similar, but most iPSC methylomes analyzed had megabase-sized loci of aberrant DNA methylation, which persist long after reprogramming and even after differentiation. Some of these loci are shared among distinct iPSC lines, suggesting that certain regions of the methylomes are susceptible to aberrant and incomplete reprogramming. There have also been reports of differences in gene expression patterns between ESCs and iPSCs, although a recent study found that most of these differences are laboratory-specific and can be attributed to microenvironment differences in growth conditions from one laboratory to the next (Newman and Cooper, 2010).

DNA sequence defects have also been described in iPSCs. Early-passage iPSCs display a range of polymorphism in copy number variant (CNV) regions compared to their parental fibroblasts (Hussein et al., 2011). As with DNA methylation, it was also found that CNVs occurred more commonly in "fragile regions" of the genome. As CNVs arise from damaged DNA improperly repaired by homologous recombination, this phenomenon suggests DNA damage and replicative stress in cells undergoing reprogramming. However, while early-passage iPSCs contain significantly more CNVs, a vast majority of these mutations put the cells at a selective disadvantage. Mid- to late-passage iPSCs therefore lose CNVs and soon approach a genomic state highly similar to ESCs. Point mutations in specific genes have also been identified in iPSCs, however unlike CNVs these display a nonrandom pattern of enrichment, with a majority occurring in proto-oncogenes and tumor suppressors (Gore et al., 2011). Half of these mutations can be traced back to the parental fibroblasts, which harbor these mutations in low frequencies; however the other half most likely arise during reprogramming and, more importantly, the subsequent selection and propagation steps. Oncogenic mutations are generalized to give a selective advantage to pluripotent cells, and although an accumulation in oncogenic mutations has also been demonstrated in ESCs, this study still establishes the need for extensive genetic testing of iPSCs before they are to be used on a clinical scale.

6. Conclusions

Although induced pluripotency as a reprogramming technique currently brings significant concerns about carcinogenicity as well as genomic and epigenomic integrity, a significant portion of the ESC research community has jumped ship in recent years in order to study iPSCs. This is because of the exciting promise these cells hold, as well as the mainstream belief that the obstacles that come with them will be overcome. With applications in a variety of fields including regenerative therapies, disease modeling, animal cloning and genetic engineering, induced pluripotency is actively transforming the stem cell community. Given how young the field is, induced pluripotency has a surprisingly well-developed body of basic research, which has already contributed enormously to our understanding on developmental biology and epigenetics, as well as given us insights on a large number of modeled genetic diseases. Taken together, the current body of literature on induced pluripotency describes why it is a very exciting time to be a part of this field.

7. References

Aasen, T., Raya, A., Barrero, M., Garreta, E., Consiglio, A., Gonzalez, F., Vassena, R., Bilic, J., Pekarik, V., Tiscornia, G., Edel, M., Boue, S. & Izpisua-Belmonte, J. (2008). Efficient and rapid generation of induced pluripotent stem cells from human keratinocytes. *Nature Biotechnology*, Vol.26, No.11, (October 2008), pp. 1276-1284, ISSN 1087-0156

Avilion, A., Nicolis, S., Pevny, L., Perez, L., Vivian, N. & Lovell-Badge, R. (2003). Multipotent cell lineages in early mouse development depend on SOX2 function. *Genes and Development*, Vol.17, No.1, (January 2003), pp. 126-140, ISSN 1549-5477

Biancotti, J., Narwani, K., Buehler, N., Mandefro, B., Golan-Lev, T., Yanuka, O., Clark, A., Hill, D., Benvenisty, N. & Lavon, N. (2010). Human embryonic stem cells as models for aneuploid chromosomal syndromes. *Stem Cells,* Vol.28, No.9, (September 2010), pp. 1530-1540, ISSN 1549-4918

Bieberich, E., Silva, J., Wang, G., Krishnamurthy, K. & Condie, B. (2004). Selective apoptosis of pluripotent mouse and human stem cells by novel ceramide analogues prevents teratoma formation and enriches for neural precursors in ES cell-derived neural transplants. *The Journal of Cell Biology*, Vol.167, No.4, (November 2004), pp. 723-734, ISSN 1540-8140

Bourchis, D., Le-Bourhis, D., Patin, D., Niveleau, A., Comizzoli, P., Renard, P. & Viegas-Pequignot, E. (2001). Delayed and incomplete reprogramming of chromosome methylation patterns in bovine cloned embryos. *Current Biology*, Vol.11, No.19, (October 2001), pp. 1542-1546, ISSN 0960-9822

Boyer, L., Lee, T., Cole, M., Johnstone, S., LEvine, S., Zucker, J., Guenther, M., Kumar, R., Murray, H., Jennifer, R., Gifford, D., Melton, D., Jaenisch, R. & Young, R. (2005). Core transcriptional regulatory circuitry in human embryonic stem cells. *Cell*, Vol.122, No.6, (September 2005), pp. 947-956, ISSN 0092-8674

Carvajal-Vergara, X., Sevilla, A., D'Souza, S.L., Ang, Y.S., Schaniel, C., Lee, D.F., Yang, L., Kaplan, A.D., Adler, E.D., Rozov, R., Ge, Y., Cohen, N., Edelmann, L.J., Chang, B., Waghray, A., Su, J., Pardo, S., Lichtenbelt, K.D., Tartaglia, M., Gelb, B.D. & Lemischka, I.R. (2010). Nature, Vol.465, No.7299, (June 2010), pp. 808-812, ISSN 0028-0836

Caspi, O., Huber, I., Kehat, I., Habib, M., Arbel, J., Gepstein, A., Yankelson, L., Aronson, D., Beyar, R. & Gepstein, L. (2007). Transplantation of Human Embryonic Stem Cell-Derived Cardiomyocytes Improves Myocardial Performance in Infarcted Rat Hearts. *Journal of the American College of Cardiology*, Vol.50, No.19, (October 2007), pp. 1884-1893, ISSN 1558-3597

Ciavatta, D., Ryan, T., Farmer, S. & Townes, T. (1995). Mouse model of human beta zero thalassemia: targeted deletion of the mouse beta maj- and beta min-globin genes in embryonic stem cells. *PNAS*, Vol.92, No.20, (September 1995), pp. 9259-9263, ISSN 0027-8424

Dimos, J.T., Rodolfa, K.T., Niakan, K.K., Weisenthal, L.M., Mitsumoto, H., Chung, W., Croft, G.F., Saphier, G., Leibel, R., Goland, R., Wichterle, H., Henderson, C.E. & Eggan, K. (2008). Science, Vol.321, No.5893, (August 2008), pp. 1218-1221, ISSN 1095-9203

Ebert, A.D., Yu, J., Rose, F.F. Jr., Mattis, V.B., Lorson, C.L., Thomson, J.A. & Svendsen, C.N. (2009). Induced pluripotent stem cells from a spinal muscular atrophy patient. Nature, Vol.457, No.7227, (January 2009), pp.277-280, ISSN 0028-0836

Eggan, K., Rode, A., Jentsch, I., Samuel, C., Hennek, T., Tintrup, H., Zevnik, B., Erwin, K., Loring, J., Jackson-Grusby, L., Speicher, M., Kuehn, R. & Jaenisch, R. (2002). Male and female mice derived from the same embryonic stem cell clone by tetraploid embryo complementation. *Nature Biotechnology*, Vol.20, No.5, (May 2002), pp. 455-459, ISSN 1087-0156

Eminli, S., Foudi, A., Stadtfield, M., Maherali, N., Ahfeldt, T., Mostoslavsky, G., Hock, H. & Hochedlinger, K. (2009). Differentiation stage determines potential of hematopoietic cells for reprogramming into induced pluripotent stem cells. *Nature Genetics*, Vol.41, No.9, (September 2009), pp. 968-976, ISSN 1546-1718

Esteban, M., Xu, J., Yang, J., Peng, M., Qin, D., Li, W., Jiang, Z., Chen, J., Deng, K., Zhong, M., Cai, J., Lai, L. & Pei, D. (2009). Generation of induced pluripotent stem cell lines from Tibetan Miniature Pig. *Journal of Biological Chemistry*, Vol.284, No.26, (June 2009), pp. 17634-17640, ISSN 0021-9258

Evans, M. & Kaufman, M. (1981). Establishment in culture of pluripotential cells from mouse embryos. *Nature*, Vol.292, No.5819, (July 1981), pp. 154-156, ISSN 0028-0836

Gangemi, R., Griffero, F., Marubbi, S., Perera, M., Capra, M., Malatesta, P., Ravetti, G., Zona, G., Daga, A. & Corte, G. (2009). SOX2 silencing in glioblastoma tumor-initiating cells causes stop of proliferation and loss of tumorigenicity. *Stem Cells*, Vol.27, No.1, (January 2009), pp. 40-48, ISSN 1549-4918

Gidekel, S., Pizov, G., Bergman, Y. & Pikarsky, E. (2003). Oct-3/4 is a dose-dependent oncogenic fate determinant. *Cancer Cell*, Vol.4, No.5, (November 2003), pp. 361-370, ISSN 1535-6108

Gordon, J., Scangos, G., Plotkin, D., Barbarosa, J. & Ruddle, F. (1980). Genetic transformation of mouse embryos by microinjection of purified DNA. *PNAS*, Vol.77, No.12, (December 1980), pp. 7380-7384, ISSN 0027-8424

Gore, A., Li, Z., Fung, H., Young, J., Agarwal, S., Antosiewicz-Bourget, J., Canto, I., Giorgetti, A., Israel, M., Kiskinis, E., Lee, J., Loh, Y., Manos, P., Montserrat, N., Panopoulos, A., Ruiz, S., Wilbert, M., Yu, J., Kirkness, E., Belmonte, J., Rossi, D., Thomson, J. & Eggan, K., Daley, G., Goldstein, L., Zhang, K. (2011). Somatic coding mutations in human induced pluripotent stem cells. *Nature*, Vol.471, No.7336, (March 2011), pp. 63-72, ISSN 1476-4687

Hakelien, A., Landsverk, H., Robl, J., Skalhegg, B. & Collas, P. (2002). Reprogramming fibroblasts to express T-cell functions using cell extracts. *Nature Biotechnology*, Vol.20, No.5, (May 2002), pp. 460-466, ISSN 1087-0156

Han, J., Sachdev, P. & Sidhu, K. (2010). A combined epigenetic and non-genetic approach for reprogramming human somatic cells. *PLoS ONE*, Vol.5, No.8, (August 2010), pp. 12297, ISSN 1932-6203

Hochedlinger, K., Yamada, Y., Beard, C. & Jaenisch, R. (2005). Ectopic expression of Oct-4 blocks progenitor-cell differentiation and causes dysplasia in epithelial tissues. *Cell*, Vol.121, No.3, (May 2005), pp. 465-477, ISSN 0092-8674

Huangfu, D., Maehr, R., Guo, W., Eijkelenboom, A., Snitow, M., Chen, A. & Melton, D. (2008a). Induction of Pluripotent stem cells by defined factors is greatly improved by small-molecule compounds. *Nature Biotehcnology*, Vol.26, No.7, (July 2008), pp. 795-797, ISSN 1546-1696

Huangfu, D., Osafune, K., Maehr, R., Guo, W., Eijkelenboom, A., Chen, S., Muhlestein, W. & Melton, D. (2008b). Induction of pluripotent stem cells from primary human fibroblasts with only Oct4 and Sox2. *Nature Biotehcnology*, Vol.26, No.11, (November 2008), pp. 1269-1275, ISSN 1546-1696

Hussein, S., Batada, N., Vuoristo, S., Ching, R., Autio, R., Narva, E., Ng, S., Sourour, M., Hamalainen, R., Olsson, C., Lundin, K., Mikkola, M., Trokovic, R., Peitz, M., Brustle, O., Bazett-Jones, D., Alitalo, K., Lahesmaa, R., Nagy, A. & Otonkoski, T. (2011). Copy number variation and selection during reprogramming to pluripotency. *Nature*, Vol.471, No.7336, (March 2011), pp., ISSN 1476-4687

Kang, L., Wang, J., Zhang, Y., Kou, Z. & Gao, S. (2009). iPS Cells Can Support Full-Term Development of Tetraploid Blastocyst-Complemented Embryos. *Cell Stem Cell*, Vol.5, No.2, (August 2009), pp. 135-138, ISSN 1875-9777

Kim, D., Gu, Y., Ishii, M., Fujimiya, M., Qi, M., Nakamura, N., Yoshikawa, T., Sumi, S. & Inoue, K. (2003). In vivo functioning and transplantable mature pancreatic islet-like cell clusters differentiated from embryonic stem cell. *Pancreas*, Vol.27, No.2, (August 2003), pp. 34-41, ISSN 1536-4828

Kim, D., Kim, C.H., Moon, J.I., Chung, Y.G., Chang, M.Y., Han, B.S., Ko, S., Yang, E., Cha, K.Y., Lanza, R. & Kim, K.S. (2009). Generation of human induced pluripotent stem cells by direct delivery of reprogramming proteins. *Cell Stem Cell*, Vol.4, No.6, (June 2009), pp. 472-476, ISSN 1875-9777

Kim, J. (2009). Oct4-induced pluripotency in adult neural stem cells. *Cell*, Vol.136, No.3, (February 2009), pp. 411-419, ISSN 1097-4172

Kim, J., Zaehres, H., Wu, G., Gentile, L., Ko, K., Sebastiano, V., Arauzo-Bravo, M., Ruau, D., Han, D., Zenke, M. & Scholer, H. (2008). Pluripotent stem cells induced from adult neural stem cells by reprogramming with two factors. *Nature*, Vol.454, No.7204, (July 2008), pp. 646-650, ISSN 1476-4687

Kim, K., Doi, A., Wen, B., Ng, K., Zhao, R., Cahan, P., Kim, J., Aryee, M., Ji, H., Ehrlich, L., Uabuuchi, A., Takeuchi, A., Cunniff, K., Hongguang, H., McKinney-Freeman, S., Naveiras, O., Yoon, T., Irizarry, R., Jung, N., Seita, J., Hanna, J., Murakami, P., Jaenisch, R., Weissleder, R., Orkin, S., Weissman, I., Feinberg, A. & Daley, G. (2010). Epigenetic memory in induced pluripotent stem cells. *Nature*, Vol.467, No.7313, (September 2010), pp. 285-290, ISSN 1476-4687

Kleinsmith, L. & Pierce, G.J. (1964). Multipotentiality of Single Embryonal Carcinoma Cells. *Cancer Res.,* Vol.24, No.1, (October 1964), pp. 1544-1551, ISSN 0008-5472

Kopp, J., Ormsbee, B., Desler, M. & Rizzino, A. (2008). Small increases in the level of Sox2 trigger the differentiation of mouse embryonic stem cells. *Stem Cells,* Vol.26, No.4, (April 2008), pp. 903-911, ISSN 1549-4918

Krizhanovsky, V. & Lowe, S. (2009). Stem cells: The promises and perils of p53. *Nature,* Vol.460, No.7259, (August 2009), pp. 1085-1086, ISN 1476-4687

Lan, J., Hua, S., Zhang, H., Song, Y., Liu, J. & Zhang, Y. (2010). Methylation patterns in 5' terminal regions of pluripotency-related genes in bovine in vitro fertilized and cloned embryos. *Journal of Genetics and Genomics,* Vol.37, No.5, (May 2010), pp. 297-304, ISSN 1673-8527

Lanza, R.P., Cibelli, J.B., Blackwell, C., Cristofalo, V.J., Francis, M.K., Baerlocher, G.M., Mak, J., Schertzer, M., Chavez, E.A., Sawyer, N., Lansdrop, P.M. & West, M.D. (2000). Extension of cell life-span and telomere length in animals cloned from senescent somatic cells. *Science,* Vol.288, No.5466, (April 2000), pp.665-669, ISSN 1095-9203.

Lee, G., Papapetrou, E.P., Kim, H., Chambers, S.M., Tomishima, M.J., Fasano, C.A., Ganat, Y.M., Menon, J., Shimizu, F., Viale, A., Tabar, V., Sadelain, M. & Studer, L. (2009). Modelling parthenogenesis and treatment of familial dysautonomia using patient-specific iPSCs. *Nature,* Vol.461, No.7262, (September 2009), pp. 402-406, ISSN 0028-0836

Lengner, C., Camargo, F., Hochedlinger, K., Welstead, G., Zaidi, S., Gokhale, S., Scholer, H., Tomilin, A. & Jaenisch, R. (2007). Oct4 Expression Is Not Required for Mouse Somatic Stem Cell Self-Renewal. *Cell Stem Cell,* Vol.1, No.4, (October 2007), pp. 403-415, ISSN 1875-9777

Li, W., Wei, W., Zhu, S., Zhu, J., Shi, Y., Lin, T., Hao, E., Hayek, A., Deng, H. & Ding, S. (2009). Generation of rat and human induced pluripotent stem cells by combining genetic reprogramming and chemical inhibitors. *Cell Stem Cell,* Vol.4, No.1, (January 2009), pp. 16-19, ISSN 1875-9777

Liao, J., Wu, Z., Wang, Y., Cheng, L., Cui, C., Gao, Y., Chen, T., Rao, L., Chen, S., Jia, N., Dai, H., Xin, S., Kang, J., Pei, G. & Xiao, L. (2008). Enhanced efficiency of generating induced pluripotent stem (iPS) cells from human somatic cells by a combination of six transcription factors. *Cell Research.,* Vol.18, No.5, (May 2008), pp. 600-603, ISSN 1748-7838

Lin, S., Chang, D., Chang-Lin, S., Lin, C., Wu, D., Chen, D. & Ying, S. (2008). Mir-302 reprograms human skin cancer cells into a pluripotent ES-cell-like state. *RNA,* Vol.14, No.10, (October 2008), pp. 2115-2124, ISSN 1469-9001

Lister, R., Pelizzola, M., Kida, Y., Hawkins, R., Nery, J., Hon, G., Antosiewcz-Bourget, J., O'Malley, R., Castanon, R., Klugman, S., Downes, M., Yu, R., Stewart, R., Ren, B., Thomson, J., Evans, R. & Ecker, J. (2011). Hotspots of aberrant epigenomic reprogramming in human induced pluripotent stem cells. *Nature,* Vol.471, No.7336, (March 2011), pp. 68-76, ISSN 1476-4687

Marion, R., Strati, K., Li, H., Tejera, A., Schoeftner, S., Ortega, S., Serrano, M. & Blasco, M. (2008). Telomeres Acquire Embryonic Stem Cell Characteristics in Induced Pluripotent Stem Cells. *Cell Stem Cell,* Vol.6, No.4, (February 2009), pp. 141-154, ISSN 1875-9777

Mintz, B. & Illmensee, K. (1975). Normal genetically mosaic mice produced from malignant teratocarcinoma cells. *PNAS*, Vol.72, No.9, (September 1975), pp. 3585-3589, ISSN 0027-8424

Mitsui, K., Tokuzawa, Y., Itoh, H., Segawa, K., Murakami, M., Takahashi, K., Maruyama, M., Maeda, M. & Yamanaka, S. (2003). The Homeoprotein Nanog Is Required for Maintenance of Pluripotency in Mouse Epiblast and ES Cells. *Cell*, Vol.113, No.5, (May 2003), pp. 631-642, ISSN 0092-8674

Moretti, A., Bellin, M., Welling, A., Jung, C.B., Lam, J.T., Bott-Flügel, L., Dorn, T., Goedel, A., Höhnke, C., Hofmann, F., Seyfarth, M., Sinnecker, D., Schömig, A. & Laugwitz, K.L. (2010). *New England Journal of Medicine*. Vol.363, No.15, (October 2010), pp. 1397-1409, ISSN 0028-4793

Moustafa, L. & Brinster, R. (1972). Induced chimaerism by transplanting embryonic stem cells into mouse blastocysts. *Journal of Experimental Zoology*, Vol.181, No.2, (August 1972), pp. 193-201, ISSN 0022-104X

Nakagawa, M., Koyanagi, M., Tanabe, K., Takahashi, K., Ichisaka, T., Aoi, T., Okita, K., Mochiduki, Y., Takizawa, N. & Yamanaka, S. (2008). Generation of induced pluripotent stem cells without Myc from mouse and human fibroblasts. *Nature Biotechnology*, Vol.26, No.1, (January 2008), pp. 101-106, ISSN 1546-1696

Newman, A. & Cooper, J. (2010). Lab-specific gene expression signatures in pluripotent stem cells. *Cell Stem Cell*, Vol.7, No.2, (August 2010), pp. 258-262, ISSN 1875-9777

Nichols, J., Zevnik, B., Anastassiadis, K., Niwa, H., Klewe-Nebenius, D., Chambers, I., Scholer, H. & Smith, A. (1998). Formation of Pluripotent Stem Cells in the Mammalian Embryo Depends on the POU Transcription Factor Oct4. *Cell*, Vol.95, No.3, (October 1998), pp. 379-391, ISSN 0092-8674

Niwa, H., Miyazaki, J. & Smith, A. (2000). Quantitative expression of Oct-3/4 defines differentiation, dedifferentiation or self-renewal of ES cells. *Nature Genetics*, Vol.24, No.4, (April 2000), pp. 372-376, ISSN 1061-4036

Okita, K., Ichisaka, T. & Yamanaka, S. (2007). Generation of germline-competent induced pluripotent stem cells. *Nature*, Vol.448, No.7151, (July 2007), pp. 313-317, ISSN 1476-4687

Okita, K., Nakagawa, M., Hyenjong, H., Ichisaka, T. & Yamanaka, S. (2008). Generation of mouse induced pluripotent stem cells without viral vectors. *Science*, Vol.322, No.5903, (November 2008), pp. 949-953, ISSN 1095-9203

Pandya, A., Talley, L., Frost, A., Fitzgerald, T., Trivedi, V., Chakravarthy, M., Chieng, D., Grizzle, W., Engler, J., Krontiras, H., Bland, K., LoBuglio, A., Lobo-Ruppert, S. & Ruppert, J. (2004). Nuclear Localization of KLF4 is Associated with an Aggressive Phenotype in Early-Stage Breast Cancer. *Clinical Cancer Research*, Vol.10, No.8, (April 2004), pp. 2709-2719, ISSN 1078-0432

Park, I., Arora, N., Huo, H., Maherali, N., Ahfeldt, T., Shimamura, A., Lensch, M., Cowan, C., Hochedlinger, K. & Daley, G. (2008a). Disease-Specific Induced Pluripotent Stem Cells. *Cell*, Vol.134, No.5, (September 2008), pp. 877-886, ISSN 1097-4172

Park, I., Zhao, R., West, J., Yabuuchi, A., Huo, H., Ince, T., Lerou, P., Lensch, M. & Daley, G. (2008b). Reprogramming of human somatic cells to pluripotency with defined factors. *Nature*, Vol.451, No.7175, (January 2008), pp. 141-146, ISSN 1476-4687

Pechloff, K., Holch, J., Ferch, U., Schweneker, M., Brunner, K., Kremer, M. & Sparwasser, T. (2010). The fusion kinase ITK-SYK mimics a T cell receptor signal and drives oncogenesis in conditional mouse models of peripheral T cell lymphoma. *Journal of Experimental Medicine*, Vol.207, No.5, (May 2010), pp. 1031-1044, ISSN 1540-9538

Pruszak, J., Sonntag, K., Aung, M., Sanchez-Pernaute, R. & Isacson, O. (2007). Markers and methods for cell sorting of human embryonic stem cell-derived neural cell populations. *Stem Cells*, Vol.25, No.9, (September 2007), pp. 2257-2268, ISSN 1549-4918

Pryzhkova, M., Peters, A. & Zambidis, A. (2010). Erythropoietic differentiation of a human embryonic stem cell line harbouring the sickle cell anaemia mutation. *Reproductive BioMedicine Online*, Vol.21, No.2, (August 2010), pp. 196-205, ISSN 1472-6491

Pulvers, J., Bryk, J., Fish, J., Brauninger, M., Arai, Y., Schreier, D., Naumann, R., Helppi, J., Habermann, B., Vogt, J., Nitsch, R., Toth, A., Enard, W., Paabo, S. & Huttner, W. (2010). Mutations in mouse Aspm (abnormal spindle-like microcephaly associated) cause not only microcephaly but also major defects in the germline. *PNAS*, Vol.107, No.38, (September 2010), pp. 16595-16600, ISSN 1091-6490

Raya, A., Rodríguez-Pizà, I., Guenechea, G., Vassena, R., Navarro, S., Barrero, M.J., Consiglio, A., Castellà, M., Río, P., Sleep, E., González, F., Tiscornia, G., Garreta, E., Aasen, T., Veiga, A., Verma, I.M., Surrallés, J., Bueren, J. & Izpisúa Belmonte, J.C. (2009). Nature, Vol.460, No.7251, (July 2009), pp.53-59, ISSN 0028-0836

Reya, T., Morrison, S., Clarke, M. & Weissman, I. (2001). Stem cells, cancer, and cancer stem cells. *Nature*, Vol.141, No.6859, (November 2001), pp. 105-111, ISSN 0028-0836

Richt, J., Kasinathon, P., Hamir, A., Castilla, J., Sathyyaseelan, T., Vargas, F., Sathiyaseelan, J., Wu, H., Matsushita, H., Koster, J., Kato, S., Ishida, I., Soto, A., Robl, J. & Kuroiwa, Y. (2006). Production of cattle lacking prion protein. *Nature Biotechnology*, Vol.25, No.1, (January 2007), pp. 132-138, ISSN 1087-0156

Schuldiner, M., Itskovitz-Eldor, J. & Benvenisty, N. (2003). Selective ablation of human embryonic stem cells expressing a "suicide" gene. *Stem Cells*, Vol.21, No.3, (May 2003), pp. 257-265, ISSN 1066-5099

Shi, Y., Desponts, C., Do, J., Hahm, H., Scholer, H. & Ding, S. (2008). Induction of Pluripotent Stem Cells from Mouse Embryonic Fibroblasts by Oct4 and Klf4 with Small-Molecule Compounds. *Cell Stem Cell*, Vol.3, No.5, (November 2008), pp. 568-574, ISSN 1875-9777

Soldner, F., Hockemeyer, D., Beard, C., Gao, Q., Bell, G., Cook, E., Hargus, G., Blak, A., Cooper, O., Mitalipova, M., Isacson, O. & Jaenisch, R. (2009). Parkinson's disease patient-derived pluripotent stem cells free of viral reprogramming factors. *Cell*, Vol.136, No.5, (March 2009), pp. 964-977, ISSN 1097-4172

Stadtfeld, M., Nagaya, M., Utikal, J., Weir, G. & Hochedlinger, K. (2008). Induced pluripotent stem cells generated without viral integration. *Science*, Vol.322, No.5903, (November 2008), pp. 945-949, ISSN 1095-9203

Sumer, H., Jones, K., Liu, J., Heffernan, C., Tat, P., Upton, K. & Verma, P. (2009). Reprogramming of somatic cells after fusion with iPS and ntES cells. *Stem Cells & Development*, Vol.19, No.2, (February 2010), pp. 239-246, ISSN 1557-8534

Takahashi, K. & Yamanaka, S. (2006). Induction of pluripotent stem cells from mouse embryonic and adult fibroblast cultures by defined factors. *Cell*, Vol.126, No.4, (August 2006), pp. 663-676, ISSN 0092-8674

Taranger, C., Noer, A., Sorensen, A., Hakeliel, A., Boquest, A. & Collas, P. (2005). Induction of dedifferentiation, genomewide transcriptional programming, and epigenetic reprogramming by extract of carcinoma and embryonic stem cells. *Molecular Biology of the Cell*, Vol.16, No.12, (December 2005), pp. 5719-5735, ISSN 1059-1524

Telugu, B., Ezashi, T. & Roberts, R. (2010). The Promise of stem cell research in pigs and other ungulate species. *Stem Cell Reviews*, Vol.6, No.1, (March 2010), pp. 31-41, ISSN 1558-6804

Thomson, J., Itskovitz-Eldor, J., Shapiro, S., Waknitz, M., Swiergiel, K., Marshall, V. & Jones, J. (1998). Embryonic stem cell lines derived from human blastocysts. *Science*, Vol.282, No.5391, (November 1998), pp. 1145-1147, ISSN 0036-8075

Thuan, N. V., Kishigami, S. & Wakayama, T. (2010). How to improve the success rate of mouse cloning technology. *Journal of Reproduction and Development*, Vol.56, No.1, (January 2010), pp 20-30, ISSN 0916-8818.

Tomioka, I., Maeda, T., Shimada, H., Kawai, K., Okada, Y., Igarashi, H., Oiwa, R., Iwasaki, T., Aoki, M., Kimura, T., Shiozawa, S., Shinohara, H., Suemizu, H., Sasaki, E. & Okano, H. (2010). Generating induced pluripotent stem cells from common marmoset (Callithrix jacchus) fetal liver cells using defined factors, including Lin28. *Genes to Cells*, Vol.15, No.9, (September 2010), pp. 959-969, ISSN 1365-2443

Tong, C., Li, P., Wu, N., Yan, Y. & Ying, Q. (2010). Production of p53 gene knockout rats by homologous recombination in embryonic stem cells. *Nature*, Vol.467, No.7312, (September 2010), pp. 211-213, ISSN 1476-4687

Warren, L., Manos, P., Ahfeldt, T., Loh, Y., Li, H., Lau, F., Ebina, W., Mandal, P., Smith, Z., Meissner, A., Daley, G., Brack, A., Collins, J., Cowan, C., Schlaeger, T. & Rossi, D. (2010). Highly efficient reprogramming to pluripotency and directed differentiation of human cells with synthetic modified mRNA. *Cell Stem Cell*, Vol.7, No.5, (November 2010), pp. 618-630, ISSN 1875-9777

Wilmut, I., Schnieke, A., McWhir, J., Kind, A. & Campbell, K. (1997). Viable offspring derived from fetal and adult mammalian cells. *Nature*, Vol.385, No.6619, (February 1997), pp. 810-813, ISSN 0028-0836

Woltjen, K., Michael, I., Mohseni, P., Desai, R., Mileikovsky, M., Cowling, R.H.R., Wang, W., Liu, P., Gertsenstein, M., Kaji, K., Sung, H. & Nagy, A. (2009). PiggyBac transposition reprograms fibroblasts to induced pluripotent stem cells. *Nature*, Vol.458, No.7239, (April 2009), pp. 766-770, ISSN 1476-4687

Wu, L., Sun, C., Ryan, T., Pawlik, K., Ren, J. & Townes, T. (2006). Correction of sickle cell disease by homologous recombination in embryonic stem cells. *Blood*, Vol.108, No.4, (August 2006), pp. 1183-1188, ISSN 0006-4971

Wu, Z., Chen, J., Ren, J., Bao, L., Liao, J., Cui, C., Rao, L., Li, H., Gu, Y., Dai, H., Zhu, H., Teng, X., Cheng, L. & Xiao, L. (2009). Generation of pig induced pluripotent stem cells with a drug-inducible system. *Journal of Molecular and Cellular Biology*, Vol.1, No.1, (October 2009), pp. 46-54, ISSN 1759-4685

Yang, D., Zhang, Z., Oldenburg, M., Ayala, M. & Zhang, S. (2008). Human Embryonic Stem Cell-Derived Dopaminergic Neurons Reverse Functional Deficit in Parkinsonian Rats. *Stem Cells*, Vol.26, No.1, (January 2008), pp. 55-63, ISSN 1549-4918

Yu, J., Vodyanik, M., Smuga-Otto, K., Antosiewicz-Bourget, J., Frane, J., Tian, S., Nie, J., Jonsdottir, G., Ruotti, V., Stewart, R., Slukvin, I. & Thomson, J. (2007). Induced pluripotent stem cell lines derived from human somatic cells. *Science*, Vol.318, No.5858, (December 2007), pp. 1917-1920, ISSN 1095-9203

Zhong, X., Li, N., Liang, S., Huang, Q., Coukos, G. & Zhang, L. (2010). Identification of mircoRNAs regulating reprogramming factor Lin28 in embryonic stem cells and cancer cells. *Journal of Biological Chemistry*, Vol.285, No.53, (December 2010), pp. 41961-41971, ISSN 1083-351X

Zhu, X., Pan, X., Wang, W., Chen, Q., Pang, R., Cai, X., Hoffman, A. & Hu, J. (2010). Transient in vitro epigenetic reprogramming of skin fibroblasts into multipotent cells. *Biomaterials*, Vol.31, No.10, (April 2010), pp. 2779-2787, ISSN 1878-5905

7

Animal Models of Angiogenesis and Lymphangiogenesis

L. D. Jensen[1,2] et al.*
*Department of Microbiology, Tumor and Cell Biology,
The Karolinska Institute, Stockholm,
*Institution of Medicine and Health, Linköping University, Linköping,
Sweden*

1. Introduction

Blood and lymphatic vessels are present in all tissues, and play important roles for their function, homeostasis and maintenance. Angiogenesis, the growth of new blood vessels, is therefore highly important during development, but is largely not observed in the adult, except for during the female reproduction cycle and during wound healing. In pathological situations, however, angiogenesis may be turned on, and in this case contribute to the onset and progression of most severe human pathologies characterized by high mortality, including cancer, diabetes, obesity and retinopathies (Carmeliet, 2003) or is insufficiently activated such as in the case of myocardial infarction and stroke (Y Cao et al, 2005). Thus, angiogenesis is one of the largest and fastest evolving areas of research today. Angiogenesis is a highly complicated process, involving many different cell types, and it is therefore highly recommended that researchers use *in vivo* animal models for their studies. Accordingly, today there are many *in vivo* models available.

The aim of this chapter is to give insights into the most commonly used *in vivo* angiogenesis models in both mice and zebrafish. We will provide detailed descriptions and discussions of the adipose tissue-, tumor-, ischemic hind limb- wound healing- and corneal micropocket angiogenesis models in mice and developmental-, tumor-, hypoxia-induced retinal- and regenerating tail fin angiogenesis models in zebrafish. We will provide a base for comparison between the different assays to quickly identify which model is best suited for a particular research focus.

1.1 Basic mechanisms of angiogenesis

Angiogenesis is a multistep process which is tightly regulated by an intimate balance between pro- and anti-angiogenic factors. Angiogenesis is stimulated by angiogenic factors the most commonly studied being members of the vascular endothelial growth factor (VEGF), fibroblast growth factor (FGF), transforming growth factor (TGF) or platelet derived growth factor (PDGF) families in the tissue. These factors either act locally, or are

* J. Honek[1], K. Hosaka[1], P. Rouhi[1], S. Lim[1], H. Ji[1], Z. Cao[2], E. M. Hedlund[1], J. Zhang[1] and Y. Cao[1,2]
1 *Department of Microbiology, Tumor and Cell Biology, Karolinska Institute, Stockholm, Sweden,*
2 *Institution of Medicine and Health, Linköping University, Linköping, Sweden.*

released into the blood stream and activate angiogenesis through binding to their corresponding receptors on pre-existing endothelial cells and/or perivascular cells (Y Cao, 2008). Upon ligand binding to the receptors, a cascade of intracellular signal transduction is triggered resulting in activation of the endothelial cell and initiation of the cellular mechanisms required for angiogenesis including basal membrane degradation, endothelial cell (EC) proliferation and migration, formation of tube like structures, and finally vascular maturation by coverage with smooth muscle cells or pericytes to ensure stability (Davis & Senger, 2005; Jensen et al, 2007).

The dynamics of angiogenesis as well as the morphology of the resulting vasculature differs depending on which factor induces the process (R Cao, 2004). In the case of the most studied angiogenic factor, VEGF, a mature vessel may be exposed to a gradient of VEGF induced by hypoxia in a nearby tissue such as a tumor. VEGF activates VEGF receptor 2 (VEGFR2) on endothelial cells leading to production of matrix-metalloproteases (MMPs) which disrupts EC-pericyte contacts, degrade the basal membrane and induce a tip-cell behavior in one cell, which is allowed to form cell processes important for cell migration such as filopodia and lammelopodia. The tip cell will via lateral inhibition signal to surrounding cells that they should not adopt this phenotype thus leading to an ordered sprouting of just one or a few neovessels from the original mother-vessel. The tip cell will start migrating up the angiogenic factor gradient, but remain in contact with the underlying and proliferating endothelial cells (stalk cells) and thus retain the connection to the original vessel. At a certain point, the tip cell will meet other tip cells, anastomose – i.e. fuse – and thus form a circulation loop. The endothelial chords will lumenize, allowing blood to flow through the vessel and the vessels will mature by recruiting perivascular mural cells which are tightly associated with the endothelial cells thus providing stability and trophic factors for the endothelial cells, leading to their maturation and re-entry into quiescence (Risau 1997).

1.2 Angiogenesis in physiology and pathology

During development the first vessels – the aorta and the cardinal vein – are formed by vasculogenesis - i.e. the de novo formation of blood vessels by aggregation and vascular morphogenesis of single endothelial progenitor cells. Following this process, the majority of the blood vessels in the body are formed by angiogenic expansion of this initial, primitive circulation loop (Risau 1997). Thus angiogenesis is very important during development. In adults however, most tissues and therefore also the vasculature have stopped growing and instead adopted a quiescent state. In some cases such as during the female reproductive cycle and during wound healing, angiogenesis is needed to regenerate the tissue, but otherwise the erroneous or insufficient induction of angiogenesis in adults is usually associated with pathology. For example, as a tumor is a growing tissue and therefore need blood vessels for supplying energy, angiogenesis is induced after the tumor has reached a certain size (approximately 1 cubic centimetre). Similarly in growing fat tissue new blood vessels are needed when the distance from the existing blood vessels are too large for efficient oxygenation. By blocking such tumor- or adipose-angiogenesis, it is possible to inhibit the growth of these tissues and therefore anti-angiogenic drugs have potential as treatment for cancer and obesity. In patients with retinopathy retinal hypoxia is a consequence of blocked or disrupted blood vessels and also here lead to induction of retinal angiogenesis. The newly formed blood vessels are however often immature and prone to rupture which leads to micro-hemorrhages and thereby drives the pathological progression to advanced states of the disease (Carmeliet, 2003; Folkman, 1995). On the other hand, in other hypoxic tissues such as the hypoxic heart tissue downstream of a blocked coronary artery, it is important to speed up the angiogenic revascularization of the tissue in order to secure sufficient oxygen for the highly metabolically active myocardium such that it may sustain its function (Y Cao, 2010).

These aspects of physiological and pathological angiogenesis are accurately modeled by the angiogenesis models we will discuss in this chapter.

1.3 Mouse models of angiogenesis

Mouse models are the primary *in vivo* tools used in biomedical research. Most research institutions have their own mouse or rodent facilities, and today several strains have been genetically purified through serial in-breeding, which allow sophisticated investigations such as transplantation and grafting experiments to be done in a wildtype animal with a fully functional immune system. Furthermore, genetically manipulated mice have been generated for most of the genes important for angiogenesis and vascular biology. Such mice include conditional or global knock-outs or mice over-expressing angiogenic factors in defined tissues. As the murine and human vasculatures are highly similar, these tools have been very valuable in defining the role of angiogenesis and distinct angiogenic factors in the onset and progression of various human diseases.

Here we present five mouse assays of pathological angiogenesis that have shown tremendous power in the study of mechanisms behind human diseases. These are the adipose-angiogenesis models used to study obesity and diabetes, the xenograft models used to study cancer biology, the ischemia models used to study myocardial infarction and stroke, the wound healing models and the cornea micropocket models used to study basic mechanisms of angiogenesis and lymphangiogenesis.

2. Angiogenesis in adipose tissue

2.1 Introduction

According to a report of the World Health Organization in 2009, there are currently more than one billion people overweight and more than 300 million individuals considered clinically obese. The escalating number of obese individuals is no longer a problem faced only in high income countries. This adverse trend has also been adopted by low and middle income countries. Body Mass Index (BMI) is a commonly used method to categorize overweight (BMI $\geq$ 25 kg*m^{-2}) and obese (BMI > 30 kg*m^{-2}) individuals (Table 1) (Kopelman, 2000). However, rather than just being a cosmetic problem or being associated with social issues, obesity is a serious pre-disease that frequently leads to the development of severe complications and metabolic disorders that often have a fatal outcome. Amongst these diseases are dyslipidemias, fatty liver, sleep apnea, cardiovascular complications, stroke, type 2 diabetes as well as certain types of cancer such as prostate, breast and colon cancer (Y Cao, 2010; Y Cao, 2007). In the affected individuals, development of metabolic disorders is frequently related to endothelial dysfunction.

BMI (kg*m^{-2})	WHO classification
< 18.5	underweight
18.5 - 24.9	-
25 - 29.9	grade 1 overweight
30 - 39.9	grade 2 overweight (obese)
> 40	grade 3 overweight (morbidly obese)

Table 1. Association of Body Mass Index (BMI) with overweight and obesity. A BMI value greater than 30 indicates obesity. Adapted from Kopelman, 2000.

The adipose tissue constantly experiences expansion and regression during growth and repair throughout adulthood. Interestingly, it is known that adipose tissue growth relies on angiogenesis. Already in embryos, the formation of the primitive fat organ is preceded by angiogenesis. Moreover, anti-angiogenic therapy can prevent the expansion and even induce regression of adipose tissue (Brakenhielm et al, 2004).

The adipose tissue vasculature plays an important role in supplying oxygen and nutrients, as well as plasma containing cytokines and growth factors. Furthermore, the vasculature provides the adipose tissue with bone marrow-derived stem cells, able to differentiate into pre-adipocytes and adipocytes, endothelial cells as well as pericytes. The blood vessels also facilitate infiltration of inflammatory cells, such as monocytes and neutrophiles, and play a role in the removal of metabolic waste products. Due to the close interaction of blood vessels and adipocytes, studying angiogenesis in the adipose tissue is a promising strategy to identify potential novel targets for anti-obesity/anti-diabetic therapy and might open new avenues in the prevention and treatment of metabolic disorders in the future.

There are two types of adipose tissue in the human adult, the white adipose tissue (WAT) and the brown adipose tissue (BAT). WAT is frequently regarded as the 'bad' fat which stores excess energy in the form of triglycerides. Indeed, this notorious WAT usually amasses in undesirable parts of the body such as in the intra-abdominal area resulting in a so called 'apple-shaped' body type or the thighs or hips leading to the so called 'pear-shaped' body type. WAT is essential to serve as heat insulation, mechanical cushion and to provide energy for the body. WATs are complex tissues consisting of different cell types such as endothelial cells, pericytes, macrophages and mesenchymal cells having close interaction with one another and collaboratively regulating processes in the adipose tissue. Being an endocrine organ, the WAT produces a myriad of cytokines and angiogenic factors including vascular endothelial growth factor (VEGF), leptin, adiponectin, resistin, interleukin-6 (IL-6), IL-8, hepatocyte growth factor (HGF), angiopoietin (Ang)-1 and -2 FGF-2, estrogen, TGF-α and -β as well as MMP-2 and -9 (Brakenhielm & Y Cao, 2008). These factors interact to regulate the survival, proliferation and differentiation of pre-adipocytes to adipocytes.

Remarkably, adipose tissue is one of the most highly vascularized tissues in the human body (Fig. 1 and 5). In BAT, blood vessel density is several folds higher than in WAT. This reflects the higher metabolic activity of BAT. Adipocytes in WAT are characterized by a large diameter, a spherical morphology and a large unilocular lipid droplet which is surrounded by a thin layer of cytoplasm whereas adipocytes originating from BAT are smaller, contain multilocular lipid droplets and higher cytoplasm content. Furthermore, brown adipocytes express uncoupling protein 1 (UCP1) which has an important role in energy metabolism.

A   B

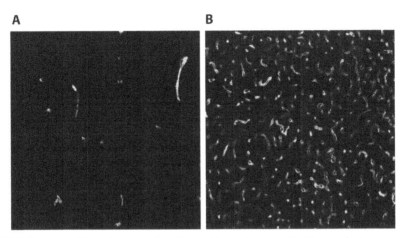

Fig. 1. Visualization of blood vessels in adipose tissue for quantitative analysis.
Fluorescence-based visualization of CD31+ endothelial cells (green) in white (A) and brown
(B) adipose tissue.

Many decades ago, BAT was discovered as a thermogenic (heat generating) tissue, active in
new born human babies, but it was thought to disappear or rather to be inactivated in adults
(Cannon & Nedergaard, 2004). BAT is densely packed with mitochondria which explains
the high metabolic status compared to white adipose tissue. In rodents, BAT is primarily
located in the interscapular region, on the dorsal side between the front limbs, with minor
amounts being found in the thymus, thorax and abdomen. The highly metabolically active,
thermogenic BAT requires a high density of blood vessels to supply oxygen and substrates
to the mitochondria and for waste removal.

In recent years, there has been accumulating evidence demonstrating the presence of active
BAT in adult humans (Cypess et al, 2009; van Marken Lichtenbelt et al, 2009; Virtanen et al,
2009). Several clinical observations have shown the presence of BAT in patients, with tumors
such as pheochromocytoma, following exposure to high levels of catecholamines or
exposure to cold. Most research utilizes the uptake of 18F-fluoro-2-deoxygucose as a tracer
in positron emission tomography (PET) and computer tomography (CT) to detect active
BAT depots in adult humans. PET-CT reveals that the distribution of the BAT depot is
located in the fascial plane in the ventral neck and thorax bilaterally instead of the
interscapular region as seen in rodents and children. The human adult possesses
approximately 10 g of BAT. If all the brown fat in the adult body was fully activated, it
would be able to burn around 4.1 kg of white fat in a year. Histological studies of human
BAT depots show high capillary density. Here, we address the possibilities of driving the
activity of BAT, of conversion of WAT to brown-like adipose tissue and of using
angiogenesis modulators to treat obesity.

2.2 Models/methods to study adipose tissue angiogenesis

To study adipose tissue angiogenesis, several mouse models including models in genetically
manipulated mice, are currently available (Xue Y et al, 2010, Nat. Protoc). These models
usually provide highly reproducible and robust results as the mice are inbred and therefore
share a highly similar genetic background. However, in humans, the cause of developing

obesity is most frequently not genetic but rather due to overeating and a lifestyle based on high caloric intake and little physical exercise. Therefore, high-fat diet fed mouse models provide a powerful tool to study non-genetically related obesity.

2.3 Genetic models
2.3.1 Ob/ob mice

In 1950, obese mice carrying the mutation obese (ob) were described for the first time (Ingalls et al, 1950). The ob mutation was later shown to be located in the gene coding for a hormone known as leptin. Leptin is important in the regulation of appetite and food intake. Leptin signaling is mediated via binding to the leptin receptor (Ob-R) and subsequent signaling to the hypothalamus. Via this pathway, food uptake, energy expenditure as well as fat and glucose metabolism are regulated (Friedman & Halaas, 1998) (Fig. 2).

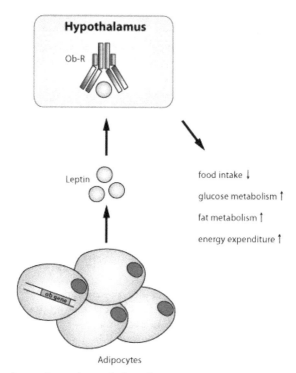

Fig. 2. Regulation of appetite and metabolism by Leptin. Leptin is the product of the *ob* gene and is expressed primarily in adipocytes. In the hypothalamus, Leptin binds to its receptor (Ob-R). It decreases food intake and increases energy expenditure, glucose as well as fat metabolism. Thereby, Leptin contributes to a satiety effect and a lean phenotype.

Due to a lack of leptin, these mice exhibit uncontrolled food intake. Constant overeating therefore results in a gain of body weight. Consequently, ob/ob mice can reach a weight that is three times higher compared to wild type littermates, and their body fat content can be elevated up to fivefold. Apart from this obvious phenotype, ob/ob mice also show decreased physical activity and energy expenditure, infertility and immune deficiencies.

Heterozygotes on the other hand do not display any phenotype as the mutation is recessive. The leptin deficient mouse can be used as an excellent model to study the role of angiogenesis in adipose tissue expansion. Obesity in these mice can be prevented by treatment with anti-angiogenic drugs (Brakenhielm et al, 2004) (this will be discussed more in depth in the Treatment section). Since these mice are comparable to morbidly obese humans regarding the obesity phenotype, using this model might be helpful to identify potential novel targets to treat obesity and obesity-related metabolic disorders in the future.

2.3.2 Db/db mice

The autosomal recessive mutation diabetes (db) was first described in 1966 in the mouse strain C57BL/KsJ (Hummel et al, 1966). These mice are deficient for the leptin receptor. Animals which are homozygous for this mutation, exhibit a phenotype that resembles human diabetes mellitus. This mutant strain is also characterized by an obese phenotype. Furthermore, homozygous mutants are infertile and hyperglycemic while heterozygotes are phenotypically indistinguishable from wild type littermates. These mice are excellent models for studying mechanisms of obesity-related diabetes and insulin insensitivity, and the role of angiogenesis in this regard.

2.4 Other mouse models of obesity

By injecting 3T3 preadipocyte cells subcutaneously into nude mice, researchers are allowed to study the close spatial and temporal correlation between neovascularization and adipose tissue development (Neels et al, 2004). 3T3 preadipocytes will differentiate into adipocytes, and start forming mature adipose tissue *in vivo*. The developing fat pad can be removed at different time points ranging from 1 to 21 days and stained for endothelial cell markers by immunohistochemistry (Fig. 3). Furthermore, the gene expression profile can be analyzed focusing on angiogenesis- as well as adipogenesis-specific genes. Due to the controlled onset and development of the fat pad in adult mice, this model also allowed for studying the origin of the cells that contribute to the formation of new blood vessels during vascularization of the growing adipose tissue.

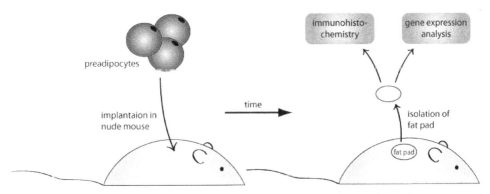

Fig. 3. A mouse model to study adipose tissue angiogenesis. In this model, 3T3 pre-adipocytes are injected into nude mice. Blood vessel formation and gene expression can be studied in the developing fat pad.

2.5 High-fat diet models

One of the first methods to obtain diet-induced obesity in rodent models was based on the so-called cafeteria diet. The cafeteria diet was first described in 1976 as a method to produce obesity in rats (Sclafani et al, 1976). Besides a nutritionally balanced diet, the rats were offered a variety of palatable food items such as cookies and chocolate, candy, salami or cheese and other foods containing high amounts of salt, sugar and fat. Interestingly, the animals ignored the nutritionally adequate chow in favor of the cafeteria food. Although this diet mimics the modern Western diet as consumed by millions of people, a major drawback of this diet is that the animals are allowed to choose independently from the available food. This self-selection may result in substantial differences in the choice of food and thereby nutrients and calories as well as the composition of protein, fat and carbohydrates consumed. Therefore, the cafeteria diet does not guarantee robust and reproducible results.

A more controlled setting to study diet-induced obesity in rodents can be achieved by providing diets in a pellet-form with a high content of fat. The typical standard chow for a laboratory mouse consists of 11.4 % of calories derived from fat, 62.8 % from carbohydrates and 25.8 % from protein, resulting in a nutritional value of 12.6 kJ/g food. A typical high-fat diet may be composed of 58 % fat, 25.6 % carbohydrate and 16.4 % protein. The high content of fat leads to a nutritional value of 23.4 kJ/g food (Winzell AM and Ahren B, 2004). Animals fed with this extreme diet elicit rapid weight gain and are prone to developing obesity.

2.6 Cold induced angiogenesis in adipose tissue

Some types of WAT in mice can acquire a BAT-like phenotype after exposure to cold temperature (4°C). Cold exposure leads to activation of the sympathetic nervous system which increases the capacity of non-shivering thermogenesis (NST) leading to increased heat production that is independent from non-productive muscle activity (shivering) (Xue et al, 2009). Acclimation of rodents to cold should be performed gradually. Mice should be adapted at 18°C for at least one week before transferring them to 4°C. The duration of adaptation is dependent on strains; genetically manipulated strains such as thermogenically incompetent UCP-1 knock-out mice require longer adaptation. Exposure of rodents to cold results in the transition of inguinal WAT to a BAT-like phenotype. Surprisingly, short term exposure (1 week) to cold is sufficient to regulate many genes involved in adipose tissue functions. For example, cold exposure results in up-regulation of BAT-related markers such as (UCP-1) and PGC-1α in the inguinal WAT. The density of blood vessels is highly correlated with the metabolic demand in the different adipose depots. During the transition from WAT to BAT-like, blood vessel density increases dramatically already after one week of cold exposure. After five weeks the WAT exhibits an even higher increase in blood vessel density. Blood vessels are constantly remodeling depending on the metabolic status of the adipose tissues. The transition of WAT to a BAT-like phenotype upon cold exposure is accompanied by the increase in pro-angiogenic factors such as VEGF. This example further demonstrates the importance of tight regulation of blood vessels and angiogenesis in adipose tissue remodeling and function.

2.7 Treatments

The major focus of biomedical research should be on interfering with physiological or pathological processes through treatment. Thus, while the models mentioned in this chapter are valuable in studying the role of the vasculature in initiation and progression of diseases, they are also suitable for evaluation of potentially clinical benefits of novel or previously uncharacterized drugs.

2.7.1 Anti-angiogenic blockade of adipose tissue angiogenesis

Following the example of cold-induced transition from WAT to BAT-like depots described above, this change occurs in parallel with increased expression of VEGF receptors 1 and 2 coupled to increased density of blood vessels in the adipose tissue. Since VEGF-A activates signaling through binding to VEGFR1 and VEGFR2, treatment with neutralizing antibodies against VEGFR1 (MF1) or VEGFR2 (DC101) would provide mechanistic insights into the roles of these receptors. Inhibition with VEGFR2 abolishes the cold-induced vascularization, demonstrating that the VEGFR2 signaling pathway is involved in the regulation of the angiogenic switch in cold. On the other hand, inhibition with VEGFR1 (MF1) resulted in further increased angiogenesis in both WAT and BAT. This suggests that VEGFR1 could be involved in the negative regulation of angiogenesis in adipose tissues (Xue, 2009).

2.7.2 Angiogenesis inhibitors to counteract obesity

As mentioned, the expansion and regression of WAT is highly dependent on angiogenesis. Indeed, treatment with angiogenesis inhibitors, angiostatin, endostatin or thalidomide, results in the reduction of body weight in obese mice (Arbiser et al, 1999). Treatment of leptin deficient obese mice and high fat diet-fed wt C57Bl mice with TNP-470, a selective angiogenic inhibitor, has resulted in reduction of body weight and adipose tissue depot masses. The vascular density in the adipose tissue in TNP-470 treated animals was significantly lower indicating that TNP-470 exerts a direct anti-angiogenic effect on adipose tissues (Brakenhielm, 2004).

Leptin is considered to be a stimulator of angiogenesis. The secretion of leptin is proportionate to the size of adipocytes and regulated by the level of oxygen in adipose tissue. In hypoxic situations such as during hyperplasia or hypertrophy of adipocytes, VEGF protein as well as leptin levels are up-regulated thereby stimulating angiogenesis to provide adequate delivery of oxygen and nutrients to the adipocytes. Co-implantation of leptin with VEGF and FGF-2 in the avascular mouse cornea revealed a remarkable synergistic angiogenic stimulation. However, the treatment of obese individuals with leptin remains controversial. Despite that leptin stimulates angiogenesis, administration of leptin to individuals with a homozygous mutation in leptin genes confer beneficial outcome in terms of body weight reduction (Frederich et al, 1995).

2.8 Methods and assays to study blood vessels in tissues

The models mentioned in this chapter are further strengthened by (usually post mortem) analysis of the tissue in which the blood vessels are growing. In pathological situations the blood vessels are usually of poor quality and functionality, which greatly contributes to progression of the disease. In this section we will discuss several histological and functional tests that can be done to gain additional insight into the structure and quality of the blood vessels.

2.8.1 Hematoxylin/Eosin staining

For Hematoxylin/Eosin (HE) staining of tissues, it is recommended that sections with paraffin-embedded tissue of a thickness of 3 – 5 µm for interscapular BAT (iBAT) and inguinal WAT (iWAT), respectively, are used. For non-adipose tissue types such as tumor, cornea, muscle or dermal tissues, a thickness 5 µm is recommended. Following deparaffinization with xylene and rehydration using 99.7 %, 95 % and 70 % solutions of ethanol, the slides are stained for 3-5 min. with hematoxylin. This results in a clear blue/purple staining of the nuclei of the cells. Eosin is then used to stain the cytoplasmic contents of the cells pink/red. Depending on the different compartments within the cell, different shades of blue to pink can be observed (Fig. 4). With this method, the adipose tissue can be studied with regard to the size of adipocytes. Blood vessels however, cannot be visualized using this method.

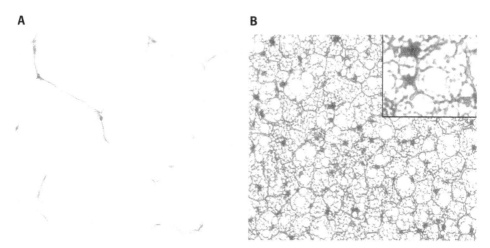

A   B

Fig. 4. Analysis of adipocytes by hematoxylin and eosin staining in white (A) and brown (B) adipose tissue. Nuclei are stained with hematoxylin (purple) and cytoplasmic components with eosin (pink).

2.8.2 Whole mount immunohistochemistry

The use of whole mount immunohistochemistry allows investigations of the vasculature in adipose and other tissues, especially in regard to its structure and functionality.

Different primary antibodies can be used to visualize endothelial cells. These antibodies target for example CD31, CD34 or isolectin all of which are expressed on endothelial cells. This provides a general overview of the vasculature in the tissue and its structure, i.e. whether the vessels are organized or disorganized, if their diameter is normal or if they are dilated and also gives information on the presence of microvessels and capillaries that might have been newly formed (Fig. 5). The tissue vascularity can be assessed by calculating the area of stained vessel signals per field. However, it should be noted that an increase of adipocyte size might imply that the number of vessels decreased even if that is not the case. This is due to the fact that with increasing adipocyte size the area per field that can be covered by vessels decreases. However, the ratio of vessels per adipocyte might not have been changed. In order to take this issue into account, it is recommended to calculate vessel number per adipocyte.

To study pericyte coverage, using an anti-NG2 antibody can provide insights into vessel maturation. Early premature vessels show a low pericyte coverage index and are therefore prone to leakiness and can be considered less functional. To gain even further knowledge regarding vessel functionality, perfusion and leakiness, tetramethyl-rhodamine dextran can be used. Perfusion is studied by injecting 2000 kDa dextran, leakiness can be investigated by injecting 70 kDa dextran via the tail vein. The tissue is then fixed in 4 % paraformaldehyde (PFA) and the vasculature is counter-stained with anti-CD31 antibody using a non-red color secondary antibody (Fig. 6).

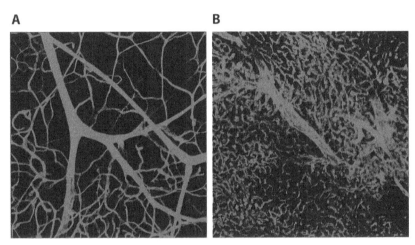

Fig. 5. Visualization of endothelial cells with the marker CD31 in whole mount immunohistochemistry. CD31+ cells (red color) are detected with a fluorochrome-linked secondary antibody in white (A) and brown (B) adipose tissue.

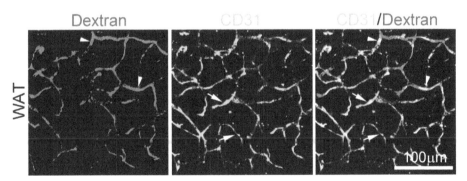

Fig. 6. Functionality test of the vasculature in adipose tissue. Dextran (red) was injected via the tail vein and perfused white adipose tissue (WAT) of a C57Bl mouse was further stained with anti-CD31 antibody (green). Perfused vessels are indicated with arrowheads and non-perfused vessels are indicated with arrows. Scale bar = 100 μm.

Whole mount staining also allows the investigation of lymphatic vessels in adipose and other tissues. Lymphatic vessel endothelial hyaluronan receptor-1 (LYVE-1), podoplanin or VEGFR-3 are common markers which can be targeted by specific antibodies to visualize lymphatic vessels (Fig. 13).

2.8.3 Immunohistochemistry on paraffin sections

While whole mount immunohistochemistry is a valuable tool to study vascular structure and functionality, different methods are advantageous to draw quantitative conclusions regarding the number of blood vessels per adipocyte. For this purpose, the use of thin sections of paraffin-embedded adipose tissue is recommended. In this approach, in contrast to whole mount immunohistochemistry where the thickness of the tissue varies as it is cut by hand, rather uniform tissue sections are prepared using a microtome. Immunohistochemistry on paraffin sections is based on the same principle as whole mount immunohistochemistry: the tissue is incubated with a primary antibody against a protein of interest. A fluorochrome or horse radish peroxidase (HRP)-labeled secondary antibody is then used to visualize the protein (Fig. 7 and 8).

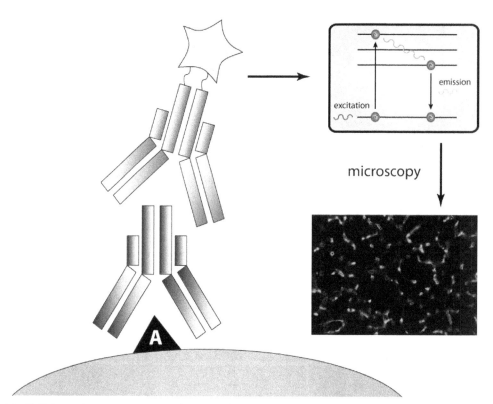

Fig. 7. Schematic representation of immunohistochemistry on paraffin sections using fluorescence-based detection of signals.

It is helpful to counterstain the tissue with DAPI, propidium iodide or Hoechst (for fluorescent stainings) or hematoxylin (for chromogenic HRP stainings) to visualize the nuclei of cells and thereby provide additional information on the structure of the tissue.

To study active angiogenesis, proliferating endothelial cells can be visualized by double staining with antibodies against an endothelial cell marker, such as CD31, and a proliferation marker, such as PCNA or Ki67. Another option is injection of 5-bromo-2'-

deoxyuridine (BrdU) into the tail vain of a mouse 1 min prior to sacrificing the animal. This synthetic thymidine analogue is incorporated into the DNA of proliferating cells. Using an anti-BrdU antibody thereby allows for detection of actively replicating cells.

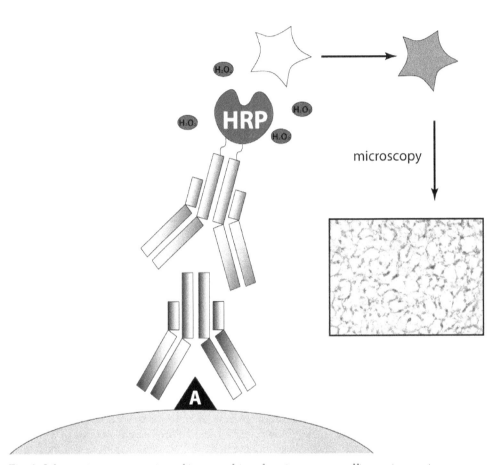

Fig. 8. Schematic representation of immunohistochemistry on paraffin sections using chromogenic detection of signals. The organic compound 3,3'-Diaminobenzidine is oxidized by HRP resulting in a brown color.

2.8.4 Immunohistochemisty on cryosections

Some antigens are masked by paraffin fixation, and in these cases staining has to be done on non-fixed, frozen tissues instead. Dissected adipose or other tissues should be embedded immediately in a plastic cryomold, snap frozen on dry ice and stored at -80°C until staining. Sectioning of cryo-embedded adipose tissue is slightly more challenging than other tissues due to its rather soft integrity. Hence, is it critical to lower the temperature of the cryotome to -30°C before sectioning cryo-embedded adipose tissue samples. Thin sections of 15 μm should be adhered on Superfrost Plus microscope slides, subsequently fixed with cold acetone and stained with specific primary antibodies. However, it is important to note, when staining for lymphatic vessels in the adipose tissue that LYVE-1 is not as widespread on the lymphatic endothelium as podoplanin in this particular tissue (Fig. 9). Here, unlike other tissues, LYVE-1 staining mainly detects inflammatory cells.

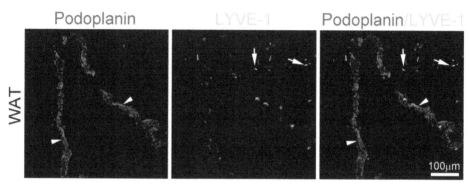

Fig. 9. Double cryosection immunostaining of the WAT of a mouse with podoplanin (red) and LYVE-1 (green). Podoplanin-positive lymphatic staining is indicated with arrowheads and LYVE-1-positive non- lymphatic staining is indicated with arrows.
Scale bar = 100 μm.

2.8.5 Hypoxia staining
Since hypoxia is one of the driving forces of angiogenesis, detection of the presence of hypoxia in tissues could also be performed. Hypoxyprobe-1 (pimonidazole hydrochloride) staining can be used to detect cell and tissue hypoxia. Pimonidazole hydrochloride has a molecular weight of 290.8 kD, ultraviolet absorbance at 324 nm and a plasma half-life of approximately 25 minutes in mice. The detection of hypoxia in tissues can be performed by intravenous or intraperitoneal injection or oral ingestion of pimonidazole hydrochloride at a dosage of 60 mg/kg, 15-90 minutes before sacrificing mice. Dissected tissues should be fixed in 4% PFA followed by paraffin embedding. Paraffin-embedded tissue sections of 3-5 μm is further stained using a peroxidase conjugated anti-pimonidazole antibodies and counterstained with nuclei staining.

3. Tumor models

3.1 Tumor models in general
Mouse tumor-angiogenesis models include xenograft/transplant models and spontaneous models in genetically engineered mice. Xenograft/transplant models are the most widely used, due to the homogeneous and fast onset and progression of the disease (Y Cao, 2005). In these models, human tumor cells are transplanted into athymic nude mice or severely compromised immunodeficient (SCID) mice (xenograft model), or murine tumor cells are transplanted into mice of the same genetic background (homologous transplantation model). Tumor cells can be injected under the skin (heterotypic transplantation, except skin

cancer) or into the same organ from which the tumors originate (orthotropic transplantation). It is important to remember when studying tumor angiogenesis that the tumor microvasculature might differ depending on the implantation sites. Orthotropic transplantation is considered to show a similar tumor vessel phenotype as the primary lesion, and thus more closely recapitulate the clinical situation. However, it is often easier and more accurate to follow the tumor growth in heterotypic models such as when the tumor grows under the skin, where it is easily visualized. Furthermore, depending on the tumor growth rate and size, the vessel structure in tumors can be different. Fast growing and big tumors (≥ 1.0 cm^3) usually have extensive necrosis in the center, which affects the vascular network in these areas. The appropriate tumor size for evaluation of tumor angiogenesis is considered to be 0.5-0.8 cm^3. Tumor cells can be modified to express high levels of angiogenic factors, such as VEGF and FGF, or reporter genes, such as green fluorescent protein (GFP), red fluorescent protein (RFP) or luciferase. These modifications allow us to study the function of specific angiogenic factors on tumor microenvironment in association with invasion and metastasis (R Cao et al, 2004). Recent technological developments have permitted us to use genetically manipulated animal models for tumor studies, in which the mice are over-expressing or deficient in angiogenesis related genes. The majority of these mice are however immunocompetent, requiring the use of murine tumor cells in such animals.

Another type of tumor-angiogenesis models are the spontaneous models. Such models are based on genetically engineered mice, where a tumor suppressor has been deleted, an oncogene is being over-expressed or both, often in a particular cell type such as the pancreatic beta-cell of the islets of Langerhans (Hanahan, 1985). These models provide a disease history that is more closely recapitulating the one in human patients, as tumors are generated from one single hyperplastic cell, and progress through steps which are highly similar to pathological progression of pre-malignant to malignant lesions in humans. In later stages of tumor development in these models, angiogenesis also becomes important and contributes to growth and metastasis of these primary lesions (Koh et al, 2010).

To study tumor angiogenesis using mouse models, control animals or tumor conditions must be incorporated in the study. If tumor cells are genetically altered, the proper control would be tumor cells which have an empty expression vector inserted in their genome instead of one coding for a particular angiogenic factor. For evaluation of treatment efficacy of anti-angiogenic drugs, the control group must be given the vehicle (solvent) in which the drug was prepared, because some vehicles can themselves affect the vascular structure in tumors. Age and genetic background of the mice must also be standardized for the different groups. Standard age for tumor experiments is 6-10 weeks. Sex is usually not important, unless the tumor cells under investigation are gender-specific such as in the case of breast or prostate cancer, but should nonetheless be the same in for example treatment and control groups.

3.2 Tumor models: Assessment of angiogenesis

Dr. Judah Folkman proposed in 1971 that all tumor growth is angiogenesis-dependent (Folkman, 1971). In general, because a limiting factor of tumor growth is the supply of sufficient levels of oxygen and nutrition, fast growing tumors are characterized by a more aggressive angiogenic phenotype which arises from the ability of the tumor cells to secrete angiogenic growth factors to support and change their microenvironments. For instance, tumor cells transduced with the gene coding for VEGF grow much faster than the empty-vector transduced control cells, due to much higher intra-tumoral blood vessel density (Eriksson et al, 2002). In these tumors however, the vasculature consists of highly irregular and immature vessels and many vascular plexuses, which does not support efficient perfusion of blood compared to a less chaotic vasculature in VEGF non-transduced tumor tissue (Fig. 10). To assess angiogenesis in mouse models, there are thus three essential points which should be addressed.

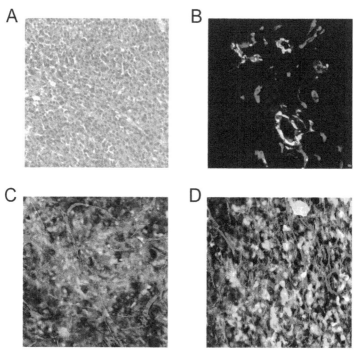

Fig. 10. Histological examinations of tumor tissue. (A) H&E staining of T241 fibrosarcoma. (B) Immunohistchemical staining of T241 fibrosarcoma visualized by fluorescence microscopy. Anti-endomucin labeling of endothelial cells is shown in red and anti-α-smooth muscle actin (SMA) labeling of pericytes/smooth muscle cells is shown in green. (C) Confocal imaging of T241 fibrosarcoma. GFP transfected tumors are shown in green, anti-CD31 labeling of endothelial cells is shown in red. (D) Confocal imaging of VEGF overexpresing T241 fibrosarcoma. GFP transfected tumors are shown in green, anti-CD31 labeling of endothelial cells is shown in red.

1. Tumor size (volume). Tumor growth depends on angiogenesis. Thus, tumor size is usually proportional to the degree of which tumor-angiogenesis is induced by the tumor. If the xenograft/transplant models are used and especially if tumors are transplanted dorsally under the skin, the size is easily measured and an accurate tumor growth curve can be generated by daily measurements of tumor width (W) and length (L) using calipers and calculating the volume (V) as $V=0.52*W^2*L$. If the tumor is not

visible because it is growing in a location inside the animal, special imaging systems, such as CT, magnetic resonance imaging (MRI) or bioluminesence imaging analysis (if tumors express luciferase) are needed to estimate the tumor size. The effect of anti-angiogenic drugs can be inferred from the reduction in tumor growth relative to vehicle treated controls.

2. Histology. Tumor tissues can be stained following the procedures listed in the section on adipose tissue angiogenesis. HE staining will give a general, structural view of the tumor, while immunohistochemical methods are used to visualize the tumor vasculature specifically. In tumors good antigens for visualizing the vasculature are for example von Willebrand Factor (vWF), CD31, CD34, endomucin, isolectin or VE-cadherin. These methods enable analysis of morphological changes and quantitative assessment of the tumor micro environment (Fig. 10). Immunohistochemical staining for multiple markers in the same sample may provide information on the localization of cells relative to each other, cell phenotypes, cell numbers, cell size, and cell conditions (i.e. apoptotic, proliferative).

3. Functional assay of tumor vessels. Tumor vasculatures are irregular and disorganized as well as leaky and poorly perfused. The stained vessels are not all functional; therefore evaluation of vessel functionality such as vessel permeability and perfusion, by using the dextran-injection method described in the adipose angiogenesis section, gives important qualitative information in all tumor models. Upon evaluation under the microscope, abluminal localization of dye indicates extravasation whereas functional vessels are characterized by retaining the dye within the vessel (Hedlund et al, 2009).

3.3 Lymphangiogenesis and lymphatic metastasis

It is well known that blood vessels can support tumor growth by providing oxygen and nutrients, and removing waste products, but the function of tumor lymphatic vessels remains poorly understood (Y Cao, 2008).

Lymphatic networks consist of lymphatic capillaries, collecting lymphatic vessels, and lymph nodes. Unlike blood vessels, lymphatic capillaries consist of one layer of lymphatic endothelial cells (LECs), discontinuous basement membrane and few vascular smooth muscle cells (VSMCs). Lymphatic vessels lack tight junctions between endothelial cells, but are instead equipped with one-way lymphatic valves which give these vessels the ability to collect fluids and macromolecules from the tissue and transport it back to the circulation. On the other hand, these features of the lymphatic endothelium also mean that malignant cells can easily enter into and disseminate via the lymphatic system, leading to lymphatic metastasis. Tumor lymphangiogenesis is therefore associated with cancer metastasis. In some common cancers, such as lung and breast cancer, lymphatic metastasis is the dominant route for tumor metastasis. Invasion of intra- or peri-tumoral lymphatics may result in dissemination of malignant cells to the lymphatic system, leading to lymphatic metastasis in regional lymph nodes (Y Cao Y, 2005; R Cao et al, 2004).

Similar to blood vessels, lymphatic vessels are quiescent in healthy individuals. The formation of lymphatic vessels in tumors is a multistep process that involves LEC proliferation, migration, tube formation and remodeling, which require up-regulation of lymphangiogenic stimulators and down-regulation of lymphangiogenic inhibitors (Y Cao, 2005). Lymphatic vessel growth may represent the imbalanced consequence between positive and negative regulators tipping toward positive regulation. Understanding the molecular mechanisms that control lymphangiogenesis is therefore an important step in the development of therapeutic agents in the prevention and treatment of cancer metastasis.

Among the list of lymphangiogenic factors, members of the VEGF family are the best characterized. VEGF-A, which binds to VEGFR-2 and VEGFR-1, PlGF and VEGF-B which bind to VEGFR-1 and especially VEGF-C and VEGF-D which bind to VEGFR-3 are lymphangiogenic factors, the -C and -D isoforms being the most potent, which regulate both physiological and pathological lymphangiogenesis. VEGF-C/VEGF-D-VEGFR-3-mediated signals are also critical for the sprouting of the first lymphatic vessel from the developing veins in the embryo. This signaling pathway is essential for differentiation of endothelial progenitor cells into the lymphatic lineage (Kukk et al, 1996; Alitalo et al, 2005).

Primary tumors produce several lymphangiogenic factors, including VEGF-A, VEGF-C, VEGF-D, Insulin-like growth factor (IGF), hepatocyte growth factor (HGF) and PDGF-B. These factors induce angiogenesis and lymphangiogenesis both in the local environment and in the regional lymph nodes. Furthermore, they play an important role in establishing the pre-metastatic niche. The term pre-metastatic niche describes the adaptations of for example the lymph nodes which are needed to allow disseminating tumor cells, arriving at a later stage, to meet optimal conditions for growth in that particular site. In addition to signals produced by the tumor cells themselves, inflammatory cells such as macrophages are recruited to tumors by a wide range of tumor cell-derived cytokines and growth factors. At the tumor site, inflammatory cells play a critical role in mediating lymphangiogenesis most likely through the secretion of several lymphangiogenic cytokines.

To detect lymphatic vessels in or around tumors, immunohistochemical staining with lymphatic specific markers, such as VEGFR-3, LYVE-1 and podoplanin is recommended (see Fig. 13). Lymphangiogenesis can be evaluated using all of the models described in this chapter, by using antibodies against one or a combination of these factors either in whole mount staining or on thin sections of frozen or paraffin embedded tissue (Fig. 13).

4. Non-tumor, xenograft models

As tumor cells often produce a multitude of factors which may cooperate in inducing angiogenesis and lymphangiogenesis, these models are considered to be relatively "dirty" and unsuitable for studying the effects of just one or a few factors. In order to overcome this problem, a xenograft model for non-tumor angiogenesis has been developed. In this model, a matrigel plug is grafted onto the mouse which can be mixed with recombinant angiogenic factors prior to grafting.

4.1 Matrigel plug assay

Matrigel consist of purified basement membrane components (collagens, proteoglycans and laminin) and, while it is liquid at temperatures just above 0 degrees, it forms a gel when it is warmed to 37 °C. Thus, the material can be cooled and then injected in the mice, where it will form a three dimensional gel, in which host blood vessels can invade. Matrigel itself is a poor inducer of angiogenesis, but it can be mixed with angiogenic growth factors and/or cells prior to injection leading to a controllable induction of blood vessel growth into the plug. Plug vessels are usually evaluated 7-21 days after implantation by gross examination/photography as well as immunohistochemical staining as described above (Akhtar et al, 2002). If the plug contains functional vessels, the blood (red) vessels can be identified from the photograph. Alternatively, by using mice which express GFP in the endothelium, immunohistochemical staining can be avoided. Also in this model, intravenous dye injection can be performed to evaluate vessel perfusion and leakiness.

5. Rat ischemic hind limb model

Most of the models mentioned above are very useful to study pathological angiogenesis, and therefore used to find novel anti-angiogenic treatment options including novel anti-angiogenic drugs, but other assays are needed to study diseases where therapy would consist of accelerated blood vessel growth, such as in the treatment of myocardial infarction (MI), stroke and wound healing/regeneration.

In MI an occluded coronary artery leads to a blockade of blood flow to a part of the cardiac muscle tissue, and thus leads to severe hypoxia (ischemia). The cardiac musculature is working constantly, has a very high metabolism and is therefore particularly sensitive to reduced oxygen (and sugar) levels. Therefore, unless new, so called collateral, arteries can be formed quickly, the affected cardiac tissue will perish, and usually so will the patient (Y Cao, 2010). Effective therapy would therefore be able to induce growth of highly functional arteries in the response to tissue hypoxia/ischemia.

An excellent model to study and manipulate the growth of blood vessels and in particular arteries in response to tissue hypoxia is the hind limb ischemia model, which can be performed in either mice or rats. In this model, all arteries supplying highly oxygenated blood to one of the back limbs of the animal are ligated in two steps, resulting in near-zero blood flow in the entire limb (Lundberg et al, 2002.) (Fig. 11). This leads to tissue

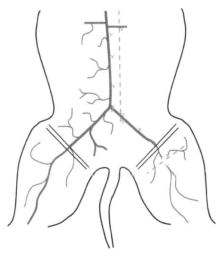

Fig. 11. Hind limb ischemia model. First, a midline incision is made (top dashed line) exposing the vessels. All branches originating from the aorta distal to the renal arteries and all branches from the left iliac artery are ligated by a small suture (blue knots). After a week a second inguinal incision is made (bottom dashed line) and the femoral artery and superficial gastric artery are ligated by a small suture (blue knots). Following this operation the left hind limb does not receive any blood and is considered ischemic.

ischemia and the induction of arteriogenesis from collateral arteries. Angiogenic factors under investigation can be injected into the limb musculature or antibodies against angiogenic or anti-angiogenic factors can be injected into circulation, thus modulating the arteriogenic response. As the hind limb is easily accessible, the circulation in the limb can be observed by Doppler-angiography and thus the same animal can be subjected to repeated investigations on how the blood flow improves over time. After euthanasia, the tissue can be excised and stained as mentioned for the adipose tissue, and the morphology of the blood vessels can therefore be studied post mortem (R Cao et al, 2003). This assay is the most commonly used assay to study therapeutic angiogenesis, and has been used in many seminal discoveries on how to therapeutically generate highly functional and stable arteries (Y Cao, 2010). However, the surgery needed to induce ischemia in the hind limb is very complicated and the assay therefore requires highly skilled and experienced surgeons. Also, there may be slight inter-individual differences in the residual blood flow in the limb after surgery – and therefore the degree of tissue hypoxia - within each experimental group.

6. Wound healing assay

The process of wound healing is divided into three stages; inflammation, new tissue formation and remodeling. Angiogenesis is critical for the formation of new tissue and vascular remodeling occurs in association with tissue remodeling at later stages. Therefore, the wound healing assay is a useful model to evaluate both angiogenesis and vascular maturation/remodelling. Wound healing models are usually performed on the skin as other accessible tissues such as the ears or the tail do not regenerate well. Usually, two circular, trans-dermal wounds are created on the back of anesthetized mice (Fig. 12) allowing one wound to serve as control while topical treatment can be administered on the other. Wound size, scar formation and re-epithelization of the wounds should be recorded daily by photography and by measuring the wound area with calipers, similar to how it is done in the tumor models. In this model, treatment given either systemically by oral administration or injection, or preferably topically on just one of the two wounds, can consist of pro- or anti-angiogenic compounds, and their effects on both the regenerative angiogenesis as well as on vessel morphology and function can be determined post mortem after the regenerated tissue has been excised, fixed and stained as mentioned for the adipose tissue. Also, transgenic or knock-out mice can be used, when available, to study the specific effects of particular genes (Xue et al, 2008).

The surgery required to create the wounds is very simple and the wounds are highly homogeneous in size and location for all animals used in the experiments. This assay is therefore very easy and robust, requires little practice and few animals as the experimental variation is relatively low. However, angiogenesis in this model occurs only in the skin, and in association with inflammation, blood clotting cascades and other highly complex biological processes, which encompass multiple cell types and a plethora of angiogenic and anti-angiogenic factors. Furthermore, regeneration of the skin is known to be quite different from regeneration of other tissues of higher clinical relevance such as the heart and nervous system, and can therefore not be expected to give much information on the role of angiogenesis in regeneration of other organs. Another drawback of this assay is that regeneration occurs by making new tissue rather than repairing/replacing damaged/dead tissue. This difference is thought to be important, as the latter is almost always the case in clinical situations where for example ischemic insults result in large patches of dead tissue that is hard to replace.

Fig. 12. Wound healing model. Two circular holes of approximately 5 mm in diameter are punched with a tissue puncher through the dorsal skin of an anesthetized wildtype C57Bl6 mouse. No bandages or cover is needed, as the skin in this region has no major blood vessels, and the wound formation leads to very little if any bleeding. Wounds close within 2 weeks and would heal completely within a month.

7. Cornea models

The cornea is an avascular tissue consisting of two, thin, transparent layers in rodents. Thus it is possible to gently cut a tiny pocket between the two layers, and in this pocket insert a pellet containing factors which are to be investigated for their angiogenic or anti-angiogenic activities *in vivo*. All vessels detected in the cornea following a few days stimulation with the implanted factors can be considered newly formed vessels, and due to the transparent nature of the cornea and the strong red color of perfused blood vessels, the angiogenic response can be followed kinetically by simply taking photographs of the eye at different time points. This makes it possible to study the effects of angiogenic factors, either alone or in combination on different processes of angiogenesis such as initial angiogenic expansion, vascular remodeling, maturation and stability in the same animal over time.

Usually angiogenic or anti-angiogenic factors are prepared in slow release pellets consisting of hydron or alumni sucrose octo-sulfate, which are dried on a nylon mesh to ensure equal size and thereby amount of factor in each pellet which is then implanted in the corneal micropocket. Using this delivery system, factors are released constantly at low amounts for weeks, thus giving rise to a continuous angiogenic stimulation or inhibition depending on the factor in question (R Cao et al, 2003). Alternatively, small pieces of tumor tissue can be implanted to study the angiogenic capacities of different tumors from different organs. Even primary human tumor samples can be investigated in this model by using immune-deficient nude or SCID mice (Jensen et al, 2009).

While blood vessels can be macroscopically studied by simply taking photographs of the cornea, more in-depth examinations on blood vessel structure and function including pericyte coverage, tip cell formation and vascular permeability require that the cornea is excised post mortem and subjected to immunohistochemical analysis such as those described for the fat tissue (Fig. 13).

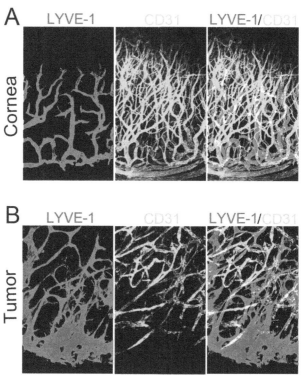

Fig. 13. Visualization of lymphangiogenesis and angiogenesis. A: Confocal analysis of mouse FGF2-induced corneal lymphatic vessels (red) stained with an anti-LYVE-1 antibody and blood vessels (green) stained with an anti-CD31 antibody. B: T241-VEGF-C tumor lymphatic vessels (red) stained with an anti-LYVE-1 antibody and blood vessels (green) stained with an anti-CD31 antibody.

Following staining with antibodies against markers specific for lymphatic endothelial cells, this assay is furthermore one of the strongest assays for investigating lymphangiogenesis. Furthermore, by treating the mice with drugs or neutralizing antibodies against particular receptors, this model can be used to parse out the specific contributions of one or a few receptors for a particular angiogenic factor in inducing angiogenesis and lymphangiogenesis.

This model is considered to be "clean" compared to the tumor models described above – i.e. the angiogenic or lymphangiogenic response can be induced by a defined factor or couple of factors. Considering all these benefits – this assay is probably the strongest *in vivo* assay to study the molecular biology of angiogenesis and lymphangiogenesis (Jensen et al, 2009). However, the assay has been criticized for being poor at describing physiologic or pathological angiogenesis as the amount of factor used in the pellet usually give rise to supra-physiological concentrations in the cornea. However, the amount of factor used per pellet can be lowered accordingly, or instead of using purified recombinant growth factors, researchers may implant tumor tissue, or a small piece of suture, which lead to corneal inflammation, which then subsequently drives the angiogenic response although via a less defined and controllable pathway. The major limitation of the assay is the technical difficulty of implanting pellets into the mouse cornea. In the beginning, researchers circumvented this problem by using larger animals such as rats or rabbits, but in order to take advantage of the growing number of transgenic or knock-out mice available today it is a major benefit to use this animal if possible.

8. Zebrafish models of angiogenesis

Zebrafish have recently gained much attention as an angiogenesis model system. Zebrafish embryos develop outside of the uterus, which greatly facilitates imaging during development. Furthermore, pigmentation can be inhibited chemically, or by using non-pigmented strains, such that the embryos maintain transparency throughout development. Many eggs are produced each breeding cycle, zebrafish are relatively easy and cheap to maintain, compared to rodents, and their embryonic development is much faster. Additional benefits of zebrafish-based models include passive uptake of chemicals added to the water, which eliminate the invasive administration procedures which are commonly needed in rodents and their small size which means that all tissues can be oxygenated by passive uptake from the water during development, thus allowing researchers to study phenotypes of embryos with severely disrupted vasculature, which are embryonically lethal and therefore difficult to study in mice (Weinstein et al, 1995).

Because of these unique characteristics of the zebrafish embryo, several genes with essential functions in the vasculature have been discovered using this system. One approach which has yielded the identification and characterization of novel genes important for developmental angiogenesis and vascular maturation is to perform unbiased screens of mutant embryos generated by random mutagenesis induced either by radiation, chemicals or insertion of small genetic fragments (Gaiano et al., 1996; Haffter et al., 1996; Knapik et al., 1996). Such screens have in the past yielded information on human congenital disorders of which the genetic background was previously unknown (Alders et al, 2009), and have – combined with the ability of the embryos to passively take up chemicals added to the water – been used to screen for, and identify novel drugs which efficiently correct the pathological vascular phenotypes (Peterson et al, 2004). Such an approach is called chemical genetics, due to the screening of novel chemicals for therapeutic effects in genetically modified zebrafish models, and is today being frequently used as a discovery tool in the pharmaceutical industry.

Recently researchers have further expanded the benefit of zebrafish-based model systems by generating many transgenic zebrafish strains which express fluorescent markers in particular cell types, organs or tissues, including endothelial cells of the vasculature (Jensen et al, 2009). By continuous observation of such transgenic embryos under the microscope, it is possible to follow the dynamics of growing vessels during zebrafish development in real time.

Such studies, which are difficult to perform in mice due to their *in utero* development, have yielded valuable insights into the process of vasculogenesis, which is the formation of the first embryonic vessels – the aorta, cardinal vein and thoracic duct – and on the origin of blood cells as well as the mechanism by which blood flow is initiated (Herbert et al, 2009; Lida et al, 2010; Yaniv et al, 2006).

27 hour old embryo

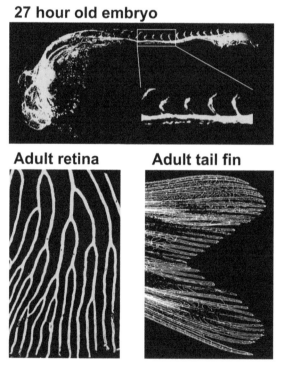

Adult retina   Adult tail fin

Fig. 14. Blood vessels in the *Fli1:EGFP* transgenic zebrafish line. In the *fli1:EGFP* transgenic zebrafish, the promoter drives EGFP expression specifically in the vasculature. The top image show EGFP positive blood vessels (green) in a 27 hours old embryo, on the bottom left is shown blood vessels in the adult retina and on the bottom right in the adult tail fin.

The zebrafish genome has been fully sequenced and annotated, which makes it easy to interfere with genes involved in vascular development by injecting specially designed synthetic RNA-like molecules called morpholinos. These morpholinos are designed to have a complementary sequence to a particular target mRNA to which they will anneal and thus specifically target that transcript for destruction. The addition of morpholic acid groups on the nucleotide backbone of morpholinos makes them stable inside the cell and they are therefore not degraded along with their mRNA target (Nasevicius & Ekker, 2000). Morpholinos, as well as mature mRNA - if over-expression rather than inhibition of a particular gene is being investigated - can be injected in the yolk of newly fertilized 1-cell stage embryos, and will then be present in all cells during development. As the cell number increases, the concentration of morpholino per cell will gradually decrease, and most morpholinos will therefore only be effective for up to 4 days after fertilization. This period is – due to the fast development of the zebrafish embryo – usually more than sufficient for angiogenesis studies, and thus does usually not pose a restriction on the study.

Morpholino-mediated disruption of a particular gene product is called knock-down, due to its transient nature compared to permanent knock-outs, and is today a widely used method to study the involvement of particular genes in angiogenesis and vascular functions during zebrafish development.

By using a zebrafish strain with enhanced green fluorescent protein expressed in endothelial cells under the *fli1* promoter (Lawson & Weinstein, 2002) (Fig. 14), knock-down of VEGF leads to a concentration-dependent inhibition of angiogenesis which can be quantified already 24 hours after fertilization (Nasevicius et al, 2000). In mice, even heterozygous deletion of VEGF leads to early embryonic lethality (Carmeliet et al, 1996) and the developmental consequence of subtle concentration differences in this critical angiogenic factor can therefore only be studied in zebrafish. Thus, zebrafish have been used recently to show that VEGF is only important for arterial growth, whereas growth of veins are practically unaffected by reduced VEGF levels.

9. Embryonic zebrafish metastasis and tumor model

Zebrafish embryos in the first few weeks of development lack an adaptive immune system, which makes them unable to reject non-zebrafish grafts and they are therefore perfect recipients for implantation of mammalian tumor cells. Due to the transparent nature of zebrafish embryos, fluorescently labeled tumor cells may furthermore be traced in a completely non-invasive fashion, leading to continuous monitoring of the behavior of the tumor cells, including their dissemination from the primary tumor mass *in vivo* (Rouhi et al, 2010).

The induction of tumor angiogenesis is a hallmark of advanced tumors, and therefore also of established tumor cell lines. Thus, mammalian tumor cells, implanted into a space around the yolk sac called the perivitelline space of transgenic *fli1:EGFP* zebrafish embryos, is a powerful way of studying early events of tumor induced angiogenesis and especially of tumor cell dissemination and metastasis via the vasculature (Lee et al, 2009; Rouhi et al, 2010). Depending on their invasive capabilities, implanted tumor cells may start disseminating from the primary site very shortly after implantation (Fig. 15). Angiogenic factors produced by the tumor cells can affect the host's existing and developing vasculature which furthermore is a good way of studying the function and morphology of vessels induced by the tumor-derived factors in question. Different kinds of manipulation e.g. knock-down or over-expression of a specific gene in the tumor cells can affect the dissemination and invasion pattern of a specific cell line implanted in the embryo. Genetic manipulation of tumor cells can be performed with help of different tools e.g. small interfering RNA (siRNA), small hairpin RNA (shRNA) or by retroviral integration of expression vectors containing angiogenic factors.

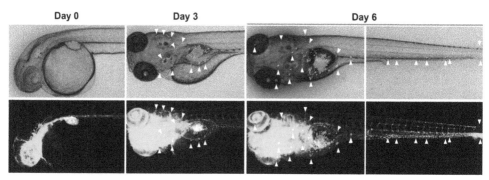

Fig. 15. Metastasis model in zebrafish embryos. DiI labeled tumor cells (red) are injected in the perivitelline space of approximately 2 day old zebrafish embryos (left panels). 3 days after injection many disseminated cells are observed (arrowheads) proximal to the tumor (middle panels). After 6 days tumor cells have spread throughout the embryo, via the vasculature (right panels). The top row shows brightfield images combined with the red-fluorescent signal from the tumor cells. The bottom row shows the EGFP-positive vasculature (green) in transgenic Fli1:EGFP zebrafish embryos combined with the red-fluorescent signal from the tumor cells.

It is also possible to study the effects of host gene expression on tumor cell dissemination and metastasis by modifying expression of host genes by injecting either morpholinos or mature mRNA immediately after fertilization, and thus prior to tumor implantation, as described above (Lee et al, 2009). This model is thus a simple system to parse out the contributions of factors derived from the tumor relative to the host on tumor angiogenesis, dissemination and metastasis.

Since disseminated tumor cells in this model will be detected by fluorescent microscopy, they should either permanently express a fluorescent protein or they can – prior to injection – be labeled with fluorescent dyes such as DiI, which can be easily distinguished from the green fluorescent color emitted from the embryo vasculature. Tumor cells are preferably implanted in the non-vascularized area in the perivitelline cavity as it can otherwise be difficult to distinguish between existing vessels, those formed by normal developmental processes and those induced by the tumor mass.

Prior to implantation, embryos will be dechorionated, anesthetized and placed on a modified agarose gel. Transplantation of tumor cells is easily carried out using a micromanipulator which is connected to a microinjector. Approximately 100 cells will be implanted into each embryo.

After microinjection, based on the purpose of the study, embryos will be immediately transferred into appropriate embryo water such as E3 water or Danieu's buffer (Fig. 15). As mentioned above, drugs may be added to the water and their effects on tumor angiogenesis, dissemination and metastasis can thus easily be evaluated.

Each tumor bearing embryo may be placed in a separate well of a multi-well plate and examined individually to monitor tumor angiogenesis, dissemination and metastasis in as high temporal resolution as required in the experiment. This model furthermore allows generation of time-lapse video sequences of the tumor cells invading into and out from the vasculature at metastatic sites, and thus examine the tumor-endothelium communications involved in tumor dissemination and metastasis in detail.

As described in more detail below, zebrafish are – compared to mice – highly amenable to studies on the physiological and pathological effects of hypoxia. In order to investigate the effects of hypoxia on tumor angiogenesis, dissemination and metastasis, tumor cell-transplanted embryos can be placed in hypoxic water in a special aquarium (Lee et al, 2009). As tumor hypoxia in mammalian models is very difficult to control and monitor, this system allows studies of the effects of highly defined oxygen concentrations and periods of hypoxia exposure. In all conditions above fish embryos will be kept at 28.5°C, which is the standard temperature for rearing zebrafish embryos and larvae. Death rate is relatively high in hypoxia experiments and in order to have enough embryos at the end to make statistically correct conclusions, a relatively high number of embryos should be implanted with tumor cells and placed inside the hypoxia chamber.

10. Spontaneous tumor models in zebrafish

As it is the case for mice, there are also several genetically engineered spontaneous tumor models available in the zebrafish. While these models may be regarded to more closely model the clinical pathogenesis of cancer, the same limitations are relevant for the zebrafish models as those described for mice, most notably the very heterogeneous tumor onset and growth both for different tumors within each fish but also between different fish.

10.1 Peripheral nerve sheath tumors

Homozygous *tp53M214K* mutant zebrafish which carry a mis-sense mutation in exon 7 of the *P53* gene, giving rise to spontaneous formation of abdominal and periocular tumors at an age of 8-16.5 months. HE staining of ocular and abdominal tumors showed that these tumors are mainly composed of spindle cell and epitheloid cell neoplasms which recapitulate characteristics of human malignant peripheral nerve sheath tumor (MPNST). Intraperitoneal implantation of extracted zMPNST cells into irradiated wild type adult zebrafish led to formation of typical zMPNST tumors after less than one month post transplantation (Berghmans et al, 2005).

10.2 Melanoma

By crossing the *tp53* mutant zebrafish with a transgenic line expressing a mutant, constitutively active form of BRAF (V600E, which is the same mutation frequently observed in human melanomas) under control of the melanocyte microphthalmia-associated transcription factor (*mitfa*) promoter, it was found that these fish develop melanomas spontaneously in preference to MPNSTs. In these fish formation of small nevi can be observed already at a young age, and these nevi continue to progress into large malignant, metastatic melanomas in multiple locations from approximately 4-6 months of age.
Malignant melanoma tumor cells from the tumor bearing fish can be transplanted into sub-lethally gamma irradiated wild type zebrafish, where they will grow and spread much like tumors in the mouse xenograft models described previously (Patton et al, 2005).

10.3 T-ALL

Aberrant expression of Myc in humans induces lymphoma and leukemia. Similarly in zebrafish a T cell acute lymphoblastic leukemia (T-ALL) zebrafish model was generated by microinjection of the mouse *c-myc* (*mMyc*) gene under control of the zebrafish *Rag2* promoter (*zRag2*). Onset of tumors varied between 30 to 131 days post injection of the *Rag2-mMyc* expression plasmid. Fish developing T-ALL were characterized by inflated abdominal cavities and infiltration of malignant cells (transformed lymphoblasts) throughout the body, under the skin, into base of the pectoral fin, olfactory region, and the retrorbital soft tissue that led to splayed eyes.

These malignant lymphoblasts were transplantable into irradiated wild type adult zebrafish giving rise to small tumors appearing already one week post transplantation (Langenau et al, 2003)

10.4 Rhabdomyocarcoma

Rhabdomyosarcoma (RMS) is a very aggressive soft tissue sarcoma with high incidence seen among children compared to other types of cancer. The zebrafish RMS model was generated by injecting a Rag2-kRASG12D construct, expressing a constitutively active isoform of RAS under the Rag2 promoter, into the zebrafish embryo at one cell stage. Visible highly invasive tumors in liver, intestine, kidney, and testes appeared already 10 days post injection, in accordance with the early onset of these tumors in human patients, distinguishing this model as one of the fastest spontaneous, vertebrate tumor models available (Langenau et al, 2007).

11. Adult zebrafish models

While embryonic zebrafish models have yielded many valuable insights into human vascular or vessel-related pathologies, most diseases strike adult patients, and accordingly, their clinical symptoms are expected to be more closely recapitulated by adult disease models (Dahl Ejby Jensen et al, 2009). To this end there are today also a growing number of adult zebrafish models available which are highly valuable in the study of angiogenesis-related disorders including retinopathy, regeneration/wound healing and cancer.

11.1 Zebrafish retinal angiogenesis models

Diabetic retinopathy and age-related macular degeneration are severely debilitating disorders in which angiogenesis is a major driving force of the pathology. During progression of these diseases, retinal hypoxia induces pathological growth of immature and fragile blood vessels in the retina, which is associated with the progression to severe states of the diseases. Mice are not convenient to use for studies on retinal hypoxia, as the traditional methods of generating tissue hypoxia in mice (vessel occlusion/ligation) are not applicable in the retina, as there are no easily reachable arteries which can be ligated without causing excessive damage to the animal. Furthermore, mice cannot be exposed to severely hypoxic environments, as their respiration system is not adapted to withstand low atmospheric oxygen levels.

Zebrafish are much more robust to environmental hypoxia, and can withstand even very low oxygen levels in the water for a long time. In contrast to mammals, by incubating zebrafish in an aquarium, where the oxygen levels in the water can be tightly controlled by regulating the perfusion of nitrogen gas, researchers are given full control on the precise degree of hypoxia, the amount of time the tissue experiences hypoxia and the possibilities of distinguishing hypoxia-effects from other effects of restricted circulation such as acidosis and accumulation of waste products.

The zebrafish retinal vasculature is furthermore particularly amenable to studies on angiogenesis due to its remarkably simple structure (Fig. 16). As in mice and humans, the retinal vasculature is supplied with blood from a major optic artery, which branches out at the center of the optic disc to form approximately 4-7 so-called grade I arteries which cover the inner surface of the retina. These arteries branch a few more times before anastomosing with protrusions from the circumferential vein. Thus, the zebrafish retina has a simple monolayer vasculature that is organized from the center to the periphery as arteries-capillaries-veins, which is in contrast to that in mice where the vasculature is multi-layered and arteries, capillaries and veins are co-localized throughout the retina (R Cao et al, 2008).

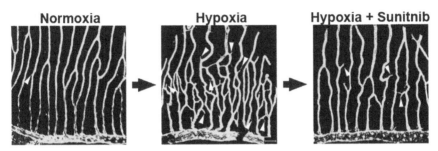

Fig. 16. Hypoxia-induced retinal angiogenesis model. Adult *Fli1:EGFP* transgenic zebrafish were incubated in 10% air-saturated water for 12 days, which induce marked angiogenic expansion of the capillary area of the retinal vasculature (middle figure) compared to fish in normoxic water (left figure). Hypoxia-induced retinal angiogenesis is dependent on VEGF signaling as the VEGF receptor inhibitor sunitnib is able to block the formation of new vessels under hypoxia in the zebrafish retina (right figure). Yellow arrowheads indicate angiogenic sprouts. Figure adapted from R Cao et al, 2008.

This retinal vasculature is highly sensitive to hypoxia. After only a few days of exposure to environmental hypoxia, a marked angiogenic expansion of the capillary area is clearly observed, which can be modulated either by the degree, or the amount of time, the fish is kept in hypoxia. An example of hypoxia-induced retinal angiogenesis in the adult zebrafish is given in Fig. 16.

In this model, adult zebrafish are put into a specifically designed aquarium where the water oxygen levels can be automatically controlled and monitored using an oxygen sensor coupled to a valve which opens for perfusion of nitrogen gas when the oxygen levels are above a set level defined by the researcher. Zebrafish can withstand very low oxygen levels better when being gradually adapted, so by slowly reducing the oxygen concentration in the hypoxia aquarium over the course of 1-2 days, the fish can be maintained in water with an oxygen concentration of only 8-10% of that in fully air-saturated water (Z Cao et al, 2010). After exposure to this level of hypoxia for a few days, retinal neovascularization is readily detected post mortem, under a fluorescent microscope. Also in this assay, it is convenient to add orally active drugs to the water during exposure to hypoxia, to study how such drugs may interfere with hypoxia-induced retinal neovascularization. Hypoxia-induced newly formed vessels can easily be distinguished due to the normally very simple and organized vasculature of the retina in adult zebrafish. Quantifying the number of sprouts, new branches and vascularization area are convenient ways of comparing the effects of the investigated drugs compared to non-treated controls (Z Cao et al, 2010). While capillary sprouting begins on day 3 after exposure to hypoxia, formation of a neovascular network becomes obvious on day 6 after exposure to hypoxia (R Cao et al, 2008).

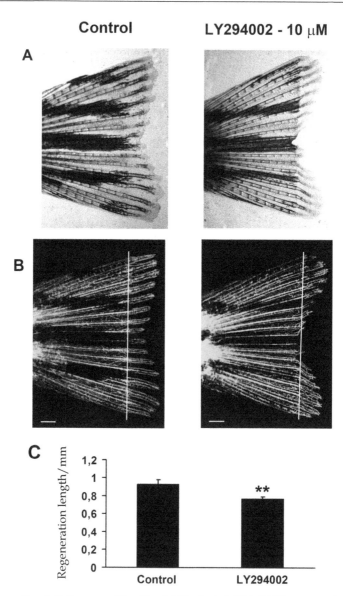

Fig. 17. Regenerating tail fin assay. The distal 1/3 of adult *Fli1:EGFP* transgenic zebrafish tail fins was amputated under anaestisia, and the fish were placed in either clean water (left panels) or water containing the phosphatidyl-inositol-3-kinase (PI3K) inhibitor LY294002 (right panels). After 6 days the fish were sacrificed, the tail fins were fixed in 4% PFA and the regeneration could be determined by bright-field microscopy (A). The vasculature in the regenerating tail fin tissue was observed by fluorescence microscopy (B). Quantification of the regeneration length is an easy and convenient way to demonstrate the effects of the tested drugs (C). Bars indicate 0.5 mm. Figure reproduced from Alvarez et al, 2009.

This experiment allows researchers to study the mechanisms of hypoxia-induced angiogenesis *in vivo*, to study neovascularization, vascular remodeling, and leakiness under pathological settings, to correlate vascular changes with disease development, to screen potential orally active therapeutic agents, and to assess the beneficial effects of known antiangiogenic agents for the treatment of retinopathy. However, there are also limitations. Unlike the clinical situation, this model does not induce local tissue hypoxia but rather rely on systemic exposure to hypoxia. Also, owing to the limitation of available antibodies against zebrafish proteins, staining of particular cell subsets in the retina may be difficult, reducing the amount of information that can be obtained from histological examinations compared to mouse models.

11.2 Angiogenesis in the regenerating zebrafish fin

Adult zebrafish have a remarkable regenerative capability. Many tissues which may not be regenerated in mammals are quickly regenerated in zebrafish. Among these are the heart, retina, maxillary barbell and fins (Poss et al, 2002; Vihtelic & Hyde, 2000; Alvarez et al, 2009). Importantly, as they regenerate, new blood and lymph vessels grow into the regenerating tissue – which enables studies on regenerative angiogenesis. One commonly used assay in the adult zebrafish, based on this principle is the regenerating tail fin. After amputation, the tail fin will re-grow and after approximately 1 month, the fin is back to its original size (Huang et al, 2003). This process of fin regeneration encompasses many of the same mechanisms as in human wound healing and regeneration, and is therefore a good model of regenerative angiogenesis. As the fin is largely transparent, and as morpholinos can be introduced by microinjection and electroporation, this assay is almost as versatile as the zebrafish developmental angiogenesis assays, the major difference being that it is performed in an adult animal. Also in the fins, the vasculature is remarkably simple (Fig. 17), and as the fin grows back various levels of vascular remodeling can be observed and therefore studied in detail (Huang et al, 2003). This regeneration model is probably the most commonly used, adult zebrafish angiogenesis model, and is considered to be complementary to the developmental angiogenesis models.

12. References

Akhtar N., Dickerson EB. & Auerbach R. (2002). The Spongue/Matrigel Angiogenesis Assay. *Angiogenesis*, 5, 1-2, pp. 75-80

Alders M., Hogan BM., Gjini E., et al. (2009). Mutations in CCBE1 cause generalized lymph vessel dysplasia in humans. *Nat Genet.*, 41, 12, pp. 1272-4

Alitalo K., Tammela T. & Petrova TV. (2005). Lymphangiogenesis in development and human disease. *Nature*, 438, 7070, pp. 946-53

Alvarez Y., Astudillo O., Jensen L., et al. (2009). Selective inhibition of retinal angiogenesis by targteting PI3 kinase. *PLoS ONE*, 4, 11, pp. e7867

Arbiser JL., Paningrathy D., Klauber N., et al., (1999). The antiangiogenic agents TNP-470 and 2-methoxyestradiol inhibit the growth of angiosarcoma in mice. *J Am Acad Dermatol.* 40, 6 Pt 1, pp. 925-9.

Berghmans S., Murphey RD., Weinholds E., et al. (2005). Tp53 mutant zebrafish develop malignant peripheral nerve sheath tumors. *Proc Natl Acad Sci U S A*, 102, 2, pp. 407-12

Brakenhielm E., Cao R., Gao B., et al. (2004). Angiogenesis inhibitor, TNP-470, prevents diet-induced and genetic obesity in mice. *Circ Res*, 94, 12, pp. 1579-88.

Brakenhielm E. & Cao Y. (2008). Angiogenesis in adipose tissue. *Methods Mol Biol*, 456, pp. 65-81

Cannon B. & Nedergaard J. (2004). Brown adipose tissue: function and physiological significance. *Physiol Rev*, 84, 1, pp. 277-359.

Cao R., Brakenhielm E., Pawliuk R., et al. (2003). Angiogenic synergism, vascular stability and improvement of hind-limb ischemia by a combination of PDGF-BB and FGF-2. *Nat. Med*, 9, 5, pp. 604-13

Cao R., Bjorndahl M., Religa, P., et al. (2004). PDGF-BB induces intratumoral lymphangiogenesis and promotes lymphatic metastasis. *Cancer Cell*. 6, 4, pp. 333-45

Cao R., Eriksson E., Kubo H., et al. (2004). Comparative evaluation of FGF-2-, VEGF-A-, and VEGF-C-induced angiogenesis, lymphangiogenesis, vascular fenestrations, and permeability. *Circ. Res*, 94, 5, pp. 664-70

Cao R., Jensen LD, Soll I., et al. (2008). Hypoxia-induced retinal angiogenesis in zebrafish as a model to study retinopathy. *PLoS ONE*, 3, 7, pp. e2748

Cao Y. (2005). Emerging mechanisms of tumour lymphangiogenesis and lymphatic metastasis. *Nat. Rev. Cancer*, 5, 9, pp. 735-43

Cao Y., Hong A., Schulten H., et al. (2005). Update on therapeutic neovascularization. *Cardiovasc Res*, 65, 3, pp. 639-48

Cao Y. (2007). Angiogenesis modulates adipogenesis and obesity. *J Clin Invest*, 117, 9, pp. 2362-8.

Cao Y. (2008). Molecular mechanisms and therapeutic development of angiogenesis inhibitors. *Adv. Cancer Res*, 100, pp. 113-31

Cao Y. (2008). Why and how do tumors stimulate lymphangiogenesis?. *Lymphat Res Biol*, 6 ,3-4, pp. 145-8

Cao Y. (2010). Therapeutic angiogenesis for ischemic disorders: what is missing for clinical benefits?. *Discov. Med*, 9, 46, pp. 179-84

Cao, Y. (2010). Adipose tissue angiogenesis as a therapeutic target for obesity and metabolic diseases. *Nat Rev Drug Discov*, 9, 2, pp. 107-15.

Cao Z., Jensen LD., Rouhi P., et al. (2010). Hypoxia-induced retinopathy model in adult zebrafish. *Nat. Protoc*, 5, 12, pp. 1903-10

Carmeliet P. (2003). Angiogenesis in health and disease. *Nat. Med*, 9, 6, pp. 653-60

Carmeliet P., Ferreira V., Breier G., et al. (1996). Abnormal blood vessel development and lethality in embryos lacking a single VEGF allele. *Nature*, 380, 6573, pp. 435-9

Cypess AM., Lehman S., Williams G., et al. (2009). Identification and importance of brown adipose tissue in adult humans. *New Engl. J Med*, 360, 15, pp. 1509-17

Dahl Ejby Jensen L., Cao R., Hedlund EM., et al. (2009). Nitric oxide permits hypoxia-induced lyphatic perfusion by controlling arterial-lymphatic conduits in zebrafish and glass catfish. *Proc Natl. Acad. Sci. U S A*, 106, 43, pp. 18408-13

Davis GE. & Senger DR. (2005). Endothelial extracellular matrix biosynthesis, remodeling, and functions during vascular morphogenesis and neovessel stabilization. *Circ. Res*, 97, 11, pp. 1093-107

Eriksson A., Cao R., Pawliuk R., et al. (2002). Placenta growth factor-1 antagonizes VEGF-induced angiogenesis and tumor growth by the formation of functionally inactive PlGF-1/VEGF heterodimers. *Cancer Cell*, 1, 1, pp- 99-108

Folkman J. (1971). Tumor angiogenesis: therapeutic implications. *N. Engl. J. Med,* 285, 21, pp. 1182-6

Folkman J. (1995). Angiogenesis in cancer, vascular, rheumatoid and other disease. *Nat Med,* 1, 1, pp. 27-31.

Frederich RC., Löllmann B., Hamann A., et al. (1995). Expression of ob mRNA and its encoded protein in rodents. Impact of nutrition and obesity. *J Clin Invest,* 96, 3. pp. 1658-63.

Friedman JM. & Halaas JL. (1998). Leptin and the regulation of body weight in mammals. *Nature,* 395, 6704, pp. 763-70.

Hanahan D. (1985). Heritable formation of pancreatic beta-cell tumors in transgenic mice expressing recombinant insulin/simian virus 40 oncogenes. *Nature,* 15, 6015, pp. 115-22

Hedlund EM., Hosaka K., Zhong Z., et al. (2009). Malignant cell-derived PlGF promotes normalization and remodeling of the tumor vasculature. *Proc. Natl. Acad. Sci. U S A,* 106, 41, pp. 17505-10

Herbert SP., Huisken J., Kim TN., et al. (2009). Arterial-venous segregation by selective cell sprouting: an alternative mode of blood vessel formation. *Science,* 326, 5950, pp. 294-8

Huang CC., Lawson ND., Weinstein BM., et al. (2003). Reg6 is required for branching morphogenesis during blood vessel regeneration in zebrafish caudal fins. *Dev. Biol,* 264, 1, pp. 263-74

Hummel KP., Dickie MM & Coleman DL. (1966). Diabetes, a new mutation in the mouse. *Science,* 153, 740, pp. 1127-8.

Ingalls JM., Dickie MM. & Snell GD. (1950). Obese, a new mutation in the house mouse. *J. Hered.,* 41, 12 pp. 317-8

Jensen LD., Hansen AJ. & Lundbaek JA. (2007). Regulation of endotheilal cell migration by amphiphiles – are changes in cell membrane physical properties involved?, *Angiogenesis,* 10, 1, pp. 13-22

Jensen LD., Cao R., & Cao Y. (2009). In vivo angiogenesis and lymphangiogenesis models. *Curr Mol Med,* 9, 8, pp. 982-91

Koh YN., Kim HZ., Hwang SI., et al. (2010). Double antiangiogenic protein, DAAP, targeting VEGF-A and angiopoietins in tumor angiogenesis, metastasis, and vascular leakage. *Cancer Cell,* 18, 2, pp. 171-84

Kopelman PG. (2000). Obesity as a medical problem. *Nature,* 404, 6778, pp. 635-43.

Kukk E., Lymboussaki A., Taira S., et al. (1995). VEGF-C receptor binding and pattern of expression with VEGFR-3 suggests a role in lymphatic vascular development. *Development,* 122, 12, pp. 3829-37

Langenau DM., Traver D., Ferrando AA., et al. (2003). Myc-induced T cell leukemia in transgenic zebrafish. *Science,* 299, 5608, pp. 887-90

Langenau DM., Keefe MD., Storer NY., et al. (2007). Effects of RAS on the genesis of embryonal rhabdomyosarcoma. *Genes Dev,* 21, 11, pp. 1382-95

Lawson ND. & Weinstein BM. (2002). In vivo imaging of embryonic vascular development using transgenic zebrafish. *Dev Biol,* 248, 2, pp. 307-18

Lee SL., Rouhi P., Dahl Jensen L., et al. (2009). Hypoxia-induced pathological angiogenesis mediates tumor cell dissemination, invasion, and metastasis in a zebrafish tumor model. *Proc. Natl. Acad. Sci. U S A,* 106, 46, pp. 19485-90

Iida A., Sakaguchi K., Sato K., et al. (2010). Metalloprotease-dependent onset of blood circulation in zebrafish. *Curr Biol*, 20, 12, pp 1110-6

Lundberg G., Luo F., Blegen H., et al. (2002). A rat model for severe limb ischemia at rest. *Eur. Surg. Res.*, 35, 5, pp. 430-8

Nasevicius A & Ekker SC. (2000). Effective targeted gene "knockdown" in zebrafish. *Nat. Genet*, 26, 2, pp. 216-20

Nasevicius A., Larson J. & Ekker SC. (2000). Distinct requirements for zebrafish angiogenesis revealsed by a VEGF-A morphant. *Yeast*, 17, 4, pp. 294-301

Neels JG., Thinnes T. & Loskutoff DJ. (2004). Angiogenesis in an in vivo model of adipose tissue development. *FASEB J*, 18, 9, pp. 983-5.

Patton EE., Widlund HR., Kutok JL., et al. (2005). BRAF mutations are sufficient to promote nevi formation and cooperate with p53 in the genesis of melanoma. *Curr. Biol*, 15, 3 pp. 249-54

Peterson RT., Shaw SY., Peterson TA., et al. (2004). Chemical suppression of a genetic mutation in a zebrafish model of aortic coarctation. *Nat Biotechnol*, 22, 5, pp. 595-9

Poss KD., Wilson LG. & Keating MT. (2002). Heart regeneration in zebrafish. *Science*, 298, 5601, pp. 2188-90

Risau W. (1997). Mechanisms of angiogenesis. *Nature*, 386, 6626, pp. 671-4

Rouhi P., Lee SL., Cao Z., et al. (2010). Pathological angiogenesis facilitates tumor cell dissemination and metastasis. *Cell Cycle*, 9, 5, pp. 913-7

Rouhi P., Jensen LD., Cao Z., et al. (2010). Hypoxia-induced metastasis model in embryonic zebrafish. *Nat Protoc*, 5, 12, pp. 1911-8

Sclafani A. & Springer D. (1976). Dietary obesity in adult rats: similarities to hypothalamic and human obesity syndromes. *Physiol Behav*, 17, 3. pp. 461-71.

van Marken Lichtenbelt WD., Vanhommerig JW., Smulders NM., et al. (2009). Cold-activated brown adipose tissue in healthy men. *N Engl J Med*, 360, 15, pp. 1500-8.

Vihtelic TS & Hyde DR. (2000). Light-induced rod and cone cell death and regeneration in the adult albino zebrafish (Danio rerio) retina. *J. Neurobiol*, 44, 3, pp. 289-307

Virtanen KA., Lidell ME., Orava J., et al. (2009). Functional brown adipose tissue in healthy adults. *N Engl J Med*, 360, 15, pp. 1518-25.

Weinstein BM., Stemple DL., Driever W., et al. (1995). Gridlock, a localized heritable vascular patterning defect in the zebrafish. *Nat. Med*, 1, 11, pp. 1143-7

Winzell MS. & Ahren B. (2004). The high-fat diet-fed mouse: a model for studying mechanisms and treatment of impaired glucose tolerance and type 2 diabetes. *Diabetes*, 53, 3, pp. S215-9

Xue Y., Religa P., Cao R., et al. (2008). Anti-VEGF agent confer survival advantages to tumor-bearing mice by improving cancer-associated systemic syndrome. *Proc. Natl. Acad. Sci. U S A*, 105, 47, pp. 18513-8

Xue, Y. Petrovic N., Cao R., et al. (2009). Hypoxia-independent angiogenesis in adipose tissues during cold acclimation. *Cell Metab*, 9, 1, pp. 99-109.

Yaniv K., Isogai S., Castranova D., et al. (2006). Live imaging of lymphatic development in the zebrafish. *Nat. Med*, 12, 6, pp. 711-6

8

Ethical and Legal Considerations in Human Biobanking: Experience of the Infectious Diseases BioBank at King's College London, UK

Zisis Kozlakidis[1,3], Robert J. S. Cason[2], Christine Mant[1] and John Cason[1,3]
[1]Department of Infectious Diseases, King's College London,
2nd Floor Borough Wing, Guy's Hospital,
[2]School of Law, Birkbeck College London,
[3]National Institute of Health Research's (NIHR) comprehensive Biomedical Research
Centre (cBRC) at Guy's and St Thomas' NHS Foundation Trust,
UK

1. Introduction

Since the dawn of time *Homo sapiens* have collected human body-parts for a variety of reasons (Lassila &Branch, 2006; Aquaron *et al.*, 2009; Daily Telegraph, 2011). Similarly, representations of pathological lesions have been collected for educational purposes for at least three hundred years (*e.g.* the Hunterian Museum in Glasgow has preserved plaster casts of diseased tissues). A biobank is a generic term to describe any collection of biological materials and may take many forms, ranging from the preservation of plant seeds (*e.g* the Svalbard Global Seed Vault, Norway) or, the storage of human materials for transplants (*e.g.* corneal biobanks). Others collect human materials for artificial insemination (sperm, eggs and embryos), for forensic investigations and animal materials to assist in the preservation of endangered species such as the Iberian lynx (Leon-Quinto *et al.*, 2009). Some biobanks only collect a single type of sample such as DNA (genebanks), whilst others archive a wide variety of clinical materials. Until recently the *modus operandi* of most medical researchers was to use fresh clinical materials to test a specific hypothesis. The premise was either proven, or not, and then the process repeated to answer subsequent questions. This approach is incredibly wasteful since materials not directly needed to test each argument were discarded. In contrast, clinical biobanks can archive and distribute complete sets of materials from patients with diseases to multiple researchers thereby maximising the benefit of every donation. They can also revolutionise the understanding of very rare conditions by gradually accumulating sufficient numbers of samples –or, by the exchange of samples between multiple biobanks (networking) - to permit statistically-significant conclusions to be derived.

These advantages of biobanks were recognized by *Time* magazine as *'one of the ten ideas that are changing the world right now'* (Park, 2009). This Chapter will be confined to those issues confronting biobanks which collect human materials for medical research. Such archives can be subdivided into those which have the aim of answering one specific research question (*e.g.* the Multiple Sclerosis Brain bank) as opposed to systematic biobanks such as the Infectious Diseases Biobank (IDB) at King's College London (KCL) (Williams *et al.*, 2009), which collects clinical materials with no specific research question in mind. The growing popularity of biobanks in medical research in recent years has inevitably raised new and important ethic and legal questions regarding how they should be managed and regulated. For example, recently, the German Ethics Council has proposed that biobanks should be regulated on the basis of five 'pillars' including the concepts of: confidentiality; open informed consent; careful ethics review; sample quality-assurance; and, a transparency of the biobank's goals (Deutscher Ethikrat, 2010). Here some of the most contentious ethico-legal issues facing clinical archives are considered, including: (i) the nature of the contract (*i.e.* informed consent) between the subject and the researcher; (ii) the concept of property or ownership rights in respect to body tissues and fluids; (iii) the duty of care of a biobank to the donor, the sample, the researcher and, society. This is contextualised against historic turning points which have led to the regulatory structures currently in force in the UK. Finally, the organization of the UK's IDB at KCL is described and proposed as a model system for facilitating research into infectious diseases.

2. Ethical considerations in human biobanking

2.1 Novel challenges associated with biobanking

The idea of a biobank to facilitate medical research would appear to be a worthwhile and commendable activity to most people. However, the establishment of such archives raise not only many of the same ethical problems that face the medical community (particularly those involved in recruiting organ donations), but also some unique questions of their own, for example:

- How can a volunteer provide fully 'informed consent' when neither they nor the recruiter have any idea about the nature of future research which will be performed on the donated sample?
- Donating a sample for no pressing medical reason could be questioned since this relatively benign procedure carries an appreciable risk of adverse events (in one study the rate was 0.59% of 89,000 blood donations: of these ~15% were haematomas and 77% vaso-vagal reflex: Garozzo *et al.*, 2010).
- Who actually owns the donated sample?
- Does a donated sample have a commercial value and can it be sold?
- What happens if research discovers that a volunteer has a potentially deadly disease?
- How can biobanks insure that they are representative of the local community?
- Should biobank samples be used for 'trivial' (*e.g.* cosmetics development) or 'controversial' (*e.g.* stem cell, cloning *etc.*) research?

2.2 An ethical framework

To answer such questions biobanks (and their regulators) must draw on contemporary ethical codes, attitudes and opinions to provide guidance to best practice. Whilst this approach can provide discussion points to it does not always produce definitive answers (Gillon, 1985). The earliest consideration of medical ethics was probably the Hippocratic Oath, which introduced the concepts of respecting patients as individuals and doing no harm (Farnell, 2004). Similar sentiments are expressed in the payer of Maimonides, originally believed to have been written by the 12th-century physician-philosopher (Friedenwald, 1917). More probably, this prayer was written by M. Herz, a physician and pupil of the Konigsberg philosopher Immanuel Kant, as print versions can only be traced back until 1793. More contemporary views on medical ethics were crystallized in a 1902 book by Dr Albert Moll on 'Ethics of the Physician:' Hahn, 1984). Two major ethical issues raised by biobanks revolve around consent and the ownership in human tissues.

The justification for consent stems from the notion of personal sovereignty; the exclusive right an individual holds over their own person. This concept is historically rooted in liberal and political thought, as noted by JS Mills: *'over himself, over his own body and mind, the individual is sovereign'* (Mills 1972). Equally though, Kant believed that the humans *'exists as an end in itself, not merely as a means to be used by this or that will at its discretion'* (Gregor, 1998). Indeed, personal sovereignty now serves as the justification for the majority of articles enshrined in the Universal Declaration of Human Rights. Whilst consent is a necessary component in everyday life and medical research it is debatable how 'informed' consent need be. On the one extreme consent procured through misleading information (or under duress) cannot be considered valid. At the other end of the scale an individual may be informed of the risks and side effects of a medical procedure, but is not expected to comprehend the full complexities of the issues. Consent is often reduced to a subtle paternalism in regards to the unequal position of patient-subject to the researcher, as well as addressing how 'informed' a *research* project can be.

Although respecting personal sovereignty is necessary, there are also ethical principles in favour of a communal duty to society. As the aim of a biobank is to facilitate medical research (which in turn will aid the development of future treatments for the general good of society), the question arises as to whether there is an obligation to assist such endeavours. As Mills states *'there are also many positive acts for the benefit of others that he may rightfully be compelled to perform: such as to give evidence in courts of justice; to bear his fair share in a common defence; or in any other joint work necessary to the interests of the society of which he enjoys the protection'* (Mills, 1972). Thus, respecting personal sovereignty does not negate the argument in favour of a public duty to assist such endeavours. Indeed, utilitarian arguments for the 'greatest good for the greatest number' (Bentham's *'felicific calculus'*: Mitchell, 1918) and Kant's transcendental deduction of a moral duty (Paton, 1948) may to some degree also imply an obligation to donate samples to a biobank.

Locke's concept of property is based on the premise that an individual owns the labour of their body, which when mixed with something in nature, confers a property right in the produced object. Indeed a Lockean justification of property rights was accepted in the US case *John Moore v The Regents of the University of California* (1990) as a foundation for a claim on a human cell-line. Thus, Lockean justification for ownership of samples in a biobank could be constructed in a similar fashion; the labour expended in collecting, preparing and storing A biobank's samples confers a right of ownership. Although a degree of ownership exists in relation to human samples it is better to conceive this as conditional ownership (or 'custodianship') rather than an absolute ownership.

2.3 Some historical precedents leading to research ethics regulation

Self-regulation of biobanks based upon general ethical principles may seem a reasonable approach to managing a few samples of blood or urine which have been willingly donated for research. However, a series of notorious cases from the 19th century up to the present day have so shocked the public that legislation of medical ethics and the storage of human body parts became inevitable: some of the most infamous cases are outlined below.

Body-snatching: The UK Murder Act of 1752 meant that the only legal source of corpses for anatomy was those of executed prisoners: however, this was insufficient to supply the demand from medical schools. Stealing a corpse was only regarded as a minor crime and thus evolved into a lucrative business. In 1827/8, the Edinburgh grave-robbers Burke and Hare realized that institutions paid more for fresh corpses and thus graduated from body-snatching to murder in order to meet this demand (Lancet, 1829; Howard & Smith, 2004). The subsequent conviction of Burke in 1829 led to the UK Anatomy Act of 1832 which stipulated that anyone practising anatomy must hold a licence and be responsible for the correct treatment of corpses. This act was repealed by the Anatomy act of 1984 which, in turn, was replaced by the Human Tissue Act of 2004 (below).

Genocide: In the 1930s/40s the National Socialist German Worker's party (NSDAP) became obsessed with the ideas of social Darwinism, eugenics and the Nietzsche concept of 'superman' (Taha, 2005). On this basis the regime initially justified killing those with congenital defects in the T4 (Tiergarten-4) euthanasia programme (Freidlander, 1995). This was criminal programme was subsequently extended to include anyone that the NSDAP deemed 'sub-human' (political opponents, Russian prisoners of war, and, notably the near genocides of European Jews and Roma: Bachrach, 2004). As part of this holocaust some victims were also subjected to non-consensual medical experiments (*e.g.* LD_{50} type testing of humans exposed to hypobaric or hypothermic conditions). Additionally, the NSDAP also assembled a collection of skeletons from euthanized prisoners for the Institute of Racial Hygiene to act as a historic record (and the basis for scientific study of) extinct human 'races'. At the end of the war the Nuremberg 'Doctors Trial' sentenced some of those responsible and resulted in the development of the Nuremberg code of practice for research involving humans (Table 1: US Government Printing Office, 1949). This is an important document since it has served as the basis of almost all subsequent refinements in medical ethics such as the most recent version of the Declaration of Helsinki (World Medical Association [WMA], 2009).

Unit 731: A less-well publicised series of medical crimes of the Second World War included those perpetrated by the Imperial Japanese Army's Unit 731. This was euphemistically named the 'Epidemic Prevention and Water Purification Department' of the Kwantung Army Group in Harbin, occupied China (Alibewk & Handelman, 1999). This unit experimented on over 10,000 humans in studies involving conscious, non-anaesthetised, *vivisections*, weapons testing (*e.g.* the effects of flamethrowers, hand grenades *etc.* upon live humans), as well as bio-weapons research (Harris, 1994; Kristof, 1995; Barrenblat, 2004).

Tuskagee syphilis study: A study of 400 poor African-American men with syphilis was initiated in 1932. To induce participation, recruits were given free medical care, meals and burials and in return provided samples of blood and cerebro-spinal fluids to researchers (Roy, 1995; Crenner, 2011). At no point were the recruits informed that they had syphilis, nor were they treated for it. The 40-year study was particularly controversial because the researchers failed to treat patients even after the discovery that penicillin was an effective cure. In 2010 it was subsequently revealed that in Guatemala the same study had been extended, between 1946-1948, to include actually infecting prisoners, soldiers, and patients in a mental hospital. A total of 696 men and women were exposed to syphilis without their informed consent. As a direct result of these revelations the US Congress passed the National Research Act in 1974 and created a commission to study construct regulations governing studies which involve human participants (Prograis, 2010).

1. The voluntary consent of the human subject is absolutely essential. This means that the person involved should have legal capacity to give consent; should be able to exercise free power of choice, without the intervention of any element of force, fraud, deceit, duress, over-reaching, or other ulterior form of coercion; and should have sufficient knowledge and comprehension of the elements of the subject matter involved as to enable him/her to make an understanding and enlightened decision. This latter element requires that before the acceptance of an affirmative decision by the experimental subject there should be made known to him the nature, duration, and purpose of the experiment; the method and means by which it is to be conducted; all inconveniences and hazards reasonable expected; and the effects upon his health which may possibly arise from participation. The duty and responsibility for ascertaining the quality of the consent rests upon each individual who initiates, directs or engages in the experiment. It is a personal duty and responsibility which may not be delegated to another with impunity.

2. The experiment should be such as to yield fruitful results for the good of society, unprocurable by other methods or means of study, and not random and unnecessary in nature.

3. The experiment should be so designed and based on the results of animal experimentation and a knowledge of the natural history of the disease or other problem under study that the anticipated results will justify the performance of the experiment.

4. The experiment should be so conducted as to avoid all unnecessary physical and mental suffering and injury.

5. No experiment should be conducted where there is a prior reason to believe that death or disabling injury will occur; except, perhaps, in those experiments where the experimental physicians also serve as subjects.

6. The degree of risk to be taken should never exceed that determined by the humanitarian importance of the problem to be solved by the experiment.

7. Proper preparations should be made and adequate facilities provided to protect the experimental subject against even remote possibilities of injury, disability, or death.

8. The experiment should be conducted only by scientifically qualified persons. The highest degree of skill and care should be required through all stages of the experiment of those who conduct or engage in the experiment.

9. During the course of the experiment the human subject should be at liberty to bring the experiment to an end if he has reached the physical or mental state where continuation of the experiment seems to him to be impossible.

10. During the course of the experiment the scientist in charge must be prepared to terminate the experiment at any stage, if he has probable cause to believe, in the exercise of the good faith, superior skill and careful judgment required of him that a continuation of the experiment is likely to result in injury, disability, or death to the experimental subject.

Table 1. The Nuremberg code for medical research involving humans.

Alder Hey hospital scandal: An investigation into the retention of hearts at hospitals in Bristol UK in the early 1990s led to a public inquiry. This subsequently found that a large number of hearts were also being held by the Alder Hey Children's Hospital and the Walton Hospital. In 2001 the Redfern Report (Royal Liverpool Hospital Children's enquiry, 2001) was published and this led to public outcry when it was revealed that Prof van Velzen had archived organs from every child subjected to a *post mortem*. Around 500,000 tissue samples were being held without any realistic likelihood of them ever being used for research. These revelations led to the creation of the Human Tissue Authority and the 2004 Human Tissue Act and in the UK.

Desecration of Alaister Cooke's remains: In 2005 it was discovered that the bones of Alaister Cooke (a distinguished BBC correspondent) and those of others had been surgically excised without permission prior to cremation by Biomedical Tissue Services Ltd. (Smit, 2008). The company then sold the treated bones for use as surgical grafts. Cooke was suffering from bone cancer when he died which would have made his tissues unsuitable for such a purpose. Reports revealed that the people involved in selling the bones altered his death certificate to hide this fact: subsequently M. Mastromarino, a former New Jersey dentist, was sentenced to between 18 and 54 years imprisonment.

The more contemporary of these cases illustrate that body-snatching is a practice which is not restricted to the dark days of the 19th century and will no doubt continue in illicit markets for the foreseeable future. Common themes linking all of these examples include: a dereliction of basic medical responsibilities by physicians; lack of compassion; complete disregard of the dignity and autonomy of the participants (and/or that of the relatives of the deceased); the *storage of body parts*; and, *the absence of consent* by the participants.

2.4 Informed consent and tissue banks

A fundamental requirement of contemporary medical ethics is that of 'informed consent' be provided by a participant before any study, or procedure, can be performed as discussed above (2.2). However, the phrase is fundamentally misleading (Kaye, 2004), since it implies a comprehension of the relevant facts and all possible outcomes of the research. However, how can non-medically qualified members of the public truly be considered to be fully 'informed'? Indeed, by definition the researchers themselves can only best-guess the possibilities (*'if we knew what we were doing, it wouldn't be called research, would it?'* Albert Einstein). This situation is exacerbated in the case of biobanks where samples may be used in future research projects that have not yet been envisioned using techniques and technologies which have yet to be developed. Indeed, one study of biobank donors found that they did not consider themselves well informed about what their samples would be used for (Hoeyer *et al.*, 2005).

This issue was been addressed by the German Ethics Council which takes the view that: *'if donors have been informed of the indefinite nature of the actual future applications, they will be aware that they are agreeing to an uncertainty. This uncertainty is not acceptable if it involves more than minimum health risk which is not the case with Biobanks'* (Deutscher Ethikrat, 2010). Thus 'open' or 'broad' consent to future usage of donated samples has been proposed as best practice for biobanks (Hansson *et al.*, 2006). Similarly, the council of Europe's biobanking recommendation acknowledges the conflict between the traditional informed consent and the needs of population genetic databases, and as a result stated that consent need not be specific, but it must be as specific as possible with regard to unforeseen uses (Council of Europe, 2006).

Such proposals are not without their critics who see equivalence between the broadening of consent and the dilution of ethics which may result in increased public distrust (Hofmann, 2009). Though others have noted that actually the reverse may be true (Lipworth *et al.*, 2009). Some commentators have gone further, suggesting that for genebanks and for population databases, informed consent should be abandoned altogether (Kaye, 2004). Furthermore, in countries such as the UK where free healthcare is provided to all by the state, there is debate as to whether there is an automatic moral obligation upon patients to automatically donate any excess clinical material taken for diagnostic purposes to medical research: *i.e.* the introduction of an 'opt-out' as opposed to the current 'opt-in' system. Such 'opt-out' genebanks are already in operation in Europe (*e.g.* the Vanderbilt DNA databank) and have driven the development of new approaches to the governance as well as innovative public education and communications strategies (Pulley *et al.*, 2010).

2.5 Inclusivity of biobanks

There are many problems facing researchers in gaining the public's confidence in donating samples to biobanks. In the case of donations to genebanks public refusal to consent (revealed by a questionnaire) was explained by a lack of personal relevance of the contribution and feelings of discomfort related to the possibility that the DNA would be used for purposes other than the original study (Melas *et al.*, 2010). The underlying concerns revolved around issues of integrity, privacy, suspiciousness, and insecurity. Interestingly though despite concerns about privacy another study of 4,569 US participants revealed that 60% would be willing to participate (Kafam *et al.*, 2009). However, the same study noted that ethnic minorities, women and those without a College degree, were concerned that the government could gain access to their personal information. Such concerns may be translated into an unwillingness not to engage with biobanks. Indeed, there are well acknowledged problems in recruiting sufficient organ donations from ethnic minorities resulting in a higher mortality amongst these communities from diseases necessitating transplants (Bratton *et al.*, 2011; Salim *et al.*, 2010). In the USA educational schemes have been introduced and appear to have partially resolved this problem (Callender *et al.*, 2010).

In industrialized nations the ethical points of reference for regulating medical research have inevitably been drawn from classical western moral philosophies and Judeo-Christian religious traditions. Relatively recently, the UK population has transformed from a predominantly Caucasian European Christian admix into a diverse multi-cultural/ethnic society as a result of immigration. Whilst more recent migrants from Eastern Europe share many of the cultural and religious traditions of the former, others from Africa, the Indian sub-continent and Asia often do not. Research biobanks (like organ donation schemes) need to be representative of their communities, consequently they must: (i) appreciate the cultural and religious sensitivities of ethnic minorities; (ii) understand historic negative perceptions of Western medical research, and, (iii) use this information to insure that ethnic minorities become fully engaged in such research projects.

An example of such cultural differences include Chinese tradition where self-determination is not a recognized phenomenon (Bowman & Hui, 2010), meaning that the family –rather than the patient- receive clinical information and make decisions to coordinate treatment. In Judaism, bodies are buried undisturbed and quickly after death as a matter of respect. Discussion within the Jewish faith about whether it is permissible to harvest organs for transplant from brain-dead persons is ongoing (Bresnahan & Mahler, 2000). Similarly, amongst Hindus and Sikhs the individual is caste-bound in its decision, and there is the concept of purification by death/rebirth axis: organ donation or contribution to a biobank might be seen to interfere with. In a systematic review of the opinions of twenty-eight major religions only one, Shintoism, was noted to be completely antagonistic to the idea of organ or tissue donation after death (United Network for Organ Sharing [UNOS], 2000). This is based on the concept that the cadaver is impure and dangerous, and injuring it is a serious crime as it damages the '*itai*' (the relationship between the dead and the bereaved).

Many ethnic minorities resident in industrialized nations also have perceptions of 'scientific imperialism' or 'bio-colonialism' (Emerson *et al.*, 2011) or scientific racism (*e.g.* the Tuskegee study 2.3 above). Not surprisingly there can be considerable distrust of western medical research. A recent example of this was the reticence of the Indonesian government to share samples of the H5N1 influenza virus with the international scientific community (Gelling, 2007). One suggestion to restore public faith is the proposal to establish a tissue trust to serve the interests of the common good (Emerson *et al.*, 2011) and would act by involving tissue donors and community members in research governance. These issues are of considerable importance to the current 'opt-in' model for biobanking in the UK. Indeed, for biobanks to be effective they must collect tissues from all of the community. Failure to do this may mean that downstream medical research using a biobanks samples may effectively result in further examples of scientific racism in that some research may be race-specific. These concerns are also a persuasive argument for locating biobanks in ethnically-diverse regions of a country.

Extending this concept of inclusivity further is the idea of harmonizing legal and ethical permissions internationally. This is important as a major ambition of biobanks world-wide is to establish networks for the international exchange of important clinical samples (Pearson, 2004). Whilst differences in national laws may complicate this process it has been suggested that if all countries simply abide by the Helsinki Declaration (WMA, 2009) any additional regulation would be counterproductive (Hansson, 2011). In contrast, others have argued in favour of a greater harmonization of ethics legislation between nations (Chalmers, 2011). Harmonization of biobank regulation is an important future goal since a survey of 126 European biobanks noted that most had currently only a very limited networking activity, and just a half having policies for cross-border sharing of samples (Zika *et al.*, 2011).

3. UK regulations and statutory bodies

In the UK there are three major governmental bodies which regulate medical research. Regulation of research ethics is by one of two types of review bodies: universities (*e.g.* the King's College London's College Research Ethics Committee) and the government's National Research Ethics Service (NRES, 2007). Whilst there is some overlap between these two bodies (*e.g.* human studies not involving NHS patients can be considered by both, investigations using NHS patient samples can only be considered by the latter), most medical researchers use the NRES's local research ethics committees (LREC) scheme. LRECs have evolved over recent years so that now specialized committees exist which are trained in issues arising from biobanking.

Storage of human tissues is regulated by a different body, the Human Tissue Authority (HTA) and premises keeping human tissues for research are required to hold a specific type of HTA licence and are subject to periodic inspections. The definition of tissues by the HTA differs considerably from the biological meaning (a collection of the same kind of cells with a common structure and function: *e.g.* muscle, skin and bone). For the HTA, a tissue is considered to be a mixture of different cells acting with common purpose (*e.g.* such as cells of the immune system). Thus, blood is considered a tissue under the HTA act, though so too are faeces and urine since they also contain a mix of immunological cells. Conversely, a cell-line derived within a week of isolation from the body is not considered a tissue on the basis of its homogeneity. Similarly, hair – not containing cellular architecture - is not regarded by the HTA as a tissue. Confusingly, if tissue architecture architecture is immediately disrupted the resultant biochemical mix is not considered a tissue by the authority (*e.g.* DNA extracted from a human tissue).

Anyone who collects, stores, uses, discloses or destroys identifiable personal information about living individuals, must also comply with the UK's 1998 Data Protection Act (DPA) and the Common Law duty of confidence. For the deceased, researchers must comply with the latter only. Anonymised personal information (as most frequently collected by biobanks) whether concerning the living or the deceased, falls outside the scope of these legal requirements. The DPA applies to `personal data´, which are data that relate to a living individual who can be identified either from those data alone or from those data taken in conjunction with other information that is available to the person who controls the data. When gathering identifiable personal information researchers should aim at all times to ensure that its processing is defensible as both `fair and lawful´. This requires as much transparency as possible about the uses to which data will be put and any risks that might be involved. The net effect of the DPA act is that centres which recruit and annoymise human data and clinical materials prior to submitting them to a biobank are subject to the provisions of the DPA, thus such personal information must be kept secure at all times. Dependent on the nature of the human samples being archived the Health and Safety Executive (HSE) will often need to be consulted by biobanks, particularly if this includes the use of clinical materials from patients infected with dangerous human pathogens (*e.g.* the IDB below).

4. The KCL infectious diseases biobank as a model system

4.1 Vision

Biological resource collections such as the Multicenter AIDS Cohort Study (Kingsley *et al.*, 1987) and the Sidney blood bank cohort (Oelrichs *et al.*, 1998) have helped drive important advances in the understanding of the pathogenesis of human immunodeficiency virus (HIV). In 2005 there were many biobanks in the UK dedicated to the collection of brains or cancer biopsies, but no equivalent facility for the collection of materials for HIV, or indeed any other infectious diseases, research. Even within Europe only three other biobanks held stocks of publicly-accessible material for HIV research, the Spanish HIV biobank (Garcia-Merino *et al.*, 2009,2010), the Sapienza University HIV biobank in Italy, and the Picardie biobank which holds only sera from infected subjects (Chaigneau *et al.*, 2007). A consultation exercise with researchers at KCL indicated that an infectious diseases tissue bank facility would be welcomed by many. This led to the establishment of a group of clinicians and scientists to develop what has now become the KCL IDB. The central issue of the IDB was to collect materials which are of significant value to researchers. For example, around this time many pathology departments were (and still are) rebranding themselves as biobanks to attract research funding. The problem with this approach is that such pathological collections are plentiful and the types of samples preserved do not always coincide with the requirements of researchers (*e.g.* materials suitable for molecular biology studies).

It was therefore established early on that the IDB would not be a genebank, but rather an archive of a broad range of clinical materials which would enable a spectrum of proteomic and genomic studies to be performed (*e.g.* containing live lymphocytes, RNA, DNA, plasma, sera, cerebro-spinal fluids *etc.*). These would be prepared and stored to a high standard and fully documented in terms of sample tracking and processing details. The initial patient cohorts selected for study were those infected with pathogens that were of significance to the local community and also to local researchers. These were patients infected with HIV, hepatitis B virus or, with bacteraemia (especially methicillin-resistant *Staphyloccocus aureus*). In the case of the major sample collection from HIV-infected subjects it was further decided to selectively recruit those with particularly interesting clinical histories. For example, those initially recruited were HIV-1 clade B infected individuals who either progressed to disease unusually quickly (rapid progressors) or, very slowly (long-term non-progressors) as these extremes are most likely to yield important answers to the determinants of pathogenesis. Importantly, none of these patients were to be receiving medication so that the natural history of the infection and disease processes could be studied.

4.2 Location and setting

The IDB is uniquely located for purpose as the local population in Lambeth and Southwark is large (~4 million) and extremely diverse. Indeed, after English the second most common spoken language is Yaruba (African) and then Portuguese (Lambeth census, 2001). This community also suffers from some of the highest rates for HIV infections, the viral hepatitides and sexually-transmitted infections in Europe. The prevalence of UK HIV infections is highest within this area and over 10% of all UK HIV cases are treated by local clinics. This is exemplified by the facts that amongst pregnant women attending St Thomas' Hospital to deliver their babies around 1% are infected with HIV and 2% with hepatitis B (Health Protection Agency, 2008). The IDB is embedded within the KCL Department of Infectious Diseases which is affiliated to King's Health Partners and Guy's And St Thomas' NHS Foundation Trust. The latter hosts an Academic Health Science Centre (AHSC) and also an NHS National Institute of Health Research (NIHR) comprehensive BioMedical Research Centre (cBRC). The latter offer considerable advantages since it has established two clinical research facilities (CRFs) that effectively comprise of two wards and resources in which to conduct clinical trials.

4.3 Ethical permissions for the IDB

The prerequisites for the IDB's ethics included the concept of the dignity and autonomy of the volunteers yet also acknowledged the uncertain future research uses of biobank samples. Thus, the patient's information sheets and consent forms were designed to make it absolutely clear about the uncertainty of future usage. They also make it transparent that their samples would probably be used for genetic research and, the possibility that they would be used both for academic or commercial research purposes anywhere in the world. It is also made clear to participants that their healthcare would be unaffected by their decision to donate a sample, that they could withdraw from the biobank project at any time (and also demand that previously donated samples be destroyed) in line with most recommendations (Gertz, 2008) and that all samples would be annonymised. An additional safeguard was that should a downstream third-party researcher make a finding that was pertinent to the health care of the volunteer, this information would be passed back up through the management chain *via* the biobank to the clinicians at the tissue collection centre (TCC) who could then break the code and advise patients accordingly (since codes linking the patient's NHS number and the biobank code are only held at TCCs). These core principles were consequently remarkably similar to those proposed by the German Research Ethics Council some four years later (Deutscher Ethikrat, 2010).

The timing of the establishment of the IDB was far from ideal since the HTA act was just being implemented. Like any such legislation it is the interpretation which sets the precedents, a process which can take some time. Currently the IDB has ethical permission from the Southampton and South West Hampshire Research Ethics Committee (B) which extends until 2014 (reference # 09/H0504/39) to collect research samples (blood, urine and faeces) and (any) residual diagnostic samples from patients (adults, children and infants) with any infectious or inflammatory disease who are attending a routine clinical appointment.

The IDB cannot recruit from patients who are prisoners or those who are incapable of providing informed consent (other than children where the parent/guardian can consent for them). There are also restrictions for researchers and IDB samples cannot be used for 'trivial' or 'controversial' research projects such as those involving: fertility, reproduction, stem cells, cosmetics or animals. The IDB can establish TCCs at any NHS location in England, Wales or Northern Ireland with the co-operation with a local medical Consultant. Local NHS R&D

offices have to be informed that a TCC is being established but can play no other role in the process. The IDB governance committee is also enabled (through devolved ethics powers) to act as an LREC and provide ethical opinions upon studies wishing to access IDB samples. The IDB stores samples under the authority of an HTA research license held by the Guy's Hospital campus (reference # 12521). This not only covers the storage of materials by the IDB, but also those researchers who remove IDB samples to other sites for the duration of their ethical permission. A diagram of the IDBs operations is provided (Figure 1).

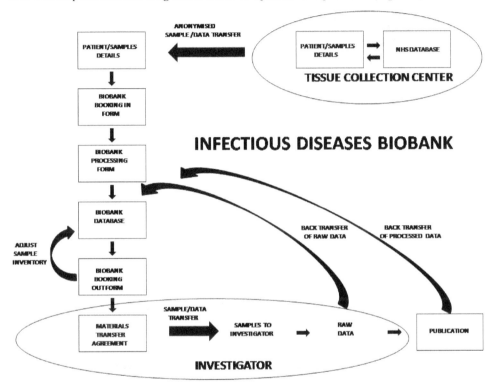

Fig. 1. Management of the IDB.

4.4 Governance

The IDB has a Governance Committee comprising of scientists, doctors, representatives of funding bodies and of the patients. This Committee is responsible for managing the IDB, strategy decisions, insuring that it conforms to current legislative requirements and also to local Medical school regulations. The IDB is regularly audited by the internal KCL Medical School representatives and by the HTA (~four times in 2010) and the results of these checks are passed back to the Governance Committee. Researchers wishing to access the IDB's samples submit a simple two-page application detailing what they propose to do and, the type and numbers of samples required. These details are scrutinized by the IDB's Governance Committee for scientific validity and also for any ethical dimensions. If successful, researchers sign a materials transfer agreement (MTA) and the samples are released.

Staff involved in recruiting volunteers must all have completed 'in house' courses on 'consent taking', 'good clinical practice' and phlebotomy. Copies of these certificates are held by the IDB. Members of the IDB staff are also encouraged to undertake an academic module in ethics, philosophy and religion (a three year 'Associate of KCL' course). The Governance Committee has also established a clear policy on charging researchers to access the IDB's material; they may either agree to pay a fixed rate for the individual samples or contribute funds towards the salaries of IDB staff, to offset the processing and storage costs incurred by the IDB. Researchers are encouraged to approach the IDB early on during the preparation of grants so that projected costs can be included in their applications.

4.5 Quality control
The IDB utilizes standardized operating procedures (SOPs) based upon EEC standards (ISO guideline 34,# 17025:2005) and works within the UNE-EN-ISO 9001:2000 guidelines to facilitate future inter-biobank networking capabilities. Samples are continuously tracked from the time of venepuncture, the time of courier collection from the TCCs, through to freezing at the IDB, with a target of processing >75% of samples within four hours of the bleed. All materials from patients with infections are processed in negative-pressure category III laboratory and stored in locked -80 °C freezers. All of the freezers are: on a protected hospital electricity supply; alarmed to the IDB's staff mobile telephones; and, checked daily for temperature fluctuations. None of the released samples from the collections have undergone a freeze-thaw cycle. Purified DNA samples are tested for the concentration of DNA and, by polymerase chain reaction (PCR) amplification of the housekeeping gene β-globin, for the absence of PCR inhibitors prior to release. In-house assessment of viral RNA viability has also revealed that viral RNA and sequences can be recovered from all plasmas of HIV patients so far tested (for those with viral loads of >350 copies *per* ml). Similarly, an independent analysis of human genomic RNA integrity has demonstrated that all of 104 samples were of high-quality (mean RNA integrity values of 9.3, on a scale where 1=degraded and 10=completely intact RNA) and were successfully used to generate DNA for transcriptome analysis (Kozlakidis *et al.*, 2011). Ultimately, the IDB aims to have all of its procedures, SOPs and operations validated by the International Standards Organisation.

In addition to merely maintaining the samples under the IDB's custiodianship we have also sought to enhance their research value. For example, for the core cohort of HIV infected patients approximately 33.3% of samples have been genotyped for their HLA class I and II alleles and their plasma viruses have had their Gag genes sequenced. The Gag region is important as protein products of this reading frame are believed to be important determinants in viral escape from the innate and adaptive arms of the immune system (Deml *et al.*, 2005). To date, several hundreds of full-length Gag genes have been cloned and sequenced (and the latter data deposited in Genebank: accession numbers FN597659-FN600533). These cloned Gag genes are available to researchers by arrangement. The intention is to obtain equivalent data for the complete HIV cohort. The other type of quality control that the IDB is actively involved in is monitoring that the samples being collected are those which are of (a) most clinical significance and (b) representative of the local community. Initial analyses of the first 200 HIV patient volunteers indicated that it was collecting a population greatly enriched for those with unusual rates of disease progression and that the ethnicity of the volunteers matches well with that of the local community (Kozlakidis *et al.*, In Press).

4.6 Transparancy

Given the multiplicity of studies that any individual sample may be used in, the Governance committee decided that logistically it would be impossible to provide individual volunteers with research feedback on their individual samples, despite reports that this is the preference of potential volunteers (Meulkamp *et al.*, 2010). However, the IDB does attempt to provide feedback in the types of studies performed and these data are displayed on the IDB's website (http://www.kcl.ac.uk/schools/medicine/research/diiid/centres /pii/ biobank/index.html). The IDB has also publicised its mission and was reviewed in *Nature Medicine* (Towie, 2006), has published in *Retrovirology* (Williams *et al.*, 2009), *Biopreservation and Biobanking* (Kozlakidis *et al.*, 2011) and has made presentations to national and international meetings (*e.g.* the 2010 Biobanking Conference; the European Virology Congress in 2008 & 2010, Nuremberg and Como). The IDB director has served on the feasibility study of Biobanking in Northern Ireland (NI) for the NI NHS R&D committee (2008), the KCL College Research ethics committee and associated sub-committees (2010-) and, is chair of the IDB Governance and ethics committees. The IDB has also consulted with patient representatives and has, as a consequence, increased its electronic footprint by establishing KCL IDB sites on the social networking sites 'Facebook' and 'Linkedin'.

4.7 Growth of the IDB

Since sample collection was initiated in January 2007 there has been a logarithmic growth in the number of patient visits to TCCs as well as the numbers of studies approved to access IDB samples (Figure 2). Indeed, currently the IDB is processing over three litres of peripheral venous blood *per* week. These examples of growth have also been matched by the expansion of the IDB into new categories of diseases. These now include those with hepatitis C virus or papillomavirus virus infections and patients with inflammatory diseases (including: as diabetes, Crohn's disease, ulcerative colitis, systemic lupus erythematosus, and, pre-multiple sclerosis syndrome). Some of the currently approved studies are listed (Table 2), and a steady stream of research publications is starting to ensue (Nath *et al.*, 2006, 2007; Alvarez *et al.*, 2008; Thorborn *et al.*, 2010). In addition, publications arising from the contract research work of the IDB are expected to increase significantly (4.8 below).

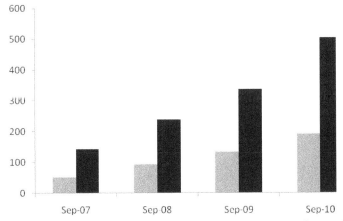

Fig. 2. Year-on-year growth of the IDB's HIV sample collection. Gray bars indicate the recruitment of new patients in total, whereas black bars indicate the number of donations per year.

T-Regulatory cells in HIV infection

Ps20 studies in HIV infections

The roles of vpu and tetherinin HIV/AIDS pathogenesis

The control of inflammation in immunity and autoimmunity

Naive B cell responses in older people

Gene expression signatures of HIV-1 infection *in vivo* and *in vitro*

ccess and study residual clinical samples made available during routine joint surgery.

Genetic variations in IL28B on the natural history of hepatitis B and C and their treatment response.

Non-infectious co-morbidities in HIV infection

Renal function and bone homeostasis in patients starting HAART

Defining the function of CD161+ CD8+ T cell subsets in HIV infection and their response to therapy

Investigating the effect of Maraviroc on Microbial Translocation in HIV infected individuals who are receiving antiretroviral therapy

The metabolic impact of Darunavir/ritonavir maintenance monotherapy after successful viral suppression with standard Atripla in HIV-1-infected patients

Table 2. Some of the types of studies currently accessing the IDB.

4.8 The IDB as a contract service

The IDB's skill and expertise is also being currently utilized to provide research support for clinical studies performed locally, these are often intervention studies and hence require independent ethical permissions from LRECs other than that of the IDB. The largest of these is currently the KCL Human Immune Response Dynamics (HIRD) study. This ground-breaking longitudinal study is investigating the response in humans to vaccination with the H1N1 ('swine flu') influenza vaccine using a protocol was approved by the Brent LREC (09/H0717/88). Briefly, the study involves the IDB collecting and archiving of peripheral venous blood samples from overnight fasted volunteers at the CRFs. Two pre-vaccination samples are harvested and, after vaccination (with PandremixTM H1N1 vaccine: GlaxoSmithKline Biologicals Ltd), a further four samples are collected until six weeks post vaccination. To date over 170 volunteers have completed the course of bleeds and vaccination.

5. Conclusions

This Chapter has summarized the development of some key ethical concepts which specifically impinge upon the newly emerging discipline of human tissue biobanking in terms of classical notions as well as historical turning points which have resulted in current UK legislation. In particular, this paper highlights the fundamental dilemma between the rights of the individual and their duty to the society and describes in practical terms how the KCL Infectious Diseases Biobank is regulated and managed. Whilst, the essential ethical keystone to any biobank is the principle of informed consent, the effective functioning of a biobank necessitate that there should be as few restrictions as possible. A major challenge not mentioned above is that of the education of the general public about biobanks as one study reported strong evidence that more people are likely to embrace the idea of biobank research if they are informed (Gaskell & Gottweis, 2011). Interestingly, they proposed the use of social networking sites to promote the concept.

Now that the IDB has been established for several years most of the governance and ethics issues have been established, nevertheless new challenges are arising. Most pressing for the immediate future is improving the IDB's IT and data processing capabilities. This arises from the facts that the IDB's MTA which requires researchers to feedback raw experimental data and with some of the initial studies drawing to a close the quantity of information is becoming overwhelming. Ultimately though this will eventually permit the IDB to amass a detailed database on the patient volunteers and subsequently permit the multivariate analyses on these cohorts to identify important pathogenic markers.

6. Acknowledgements

The IDB is grateful for generous funding from Guy's and St Thomas' Charity and from the NIHR cBRC. In addition the participation of the volunteers and enthusiasm of the IDB staff is greatly appreciated.

7. References

Alibk K, Handelman S. Biohazzard: The chilling story of the largest covert biological weapons programme in the world – told from inside by the man who ran it. 1999. Delta (2000) ISBN 0-385-33496-6

Alvarez R, Reading J, King DF, Hayes M, Easterbrook P, Farzaneh F, Ressler S, Yang F, Rowley D, Vyakarnam A. WFDC1/ps20 is a novel innate immunomodulatory signature protein of human immunodeficiency virus (HIV)-permissive CD4+ CD45RO+ memory T cells that promotes infection by upregulating CD54 integrin expression and is elevated in HIV type 1 infection. *J Virol.* 2008; 82: 471-486.

Aquaron R, Djatou M, Kamdem L. Sociocultural aspects of albinism in Sub-Saharan Africa: mutilations and ritual murders committed in east Africa (Burundi and Tanzania) *Med Trop* 2009; 69: 449-453.

Bachrach S. In the name of public health-Nazi racial hygiene. *N.Eng J Med* 2004; 351: 417-420.

Barenblat D. A plague upon humanity: the secret genocide of Axis Japan's germ warfare operation. Harper-Collins, 2004. ISBN 0-06-018625-9

Bowman KW, Hui EC. Bioethics for clinicians. 20. Chinese bioethics. *CMAJ.* 2000, 163.1481-1485.

Bratton C, Chavin K, Baliga P.Racial disparities in organ donation and why. *Curr Opin Organ Transplant.* 2011; 16: 243-249.

Bresnahan MJ, Mahler K. Ethical debate over organ donation in the context of brain death. *Bioethics* 2010; 24: 54-60.

Callender CO, Miles PV. Minority organ donation: the power of an educated community. *J Am Coll Surg.* 2010; 210: 708-715.

Chaigneau C, Cabioch T, Beaumont K, Betsou F. Serum biobank certification and the establishment of quality controls for biological fluids: examples of serum biomarker stability after temperature variation. *Clin Chem Lab Med.* 2007; 45: 1390-1395.

Chalmers D. Genetic research and biobanks. *Methods Mol Biol.* 2011; 675: 1-37.

Council of Europe, 2006. https://wcd.coe.int/wcd/ViewDoc.jsp?id=977859

Crenner C. The Tuskegee Syphilis Study and the Scientific Concept of Racial Nervous Resistance. *J Hist Med Allied Sci.* 2011 Feb 12. [Epub ahead of print]

Daily Telegraph, 2011. http://www.telegraph.co.uk/culture/culturenews/5319434Museum-returns-old-Aboriginal-skull-to-Australia.html

Deml L, Speth C, Dierich MP, Wolf H, Wagner R.Recombinant HIV-1 Pr55gag virus-like particles: potent stimulators of innate and acquired immune responses. *Mol Immunol.* 2005; 42: 259-277.

Deutscher Ethikrat, 'Human biobanks for research'. Press release: 05/2010 http://www.thikrat.org/press/press-releases/2010/press-release-05-2010

Emerson CI, Singer PA, Upshur RE. Access and use of human tissues from the developing world: ethical challenges and a way forward using a tissue trust. *BMC Med Ethics.* 2011; 12:2.

Farnell LR. 2004. Chapter 10. Greek hero cults and ideas of immortality. Kessinger Publishing. Pp234-279. ISBN 978-1417921348. P269. The famous Hippocratean oath may not be an authentic deliverance of the great master, but is an ancient formula current in his school.

Freidenwald H. 1917. Prayer of Mimondes. *Johns Hopkins Hospital Bulletin*, August 1917.

Freidlander H. The origins of the Nazi genocide: from euthanasia to the final solution. Chapel Hill: University of North Carolina Press, 1995.

García-Merino I, de Las Cuevas N, Jiménez JL, Gallego J, Gómez C, Prieto C, Serramía MJ, Lorente R, Muñoz-Fernández MA; Spanish HIV BioBank. The Spanish HIV BioBank: a model of cooperative HIV research. *Retrovirology* 2009; 6:27.

García-Merino I, de Las Cuevas N, Jiménez JL, García A, Gallego J, Gómez C, García D, Muñoz-Fernández MA. Pediatric HIV BioBank: a new role of the Spanish HIV BioBank in pediatric HIV research. *AIDS Res Hum Retroviruses* 2010; 26: 241-244.

Garozzo G, Crocco I, Giussani B, Martinucci A, Monacelli S, Randi V. Adverse reactions to blood donations: the READ project. *Blood Transfus.* 2010; 8: 49-62.

Gaskell G, Gottweis H. Biobanks need publicity. *Nature* 2011; 471: 159-160.

Gelling P. 2007. http://www.nytimes.com/2007/03/26/world/asia/26cnd-flu.html?_r=1&scp=7&sq=indonesia %20h5n1&st=cse

Gertz R.Withdrawing from participating in a biobank--a comparative study. *Eur J Health Law* 2008; 15: 381-389.

Gillon R."It's all too subjective": scepticism about the possibility or use of philosophical medical ethics. *Br Med J (Clin Res Ed).* 1985; 290: 1574-1575.

Gregor MJ. 1998. *Groundwork of the metaphysics of morals.* Cambridge University Press. Cambridge, U.K., New York. ISBN 0521622352.

Hahn SZ. Medical ethics in the life of a physician of the soul- reflections on the medical ethics concept of Albert Moll (1862-1936). *Gesamte Inn Med* 1984; 39: 558-561.

Hansson MG. The need to downregulate: a minimal ethical framework for biobank research. *Methods Mol Biol.* 2011; 675: 39-59.

Hansson MG, Dillner J, Bartram CR, Carlson JA, Helgesson G.Should donors be allowed to give broad consent to future biobank research? *Lancet Oncol.* 2006; 7: 266-269.

Harris SH. Factories of death: Japanese biological warfare 1932-1945 and the American cover-up. Routledge, 1984. ISBN 0-415-09105-5

Health Protection Agency, 2008. http://www.hpa.org.uk/web/HPAwebfile/HPAweb_C/1200471685793

Hoeyer K. The role of ethics in commercial genetic research: notes on the notion of commodification. *Med Anthropol.* 2005; 24: 45-70.

Hoeyer K, Olofsson BO, Mjörndal T, Lynöe N.The ethics of research using biobanks: reason to question the importance attributed to informed consent. *Arch Intern Med.* 2005; 165: 97-100.

Hofmann B. Broadening consent--and diluting ethics? *J Med Ethics* 2009; 35: 125-129.

Howard A, Smith M. 2004. William Burke and William Hare. River of blood: serial killers and their victims. Universal. p 50. ISBN 1581125186

Kaufman DJ, Murphy-Bollinger J, Scott J, Hudson KL.Public opinion about the importance of privacy in biobank research. *Am J Hum Genet.* 2009 Nov;85(5):643-54. Epub 2009 Oct 29.

Kaye J. Abandoning informed consent: the case of genetic research in population collections –genetic databases: socio-ethical issues in the collection and use of DNA. Tutton R, Corrigan O (ed.) London, New York: Routledge, 2004.

Kingsley LA, Detels R, Kaslow R, Polk BF, Rinaldo CR Jr, Chmiel J, Detre K, Kelsey SF, Odaka N, Ostrow D, et al.Risk factors for seroconversion to human immunodeficiency virus among male homosexuals. Results from the Multicenter AIDS Cohort Study. *Lancet* 1987; 1(8529): 345-349.

Kozlakidis Z, Mant C, Abdinur F, Cope A, Steiner S, Peakman M, Hayday A, Cason J. Variation of peripheral blood mononuclear cell RNA quality in archived samples. *Biopreservation and Biobanking* 2011: In the press.

Kozlakidis Z, Mant C, Peters B, Post F, Fox J, Philpott-Howard J, Tong W C-Y, Edgeworth J, Peakman M, Malim M, Cason J. How representative are research tissue banks of the local population? Experience of the Infectious Diseases BioBank at King's College London. *Biopreservation and Biobanking:* In press.

Kristof ND. Unmasking horror. (March 17, 1995). New York Times. A special report. Japan confronting gruesome war atrocity.

Lambeth census, 2001. www.lambeth.gov.uk/services/councildemocracy/statisticscensus information *Lancet* editorial, 1828-9 (1) pp 818-821. 28th March 1829.

Lassila KD, Branch MA. Whosse skull and bones? Yale Alumni Magazine (May/June 2006). http://www.yalealumnimagazine.com/issues/2006_05/notebook.html

Leon-Quinto T, Simon MA, Cadenas R, Jones J, Martinez-Hernandez FJ, Moreno JM, Vargas A, Martinez F, Soria B.Developing biological resource banks as a supporting tool for wildlife reproduction and conservation The Iberian lynx bank as a model for other endangered species. *Anim Reprod Sci.* 2009; 112: 347-361.

Lipworth W, Morrell B, Irvine R, Kerridge I.An empirical reappraisal of public trust in biobanking research: rethinking restrictive consent requirements. *J Law Med.* 2009; 17: 119-132.

Melas PA, Sjöholm LK, Forsner T, Edhborg M, Juth N, Forsell Y, Lavebratt C.Examining the public refusal to consent to DNA biobanking: empirical data from a Swedish population-based study. *J Med Ethics.* 2010; 36: 93-98.

Meulenkamp TM, Gevers SK, Bovenberg JA, Koppelman GH, van Hylckama Vlieg A, Smets EM.Communication of biobanks' research results: what do (potential) participants want? *Am J Med Genet A.* 2010; 152A: 2482-2492.

Mills JS. 1972. On liberty in utilitarianism, liberty and representative government, 73, Everyman library. London : J.M. Dent (ed.)

Mitchell WC.. Bentham's *Felicific Calculus. Political Science Quarterly* 1918; 33: 161-183.

Moore v. Regents of the University of California (1990) 51 Cal. 3d 120; 271 Cal. Rptr. 146; 793 P.2d 479.

Nath R, Mant CA, Kell B, Cason J, Bible JM.Analyses of variant human papillomavirus type-16 E5 proteins for their ability to induce mitogenesis of murine fibroblasts. *Cancer Cell Int.* 2006; 6: 19.

Nath R, Mant C, Luxton J, Hughes G, Raju KS, Shepherd P, Cason J.High risk of human papillomavirus type 16 infections and of development of cervical squamous intraepithelial lesions in systemic lupus erythematosus patients. *Arthritis Rheum.* 2007; 57: 619-625.

NRES, 2007. http://www.nres.npsa.nhs.uk

Oelrichs R, Tsykin A, Rhodes D, Solomon A, Ellett A, McPhee D, Deacon N.Genomic sequence of HIV type 1 from four members of the Sydney Blood Bank Cohort of long-term nonprogressors. *AIDS Res Hum Retroviruses* 1998; 14: 811-814.

Park A. 2009. http://www.time.com/time/specials/packages/article/0,28804,1884779_1884 72_1884766,00.html

Paton HJ. The categorical imperative: A study of Kant's moral philosophy. The University of Chicago Press. Chicago, Illinois, 1948, p3.

Pearson H. Summit calls for clear view of deposits in all biobanks. *Nature* 2004; 432: 426.

Prograis LJ Jr.Tuskegee Bioethics Center 10th anniversary presentation: "Commemorating 10 years: ethical perspectives on origin and destiny". *J Health Care Poor Underserved.* 2010; 21(3 Suppl): 21-25.

Pulley J, Clayton E, Bernard GR, Roden DM, Masys DR. Principles of human subjects protections applied in an opt-out, de-identified biobank. *Clin Transl Sci.* 2010; 3: 42-48.

Roy B. The Tuskegee Syphilis Experiment: biotechnology and the administrative state. *J Natl Med Assoc.* 1995; 87: 56-67.

Royal Liverpool Children's enquiry, 2001. Report ordered by the House of Commons. The Stationary Office. http://www.rlcinquiry.org.uk/

Salim A, Berry C, Ley EJ, Schulman D, Desai C, Navarro S, Malinoski D. The impact of race on organ donation rates in Southern California. *J Am Coll Surg.* 2010; 211: 596-600.

Smit M. 2008. http://www.telegraph.co.uk/news/worldnews/1582092/Alaistair-Cooke-body-snatch-leader-pleads-guilty.html

Taha A. Nietzsche, Prophet of Nazism: the cult of the superman-unveiling the Nazi secret doctorine. Author House, Bloomingto Indianna, 2005.

Thorborn G, Pomeroy L, Isohanni H, Perry M, Peters B, Vyakarnam A. Increased sensitivity of CD4+ T-effector cells to CD4+CD25+ Treg suppression compensates for reduced Treg number in asymptomatic HIV-1 infection. *PLoS One* 2010; 5: e9254.

Towie N. London hospital launches infectious disease 'biobank'. *Nat Med.* 2007; 13: 653.

UNOS. Organ and tissue donation. A reference for clergy, 4th ed., 2000. Cooper ML, Taylor GJ (eds.) Richmond, Virginia.

US Government Printing Office. Trials of war criminals before the Nuremberg Military Tribunals under Control Council Law No. 10, Vol. 2, pp. 181-182. Washington D.C.: 1949.

Williams R, Mant C, Cason J. The Infectious Diseases BioBank at King's College London: archiving samples from patients infected with HIV to facilitate translational research. *Retrovirology* 2009; 6: 98.

World Medical association Inc. Declaration of Helsinki. Ethical principles for medical research involving human subjects. *J Indian Med Assoc.* 2009; 107: 403-405.

Zika E, Paci D, Braun A, Rijkers-Defrasne S, Deschênes M, Fortier I, Laage-Hellman J, Scerri CA, Ibarreta D. A European survey on biobanks: trends and issues. *Public Health Genomics* 2011; 14: 96-103.

Part 2

Physiological Systems Engineering in Medical Assessment

9

Renal Physiological Engineering – Optimization Aspects

David Chee-Eng Ng[1] and Dhanjoo N. Ghista[2]

[1]Department of Nuclear Medicine and PET, Singapore General Hospital,
[2]Department of Graduate and Continuing Education, Framingham State University,
Framingham, Massachusetts,
[1]Singapore
[2]USA

1. Introduction

Renal overall functional performance is characterized by excretory function of major end products of protein metabolism, regulation of ionic processes, maintenance of fluid balance and blood volume regulation. Minor functions include hormonal regulation of red cell production and stabilization of blood pressure. Although the kidneys comprise less than 0.5% of total body weight, they receive approximately 20% of the total cardiac output [1]. This consideration underscore the important role played by the kidneys.

The renal circulation has a unique sequence of vascular elements: a high-resistance afferent arteriole, a high-pressure glomerular filtration capillary structure, another high-resistance efferent arteriole and a series of tubular structures with unique absorption/excretion properties. The basic functional unit is the nephron. The nephron consists of a glomerular filtration structure and a tubular system, with its associated vascular elements.

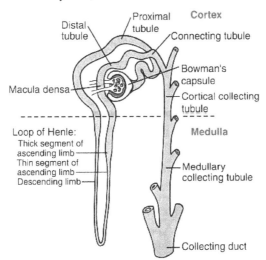

Fig. 1. Diagram of a nephron unit (adapted from [1]).

Structurally, there are 500,000 – 1,000,000 nephron units in the kidney [1,2]. Each nephron has a length of 40 mm and tubular diameter of 50 um. Microscopic puncture and perfusion techniques make it possible to measure single-nephron filtration rates, absorption and secretion rates. The single-nephron glomerular filtration rate (SNGFR) is approximately 30 nl/min.

Approximately 1200 ml/min of blood flow to the kidneys which represents 20% of total cardiac output. In the filtration mechanism, a total filtration surface of 1 m² is found. A global renal fitrate of 120 ml/min (180 L/day) is produced by both kidneys, of which 99.4% of water is reabsorbed to yield about 1 liter of urine/day. The concentration ability of the renal system is largely resides in the tubular system embedded within the renal medulla. Figure 1 demonstrates the disparate functionality of the renal nephron unit.

In this paper we will analyse the operating characteristics of the renal system and examine the optimal features of several aspects of renal physiological engineering mechanisms, particularly, the countercurrent multiplier mechanism for urine concentration and how optimal renal function in terms of renal clearance is maintained.

2. Kidney functional analysis

2.1 Countercurrent mechanisms and modelling of urine concentration

The concentration ability of the kidney is provided by a highly hyperosmotic renal medulla, which draws H_2O from the urinary filtrate within the collecting duct. The mechanism for providing and maintaining this highly hyperosmotic environment is found to be due to several mechanisms. The countercurrent multiplier process in the loop of Henle provides one of the most important mechanisms for this purpose. If the renal tubule is straight, it can provide a osmotic gradient within the renal medulla by about 300 mOsm/L, through an active Na^+ transport. However, for this gradient to continue, it requires rapid replacement against washout or dissipation of the osmotic environment in the renal medulla.

By looping the tubule in a parallel configuration and iterating the single effect of concentration, the kidney can generate and maintain the medullary concentration gradient up to 1200-1400 mOsm/L with lower energy costs. The two parallel tubes of the descending limb and the ascending limb of the loop of Henle are looped in close proximity in the hair-pin configuration (figure 1). The production of a chemical osmotic gradient is based on the active sodium reabsorption (requiring ATP) from the ascending loop; the maintenance of this gradient is crucial. The osmolality of interstitium in almost all parts of the body is about 300 mOsm/L. However, in the renal interstitium, the countercurrent mechanism provides increases up to 4 times from 300 mOsm/L to almost 1400 mOsm/L.

The hyperosmotic gradient provides the osmotic pressure to draw passive diffusion of water from the descending limb and the collecting duct. The countercurrent parallel design allows the descending limb to feedback to the ascending limb, forming a closed stable system of hyperosmolar environment and gradient, as shown in figure 2. It is the preservation of this hyperosmolar environment in steady state conditions that allows the urinary filtrate to be concentrated rapidly and efficiently.

There is also a parallel system of renal vascular network of the vasa recta to prevent this hyperosmolar environment from dissipation. The contribution of the vasa recta into the concentrating mechanism is shown as follows: assume the interstitial tissue of cortex and glomerular filtrate are iso-osmotic to plasma. The tubular fluid entering the descending loop of Henle is also iso-osmotic to plasma. This fluid becomes progressively concentrated towards the bend. In the ascending loop it becomes less concentrated as it reaches the cortex. Under the influence of anti-diuretic hormone, blood flow through the vasa recta is decreased and osmotic equilibration of blood in the vasa with medullary interstitium is enhanced. In brief, the anatomical configuration of the vasa recta minimises but does not prevent solute loss from the medulla via the blood.

Other mechanisms in concentration include the role of urea re-circulated from the collecting duct, the role of vasopressin (antidiuretic hormone or ADH) acting on water transport cellular membrane water channel proteins aquasporin 1, 2, 3 and 4 regulating water permeability. Mutations of several aquasporin genes lead to loss of function and marked abnormalities of water balance, as documented in several reports involving AQP1 knock-out animals.

The contribution of urea to the concentration gradient in the renal medulla is also an important consideration. Diffusion of H_2O occurs from the tubular lumen into the interstitium. Active transport of Na^+ occurs from the tubular fluid. The withdrawal of H_2O from the collecting tubule leads to increased concentration of urea in the collecting tubule, causing a high gradient across the duct membrane, which favours diffusion of urea from the collecting duct into the medulla. From there, the urea diffuses into the descending loop of Henle and is re-circulated into the collecting duct. This contributes to the high urea content and osmolality of the medulla in the concentrating kidney.

2.1.1 Counter-current multiplier mechanism in the loop of Henle

The countercurrent mechanism in the loop of Henle is illustrated in figure 2:

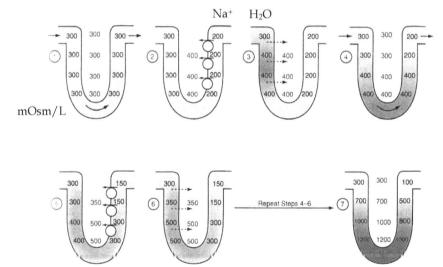

Fig. 2. Countercurrent multiplier process in the loop of Henle, in creating osmotic gradient and urine concentration (adapted from [1]). Note Na^+ absorption from the ascending limb, and passive diffusion of H_2O from the descending limb.

For the loop of Henle shown in figure 2, as a start, it is assumed that the loop of Henle is filled with a fluid with a concentration of 300 mOsm/L. First, the sodium transport from the lumen of the ascending limb to the interstitium, which instantaneously equilibrates with the descending limb. The osmolality in the ascending limb decreases. Because of the higher osmolar interstitium, the fluid in the descending limb increases in osmolality as water is shifted out by passive osmosis. However, the hair-pin structure causes the flow of hyperosmolar fluid in the descending limb to enter the ascending limb. The steps are repeated over and over, with the effect that the process gradually traps sodium in the medulla and multiplies the concentration gradient until the osmolality of the fluid in the loop of Henle and the interstitium reaches 1200 to 1400 mOsm/L.

2.1.2 Concentration of urine in the medullary interstitium

As discussed, tubular fluid entering descending loop of Henle is iso-osmotic to plasma. This tubular fluid becomes progressively concentrated towards the bend. In the ascending loop, it becomes less concentrated as it rises to the cortex (decreasing from over 1000 to 100).

Within the medullary interstitium, other mechanisms co-operate to concentrate the urine. Under the influence of ADH (anti-diuretic hormone) blood flow through the vasa recta is decreased, and osmotic equilibration of blood in the vasa recta with medullary interstitium is enhanced. Solutes such as sodium, chloride and urea enter the descending blood vessels as they pass through the progressively higher osmolality of the interstitium and H_2O leaves the vessels. In the ascending limb, the opposite events take place and H_2O is reabsorbed into the blood vessels.

In brief, the anatomical configuration of the vasa recta minimises but does not prevent solute loss from the medulla via the blood supply. Because of diffusion of H_2O from the tubular lumen into the interstitium, there is equilibrium between fluid in the collecting tubule and that in the interstitium. The withdrawal of H_2O from the collecting tubule leads to increase in the concentration of urea in the collecting tubule causing a high gradient across the duct membrane, which favours diffusion of urea from the collecting duct into the interstitium. From there, urea diffuses into the descending limb of the loop of Henle and is recirculated into the ascending limb and back into the collecting duct, contributing to the high urea concentration in the medulla in the concentrating kidney.

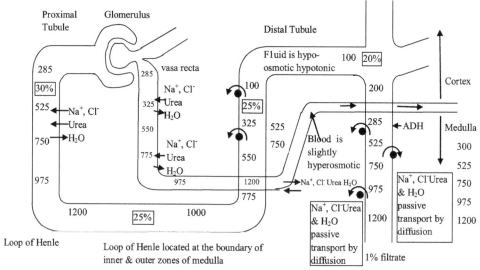

Fig. 3. Schematic diagram showing both the countercurrent multiplier process in the loop of Henle and the vasa recta producing the osmotic gradient. Depicted in the figure are: (i) the passive and active exchanges of water and ions, (ii) concentrations of tubular urine and peritubular fluid in millimetres per litre, (iii) percentages of glomerular filtrate within the tubule at various levels.

In summary (please refer to figure 3),
1. In descending limb, Na⁺ transports passively, Cl⁻ follows and H_2O transport by osmosis because the medullary region is hyperosmotic.
2. In ascending limb, Na⁺ transports actively, Cl⁻ follows.
3. In distal tubule and collecting duct, presence of ADH makes water flow out osmotically (therefore the urine becomes very concentrated).

2.1.3 Linear coupled system of the loop of Henle and analytical solutions

In this model, the driving force for increasing the osmolality (largely contributed by the concentration of Na⁺) of the descending limb is proportional to the difference in the osmotic gradient between it and renal interstitium. The renal interstitium itself has a osmotic concentration proportional to the ascending limb of the loop of Henle, due to active transport of Na⁺ out of the ascending limb.

The osmolality (largely due to the concentration of Na⁺) of the ascending limb is modelled on active sodium transport and hence the rate of fall is only related to its own concentration/osmolality.

The concentration of the interstitium is largely identical to the concentration in the descending tubule as there is passive movement of water through the descending tubule.

The schematic figure 4 illustrates the model of the renal tubule.

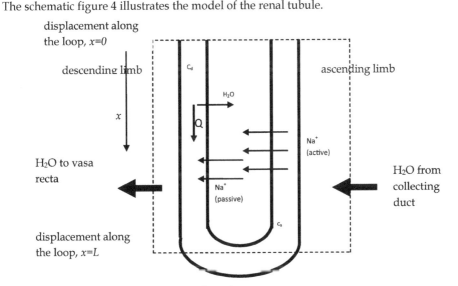

Fig. 4. Schematic diagram of the loop of Henle.

x the distance along the loop of Henle measured from the origin of the descending limb (in mm)

L the total actual length of the loop of Henle measured along one of the limb (in mm)

C_d the concentration of Na⁺ in the descending limb (in mOsm/L)

C_a the concentration of Na⁺ in the ascending limb (in mOsm/L)

C_i the concentration of Na⁺ in the interstitium (assumed proportional to C_a i.e $= k_0 C_a$) (in mOsm/L)

k_d the transport coefficient of Na⁺ ions into the descending limb (in ml/min.mm)

k_a the active transport coefficient of Na⁺ out of the ascending limb due to Na⁺ pump

Q is the tubular flow rate, assumed to be fairly constant in the first approximation (in mL/min)

The governing equations for the descending and ascending limbs of the loop of Henle are shown as a coupled system of linear first-order ODEs, with C_d and C_a the concentration/osmolality of the descending and ascending limbs of the loop of Henle respectively. In the descending limb, the change of concentration of Na$^+$ is modelled as proportional to the concentration difference between the interstitium and the descending limb. In the ascending limb, the change of concentration of Na$^+$ is modelled as directly proportional to the concentration in the ascending limb itself through active removal of Na$^+$ by the Na$^+$ pump. This leads to the following linear coupled system:

Na+ in the descending limb:
$$\frac{d(QC_d)}{dx} = k_d(C_i - C_d) = k_d k_0 C_a - k_d C_d$$
$$\frac{dC_d}{dx} = \frac{k_d}{Q}(C_i - C_d) = \frac{k_d k_0}{Q} C_a - \frac{k_d}{Q} C_d \tag{1}$$

Na$^+$ in the ascending limb:
$$Q \frac{dC_a}{dx} = k_a C_a \tag{2}$$

with $k_d, k_a > 0$. The flow rate Q in the renal tubule is taken as constant in the first approximation. Expressed as matrix equation with upper triangular matrix,

$$\begin{pmatrix} C_d' \\ C_a' \end{pmatrix} = \begin{pmatrix} -\dfrac{k_d}{Q} & \dfrac{k_d k_0}{Q} \\ 0 & \dfrac{k_a}{Q} \end{pmatrix} \begin{pmatrix} C_d \\ C_a \end{pmatrix} \tag{3}$$

The eigenvalues are $-\dfrac{k_d}{Q}$ and $\dfrac{k_a}{Q}$ and the eigenvectors are $\begin{bmatrix} 1 \\ 0 \end{bmatrix}$ and $\begin{bmatrix} k_d k_0 \\ k_d + k_a \end{bmatrix}$. The general solution of this system is given by:

$$\begin{pmatrix} C_d \\ C_a \end{pmatrix} = H_1 \begin{bmatrix} 1 \\ 0 \end{bmatrix} e^{-\frac{k_d}{Q}x} + H_2 \begin{bmatrix} k_d k_0 \\ k_d + k_a \end{bmatrix} e^{\frac{k_a}{Q}x} \tag{4}$$

where H_1 and H_2 are constants of the solution.

Analytically in phase space, since the 2 eigenvalues are real and opposite in sign, the origin of the linear system is a saddle point, asymptotically unstable. Hence, it is unlikely that the system will remain in the state of zero concentration in the ascending and descending limbs. In fact, the solution (4) shows that the system will tend towards a state where an increasing concentration exists in the loop of Henle, because of the positive eigenvalue k_a / Q (representing active sodium transport in the ascending limb) for large x. This is consistent with the observation that it is the active sodium transport in the ascending limb that drives the production of the concentration gradient within the interstitium of the renal medulla and keeps the countercurrent mechanism operational, rather than the passive osmotic gradient as governed by $-k_d / Q$ which tends to dissipate the osmotic gradient.

At the loop end of the loop of Henle, the concentration/osmolality can reach extremely high levels, driven by active sodium transport. Indeed if the active Na$^+$ transport $k_a = 0$, then the system decays to a baseline value through the exponential term associated with k_d. This shows that without active transport, the concentration gradient and the countercurrent multiplier mechanism will dissipate within the renal medulla.

If we take the boundary conditions provided by empirical data in figure 2:

$$C_d(0) = 300 \text{ mOsm/L}$$

$$C_a(0) = 100 \text{ mOsm/L}$$

and assuming trial values of $\dfrac{k_d}{Q} = \dfrac{k_a}{Q} = 1 / \text{mm}$, the concentration within the ascending and descending limbs are obtained as:

$$C_d = 250e^{-x} + 50e^{x}$$
$$C_a = 100e^{x}$$

This is plotted in the following figures:

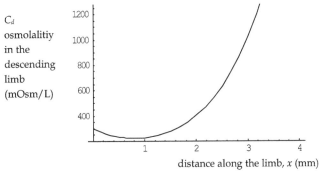

C_d osmolalitiy in the descending limb (mOsm/L)

distance along the limb, x (mm)

Top of descending limb   Bottom of descending limb

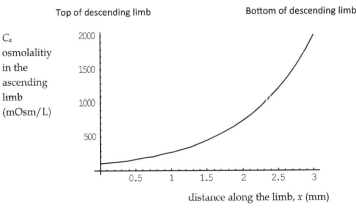

C_a osmolalitiy in the ascending limb (mOsm/L)

distance along the limb, x (mm)

Top of ascending limb   Bottom of ascending limb

Fig. 5. Variation of Fluid Osmolality in the Descending and Ascending limbs of the Loop of Henle.

Although the values of kinetic transport coefficients are trial values, we can see how the model predicts the shape of the osmolality profile within the descending and ascending loops. The graphical representations in figure 5 demonstrate the highest level of osmolality achieved at the bottom end of the loop is about 1500 mOsm/L, reasonably consistent with empirical data. This model provides analytical solutions beyond that of Keener and Sneyd [3].

The limitations of this model is in the assumption of the values of transport coefficients of the descending and ascending tubules in the appropriate units, its linearity assumption and the disregard of its interaction with other modes or mechanisms of osmotic concentration. However, it can be seen that the analytical solutions provide a reasonable qualitative profile of the urinary concentration within the loop of Henle, under the assumptions made of its properties of its different segments and its "iterative" or "multiplier optimal design".

2.2 Single compartmental model of renal clearance kinetics – (1) single input

One of the important function of the kidney is excretion of metabolic waste products. How the kidney handles this excretory function has direct implications on clinical or physiological function. It is thus of interest to analyse the behaviour of the kidney as an excretory system.

Most accessible to analysis is the renal response to a single bolus of a metabolic substrate. Most renal clearance kinetics analyse the behaviour of the excretory function of the kidney in respect to endogenous or exogenous substrates. In some assessments, it involves the administration of a single bolus dose of an exogeneous substance into the blood circulation.

When a single bolus of such a substance is introduced into the human body system though an intravenous injection, the substance will initially spread out in the circulatory system and distribute into the extravascular body-fluid compartments of the body, while it is at the same time being removed by the kidney. Hence, if we represent the human body as a two-compartment system, then there will be 2 phases of decrease of the plasma concentration of this substance. The first phase represents the fall due to rapid distribution of the substance within the body from the blood circulation into the equilibrium body fluid compartments of the body, while it is at the same time being removed by the kidney. The second phase represents the fall due largely to the renal excretion of this substance.

However, in most cases, the first phase can be ignored and corrected for by empirical approximation so that only the second slower phase needs to be measured. Hence, a single late exponential function can be used to describe the fall in the plasma concentration of the substance. This principle is used in the physiological measurement of renal clearance or glomerular filtration rate (GFR) in human subjects.

2.2.1 Renal clearance analysis using a single-bolus model of renal tracer or substrate

Assume the amount of the tracer in the entire compartment is A (in mg or mmols). Let the concentration of the tracer in the compartment at time t be C_t (in mg/L or mmol/L) and the clearance be annotated as g (in L/min). By definition, $C = A/V$, where V is the total plasma volume or distribution volume, reasonably assumed constant in the body.

By the principle of mass conservation,

$$\frac{dA}{dt} = -g \cdot C_t \tag{5}$$

This is the governing first-order linear differential equation representing the kinetics of a one-compartment system.

Integrating over all time, the total tracer dose injected (D) is given by:

$$\int_0^\infty \frac{dA}{dt} dt = D = -\int_0^\infty g C_t dt = -g \int_0^\infty C_t dt \tag{6}$$

The absolute magnitude of the renal clearance g is:

$$|g| = \left| \frac{D}{\int_0^\infty C_t dt} \right| = \frac{\text{total dose of tracer injected}}{\text{area under the tracer concentration-time curve}} \tag{7}$$

We can show that the concentration of tracer in this compartment follows an exponential variation, by rewriting equation (5) as:

$$V\frac{dC_t}{dt} = -g \cdot C_t \tag{8}$$

Separating variables, we get:

$$\int_{C_0}^{C_t} \frac{dC_t}{C_t} = -\frac{g}{V} \int_{t_0}^{t} dt$$

We have a mono-exponential clearance scheme, as follows:

$$C_t = C_0 e^{-\frac{g}{V}(t-t_0)} \tag{9}$$

By taking logarithms of both sides, we get a linear relationship on the "semi-log" scale as:

$$\ln C_t = \ln C_0 - \frac{g}{V}(t - t_0) \tag{10}$$

Equation (10) is the basis of plotting the tracer concentration against time as a semi-log graph, so that (i) the absolute value of the gradient of the slope will be given by (renal clearance)/V, which is also called the clearance constant λ, and (ii) the y-intercept will be given by C_0 which is D/V.

So the initial volume of distribution, V, will be given by

$$V = \frac{D}{C_0} \tag{11}$$

Hence,

$$\text{Renal clearance} = V \times \lambda = \frac{D}{C_0} \times \lambda \qquad (12)$$

Or, the estimated renal clearance is the

$$\text{Distribution volume} \times \text{Clearance constant} = \frac{\text{total dose injected}}{(\text{y-intercept of } \ln C \text{ vs t curve})} \times (\text{gradient of } \ln C \text{ vs t curve})$$

Historically, this methodology is often known as the indicator-dilution method or the Stewart-Hamilton method, although the origins of this method antedate the work of Stewart and Hamilton [4,5].

2.2.2 Physiological measurement of the Glomerular Filtration Rate (GFR)

If we can insert a microneedle with a flow gauge into each glomerulus in the kidney and measure the flow rate experimentally, this would constitute one way of measuring the GFR. This can be done in-vitro with micropuncture and microperfusion techniques. On a body system level, the global renal clearance has to be obtained by other ways.

Typically, GFR is deduced by measuring the renal filtered loss or clearance of a suitable substance from the plasma. In physiological tests, creatinine clearance and Cr-51 EDTA clearance are typically used. Dynamic renogram modelling of impulse tracer kinetics through the kidney had previously been performed [6].

In order to measure GFR accurately, the following properties must apply to the particular substance used to measure GFR :
1. the substance must be freely filtered through the glomerulus
2. it must not undergo renal tubular secretion or absorption
3. it must not bind to plasma proteins
4. it must not be lost through any other methods from the body
5. it must not be metabolised or changed chemically in the body.

These are severe restrictions and there are only some possible candidates for this substance, including :
1. endogenous creatinine, but this is less accurate in children
2. inulin
3. chromium-51 EDTA
4. technetium99m-DTPA

If the loss of these substances from the body can be measured, one can get a quantitative reflection of the excretory function of the kidney.

Of the four substances mentioned, only endogenous creatinine is found within the human body. The other substances have to be introduced into the human body. Inulin is a polysaccharide molecule with a molecular weight of 5200. Creatinine itself is a by-product of skeletal muscle metabolism, and it is present in the plasma at a relatively constant concentration and does not require intravenous infusion into the patient. However, creatinine is not a perfect marker for GFR because a small amount of it is secreted by the tubules and hence it is not a pure glomerular agent and tends to overestimate the GFR.

Incidentally, there is an agent, para-aminohippuric acid (PAH), which is not only filtered but also secreted to a large extent, so that it can be used to measure not the GFR but the effective renal plasma flow rate. Chromium-51 EDTA is a radiolabelled EDTA with the gamma emitter, chromium-51. This agent is very close to a purely glomerular filtered agent. Tc99m-DTPA is also a glomerular filtered substance, radiolabelled to the gamma emitter, Tc99m.

2.2.3 Continuous input of substrate model – relationship between steady-state serum creatinine concentration in the body and renal clearance

2.2.3.1 Theory and application

As opposed to the single-bolus renal kinetics, in the body, endogenous metabolic substrates are introduced into the blood circulation in a continuous way. Thus the model of renal clearance kinetics given above will have to be modified to take this continuous input into account. Analysing this continuous input model will be useful to evaluate the optimal and crucial renal handling of endogenous waste products in a typical human body.

The result of such a continuous input and renal excretion gives rise to a steady-state concentration of a renal-excreted substrate in the body. A typical endogenous substrate produced in a continuous fashion in the body and excreted by the renal route is creatinine. Figure 6 shows an inverse relationship between plasma creatinine concentration and GFR. The lower the renal clearance, the higher is the steady-state blood concentration of the substrate. However, this relationship is not linear but largely inverse rectangular hyperbolic (see figure 6).

To analyse this empirical relationship, further analysis can be performed using the single-compartment model but introducing a continuous input of substrate. This analysis follows from and extends the results obtained by Mazumdar [7].

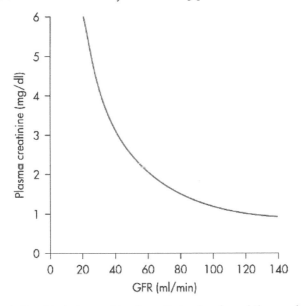

Fig. 6. Empirical relationship between blood creatinine levels and the renal clearance rate (adapted from [2]).

If instead of a single dose of tracer or substrate given as discussed in the previous section, constant doses of creatinine (substrate) are given at equal intervals of time ie, at intervals of period T, then using the single-bolus equation (7), the concentration of the substrate immediately after the second dose is given by:

$$C_1 = C_0 + \left(C_0 e^{-\frac{g}{V}T} \right) \tag{13}$$

Immediately after the third dose, the substrate concentration is given by:

$$C_2 = C_0 + \left[C_0 + \left(C_0 e^{-\frac{g}{V}T} \right) \right] e^{-\frac{g}{V}T} \tag{14}$$

$$= C_0 + C_0 e^{-\frac{g}{V}T} + C_0 e^{-\frac{2g}{V}T}$$

Immediately after the nth dose, the substrate concentration is given by:

$$C_{n-1} = C_0 + C_0 e^{-\frac{g}{V}T} + \ldots + C_0 e^{-(n-1)\frac{g}{V}T}$$

$$= C_0 \left[1 + e^{-\frac{g}{V}T} + \ldots + e^{-(n-1)\frac{g}{V}T} \right] \tag{15}$$

$$= C_0 \frac{1 - e^{-n\frac{g}{V}T}}{1 - e^{-\frac{g}{V}T}}$$

As n tends to infinity, the creatinine concentration approaches an equilibrium value, given by:

$$C_\infty = \frac{C_0}{1 - e^{-\frac{g}{V}T}} \tag{16}$$

Linearization by Taylor's series approximation gives the equilibrium concentration of the creatinine in the blood to first order, as:

$$C_\infty = \frac{C_0}{\frac{g}{V}T - \frac{1}{2}\left(\frac{g}{V}T\right)^2 + \ldots} \approx \frac{(C_0/T)V}{g} \tag{17}$$

where $(C_0/T)V$ is the amount of the creatinine introduced per unit time. Hence, using the parameter values in Table 1,

$$C_\infty \approx \frac{\text{amount of renal substrate introduced per unit time}}{\text{renal clearance}} = \frac{10.1}{\text{renal clearance}} \tag{18}$$

C_∞ is the steady-state concentration of creatinine, as it is produced and excreted continuously in the body. The relationship derived is the equilibrium or steady-state concentration of the substrate that is produced continuously in the body and excreted renally. The relationship is plotted in figure 6.

Rate of body production of creatinine metabolite, A	20-25 mg/kg body weight per day (approximately 1.5 g/day in a 70 kg man). In SI units, this would be 10.1 umol/min.
Human body renal clearance, g	120 ml/min (approximate).
Total volume of distribution of creatinine in a typical human of weight 70 kg	50,000 ml (approximate)

Table 1. Important physiological renal parameters (data from [2]).

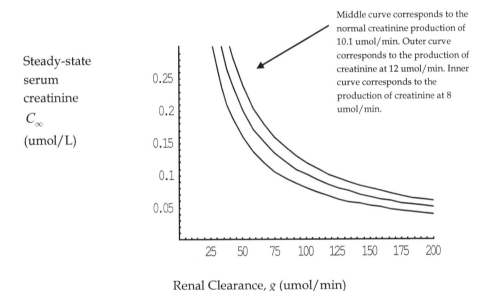

Steady-state serum creatinine C_∞ (umol/L)

Middle curve corresponds to the normal creatinine production of 10.1 umol/min. Outer curve corresponds to the production of creatinine at 12 umol/min. Inner curve corresponds to the production of creatinine at 8 umol/min.

Renal Clearance, g (umol/min)

Fig. 7. Model prediction of the relationship between blood creatinine levels and the renal clearance rate.

Graphically, equation (17) predicts a very close approximation to the empirical curve, which is an inverse rectangular hyperbolic relationship between serum creatinine levels and the renal clearance as shown in figure 7.

Depending on the rate of production of the metabolite creatinine in the human body, equation (17) also demonstrates that there is a series of iso-dose curves of renal clearance vs blood levels of renal substrate, similar to isothermal curves in ideal gas thermodynamics.

The application to human physiology can be seen as follows. It is well known that the serum creatinine levels in women is lower than in man. A typical muscular man producing a larger quantity of creatinine substrate due to muscle breakdown will, for the same renal clearance, demonstrate a higher blood concentration of creatinine for the same degree of renal clearance.

2.3 Renal clearance – convolution analysis

In general, the total amount of substrate in the body at time t is given by the convolution of the amount produced by the body per unit time, $A(t)$, which is a function of time and the biological clearance of that substance. In the case of pure renally-excreted substrate, such as creatinine, assuming a single-compartment clearance-kinetics as previously discussed, we have as follows:

Total amount of substrate in the body at time t: $A(t) * e^{-\frac{g}{V}t}$ (19)

If as above, the amount of substrate introduced into the blood compartment per unit time is constant, A, then the total amount of substrate at time t (accounting for renal clearance) is given by:

Total amount of substrate in the body at time t: $A * e^{-\frac{g}{V}t} = \int_{0}^{t} A du \cdot e^{-\frac{g}{V}(t-u)}$ (20)

The result takes a useful form for physical interpretation. Total amount of substrate in the body at time t is given by:

$$A * e^{-\frac{g}{V}t} = \int_{0}^{t} A du \cdot e^{-\frac{g}{V}(t-u)} = \frac{AV}{g}\left(1 - e^{-\frac{g}{V}t}\right)$$ (21)

Schematically, this relation is shown in figure 8, for blood creatinine levels:

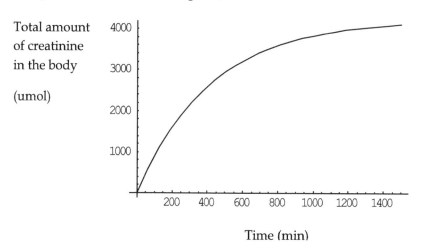

Fig. 8. Asymptotic steady-state concentration of blood creatinine levels, based on convolution analysis.

At $t \rightarrow \infty$, the equilibrium concentration of substrate in the blood compartment is:

$$C_{\infty} = \frac{1}{V}\left(\frac{AV}{g}\right) = \frac{\text{amount of substrate introduced per unit time}}{\text{renal clearance}} \qquad (22)$$

consistent with the previous equation (18).

Application of the formula can be done using the physiological values in Table 1.

The body produces creatinine at the rate of 20-25 mg/kg body weight per day, which is approximately 1.5 g/day in a 70 kg man. In SI units, this is 10.1 umol/min. The renal clearance is approximately 120 ml/min. Hence, based on equation (22), the estimated steady-state serum creatinine level in the body based on the model would be predicted to be in the order of

10.1/120 = 0.084 umol/ml or 84 umol/L, as expected empirically.

Direct correlation with physiological parameters shows this convolution analysis to give reasonably close results.

3. Conclusion

The analytical model of the loop of Henle and the renal handling of metabolic substrates is aimed at showing to some extent how the renal system is optimized for filtration and regulation of urine concentration by the countercurrent mechanism in the loop of Henle and its medullary environment, which is largely physiologically engineered to increase and maintain at steady-state the high osmolality of the urine fluid to as high as 4 times normal blood osmolality.

The renal clearance of substrates is modelled as a single-compartment kinetic model, with single-input and more physiologically as continuous-input. The analytical solutions obtained from the continuous input of creatinine predict the body creatinine level to tend to asymptotically steady-state substrate blood concentration with time, in a relationship that is an inverse rectangular hyperbolic function to the renal clearance. This is close to the relationship found empirically. The relationship is found to be related to the amount substrate input per unit time divided by renal clearance. The same conclusion is obtained from convolution analysis of renal clearance. The formula predicts reasonable estimates for the actual serum creatinine levels in the body based on renal clearance and substrate input parameters.

4. References

[1] Guyton A C, Hall J E. Textbook of Medical Physiology. Tenth edition. W B Saunders 2000.
[2] Boron W F, Boulpaep E L. Medical Physiology. Updated Edition. Elsevier Saunders 2005.
[3] James Keener, Sneyd J. Mathematical Physiology. Springer-Verlag 1998.
[4] William Simon. Mathematical Techniques for Biology and Medicine. Dover Publications 1986.
[5] Rubinow, S I. Introduction to Mathematical Biology. Dover Publications 1975.

[6] Loh K M, Ng David, Ghista D N, Ridolph H. Quantitation of renal function based on two-compartmental modeling of renal pelvis. IEEE Conference 2005.

[7] Mazumdar J. An Introduction to Mathematical Physiology & Biology. Cambridge University Press. Second edition 1999.

10

Cardiac Myocardial Disease States Cause Left Ventricular Remodeling with Decreased Contractility and Lead to Heart Failure; Interventions by Coronary Arterial Bypass Grafting and Surgical Ventricular Restoration Can Reverse LV Remodeling with Improved Contractility

Dhanjoo N. Ghista[1], Liang Zhong[2], Leok Poh Chua[3],
Ghassan S. Kassab[4], Yi Su[5] and Ru San Tan[2]
[1]Department of Graduate and Continuing Education, Framingham State University,
Framingham, Massachusetts,
[2]Department of Cardiology, National Heart Centre,
[3]School of Mechanical and Aerospace Engineering, Nanyang Technological University,
[4]Departments of Biomedical Engineering, Surgery, Cellular and Integrative Physiology,
Indiana University-Purdue University Indianapolis, Indianapolis, Indiana,
[5]Institute of High Performance Computing, Agency for Science, Technology and Research,
[1,4]USA
[2,3,5]Singapore

1. Introduction

1.1 Theme and scope

In this chapter, we are studying the course (i) of cardiomyopathy diseased LVs (with myocardial infarcts) progressing to heart failure (HF) through LV remodelling and decreased LV contractility, and (ii) their recovery through surgical therapeutic interventions of CABG and Surgical Ventricular Restoration (SVR), by restoration of myocardial ischemic segments, reversal of LV remodeling and improvement in LV contractility.

For this purpose, we first provide the methodology for detecting myocardial infarcts. Then, we characterize LV remodeling of cardiomyopathy diseased LVs (with myocardial infarcts) in terms of reduced change in curvedness from end-diastole to end-systole. In these LVs, there is also reduced contractility; so we provide an index for cardiac contractility, in terms of maximal rate-of-change of normalized wall stress, do^*/dt_{max}, and its decrease in an infarcted LV progressing to heart failure. We provide clinical studies of remodeled cardiomyopathy diseased LVs, in terms of reduced values of their curvedness index and contractility index.

By way of therapeutic interventions, we have presented the hemodynamic flow simulation of the CABG (carried out to enhance myocardial perfusion in the region around the blocked coronary artery), and pointed out certain factors and sites of wall shear stresses that cause intimal damage of vessels and hyperplasia, as potential causes for decreased patency. We have shown that surgical ventricular restoration (SVR), in conjunction with CABG, is seen to benefit the ischemic-infarcted heart, by (i) restoration of cardiac remodeling index of 'end-diastolic to end-systolic curvedness change', (ii) reduction of regional wall stresses, and (iii) augmentation of the cardiac contractility index value.

2. Myocardial infarction: What it entails

In Cardiology, the etiology of congestive heart failure (CHF) is coronary artery disease in approximately two-thirds of cases. The majority of these patients have hearts with myocardial infarcted segments. This infarcted myocardial wall mitigates adequate contraction of the wall. So, the end-result of an infarcted left ventricle (LV) is poor intra-LV velocity distribution and pressure-gradient distribution, causing impaired outflow from the LV into the aorta.

In the infarcted myocardial segments, the myocardial infrastructure of actin and myosin filaments (and their cross – bridges) is disrupted, and hence there is no contraction within these infarcted myocardial segments. The below figure 1 illustrates a myocardial sarcomere segment's bioengineering model, composed of two symmetrical myocardial structural units (MSUs). In these MSU(s), the contractile elements represent the actin-myosin contractile components of the sarcomere segment. The disruptions of these contractile elements impair the contractile capability of that sarcomere segment. Hence, a LV with infarcted myocardial segments will have diminished contractility, inadequate and improper intra-LV flow, and poor ejection.

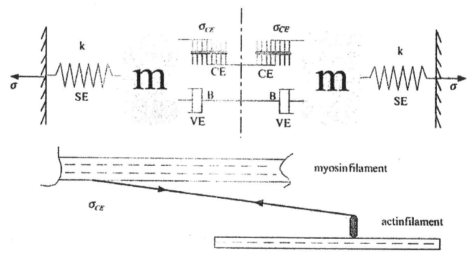

Fig. 1. Based on the conventional Hill three-element model and Huxley cross bridge theory, we have developed a myocardial model involving the LV myocardial mass, series-elastic element (CE),. In this figure we have linked the anatomical associations of these myocardial model elements with microscopic structure of the heart muscle. This figure illustrates the sarcomere element contractile model, involving: the effective mass (m) of the muscle tissue that is accelerated: elastic parameter k of the series element stress σ_{SE} (k = elastic modulus of the sarcomere), viscous damping parameter B of the stress σ_{VE} in the parallel viscous element VE, the generated contractile stress σ_{CE} between myosin (thick) and actin (thin) filaments.

3. Detection of myocardial infarcted segments

Now, infarcted myocardial segments can be detected as highly reflectile echo zones (HREZs) in 2-dimensional B-scan echocardiograms. In this context, we have shown how infarcted myocardial segments can be detected (in shape and size), by echo-texture analysis, as highly reflectile echo zones or HREZs. Each myocardial tissue component of the heart generates a grey scale pattern or texture related to the tissue density and fibrous content. In diseased states (such as myocardial ischemia, myocardial fibrosis, and infiltrative diseases), changes in myocardial tissue density have been recognised by employing echo intensity and mean grey level of pixel as the basis for recognition of such myocardial disorders. It is found that hyper-reflectile echoes (HREs) correlated well with diseased cardiac muscle, and that myocardial tissue containing HREs corresponded with foci of sub-endorcardial necrosis and even calcification.

In our study [1], in order to determine highly reflectile echo zones (HREZs), echocardiograms were recorded, and each image was made up of 256 x 256 pixels, with each pixel having a resolution of 0-256 grey scales. The echocardiographic images were digitised into 256 grey scales. Then, echo intensity levels from normal infants were used to delineate the range of echo intensities for normal tissues. The upper bound of the echo intensity was set to 100 per cent in each normal infant, and the intensities from the rest of the image were referenced to this level. Normally, pericardium had the highest intensity level. It was found that the upper-bound of the echo intensity value for healthy tissue (expressed as a percentage of the pericardial echo intensity value) was 54.2.

Patient (Sex)	Region A	Region B	Region C	Region D	HRE and its location
B (M)	M: 167.44	54.76	51.02	82.20	105.74
	SD: 25.00	28.2	17.71	24.68	30.88
	N: 65	84	75	31	65
	P: 100	32.7	30.5	49.1	63.1 Septum
P (F)	148.76	61.73	79.81	61.7	108.18
	26.78	23.02	22.05	24.2	13.03
	50	75	47	49	40
	100	41.5	53.8	41.5	72.6 Septum
Br (M)	141.65	68.3	69.3	33.93	89.412
	29.56	26.8	24.8	24.4	28.0
	40	40	49	44	79
	100	41.5	53.8	41.5	73.1 Septum
F (F)	157.34	50.1	60.8	53.8	112.1
	30.0	29.5	18.8	22.7	10.3
	35	45	49	44	31
	100	31.8	38.6	34.2	71.2 R. ventricle
III (M)	168.1	54.7	58.3	61.4	96.4
	21.35	21.8	16.9	20.0	14.7
	47	36	35	37	49
	100	32.5	34.6	37.1	57.3 L. ventricle
G (M)	117.7	46.9	45.5	42.7	85.3
	20.6	19.0	20.6	19.1	22.6
	45	44	40	49	37
	100	39.8	38.7	36.2	72.5 R. ventricle

A = Posterior Pericardium, B = Anterior Myocardium, C = Posterior Myocardium, D = Septum

Table 1. Echo intensity values for various anatomic regions of diseased pediatric hearts (based on long axis view). The numbers in the four rows represent Mean (M), Standard Deviation (SD), Number of Pixels (N), Percentage of Posterior Pericardial Intensity (P). This figure is adopted from Ref [1].

For patients whose echo-texture analysis showed presence of HREs, it was found that the echocardiographic intensities of the HREs from these patients intensities were distinctly higher than the echo intensity range of normal tissue (as depicted in Table 1).
Myocardial tissue pixels having echo-intensity values greater than 200 are generally noted to be infarcted. This region's echo-intensity values can remain unaffected by administration of a myocardial perfusing agent. This infarcted sub-region is seen to be surrounded by an ischemic sub-region whose pixels are noted to have echo intensity values between 100 and 200. This region's echo-intensity can be reduced by the administration of a myocardial perfusing agent. The surrounding healthy tissue has echo intensity less than 100. Figure 2(a) depicts an echo image of an infant with visible scars regions 1 and 2, while figure 2(b) depicts printouts of the echo intensities from these two regions, wherein the infarcted segments are depicted in dark colour and the surrounding ischemic segments are depicted in a lighter shade.

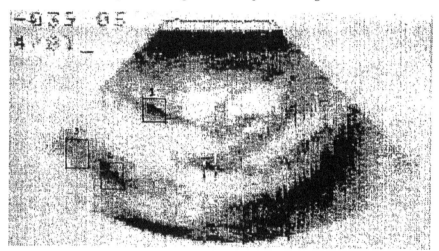

Fig. 2. (a) Long axis view of a pediatric patient's heart showing HRE regions 1 and 2 and a healthy region 3 [1].

Y/X	99	100	101	102	103	104	105	106	107	108	109	110	111	112	113	114
98	79	78	88	90	99	96	102	108	91	77	92	86	135	122	73	55
99	114	115	101	114	126	128	114	116	119	126	82	68	84	103	78	57
100	151	137	125	128	136	135	133	134	149	137	91	75	74	73	82	83
101	175	177	171	151	144	143	154	147	138	142	139	139	126	64	76	71
102	202	196	174	125	192	193	183	164	131	131	125	132	92	89	81	116
103	139	143	183	193	206	217	233	248	209	146	116	102	111	113	117	116
104	147	136	143	178	203	251	250	255	229	201	75	71	92	82	88	95
105	108	110	132	151	210	223	227	249	255	255	230	210	104	87	81	112
106	84	104	88	121	147	184	227	239	255	255	252	247	230	125	76	70
107	83	110	108	122	135	175	194	187	206	228	211	255	255	184	141	131
108	68	92	122	131	145	147	149	151	217	181	189	222	241	178	190	167
109	56	76	81	122	132	137	145	143	154	150	156	158	195	190	206	190
110	76	63	96	96	82	83	103	120	142	128	133	141	153	181	192	194
111	59	57	63	66	70	103	106	118	96	94	86	110	129	150	96	66
112	58	60	59	57	58	61	71	77	106	89	91	92	100	147	97	85
113	74	71	78	60	56	58	57	62	71	70	79	83	78	92	67	76
114	57	57	65	63	57	56	63	56	51	56	58	80	85	78	67	55
115	51	60	63	63	58	67	56	57	54	59	57	58	59	76	68	81

Fig. 2. (b) Pixel values correcponding to highly reflectile echo region 1. The central region having echo-intensity values greater than 200 is infarcted, while its immediately surrounding region shown in lighter shade is ischemic.

In this way, in each highly reflectile echo zone (HREZ) made up of , say, N number of pixels, we can determine the number (I) of infarcted pixels. The ratio I/N represents the infarcted potion of that HREZ myocardial segment. The total number of all the infarcted pixels in all the HREZs provides an indication of the amount of infarcted myocardium of the heart or of the LV.

4. Cardiac remodeling following myocardial infarction and progression to heart failure

Myocardial infarction (MI) reduces the amount of ventricular contractile myocardium, which in turn reduces the left ventricular (LV) contractile capacity for pumping out adequate cardiac output (CO). As MI extends, it progresses into heart failure (HF). A manifestation of MI and its decreased contractile capacity is the inability of the LV to retain its compact systolic curved shape. In other words, in HF resulting from MI, the left ventricle undergoes remodeling and its curvedness index decreases. As MI progresses into HF, LV size increases, LV function deteriorates, and symptoms of HF become evident. Thus, cardiac remodeling (expressed in terms of regional curvedness of the LV) constitutes a measure of the progression of HF after MI, and interventions causing some reversal of this LV remodeling process have been shown to improve mortality in patients with HF. Therefore, reversal of LV remodeling (through medical or surgical treatment) has been emerging as a therapeutic target in HF of all etiologies. The challenge is to develop specific measures of LV remodeling that can be incorporated into the clinical management pathway. For patients with HF after MI, the LV shape is more spherical in terms of the global sphericity index. However, focus on global sphericity index may be misleading, since the simple plane ratio reflects a linear alteration in the two axes of the LV chamber. Hence, regional curvedness index, determined from 3D magnetic resonance imaging (MRI)-based LV model, is proposed as a measure of LV shape. The curvedness value describes the magnitude of the curvature at a surface point, i.e., a measure of degree of curvature a point; the regional curvedness describes the curvedness of the segment of the LV. In MI patients, the regional curvedness index value does not increase significantly from end-diastole to end-systole (due to decreased LV contractile capacity), as in the case of normal patients.

In the case of patients with HF following MI, the LV cannot generate adequate contractile force, as explained earlier in section 1. As a result, the overall LV contractility index of maximal rate-of-change of normalized systolic wall stress $d\sigma^*/dt_{max}$ is decreased, and the LV ejection force is also diminished. Now, LV regional wall stress is proportional to the wall surface radius-of-curvature and inversely proportional to the wall thickness. Hence, for patients with HF after MI, the inability of the remaining viable myocardium to compensate for the increased wall stress associated with LV dilatation and thinning is a trigger for LV enlargement.

4.1 Remodeling quantification in terms of local curvedness indices and regional curvedness index
4.1.1 Overall approach
The remodeling is characterized in terms of the *Regional Curvedness* index of the 16 segments of the LV endocardial surface. The method entails generation of the LV endocardial shape and its compartmentalization into 16 segments or regions (as explained later on). Each segment is discretized into triangular meshes, so that a point on the endocardial surface is a vertex of a triangular mesh. At each vertex point, we determine the *Local Curvedness* index

[2], by employing its surrounding neighboring vertices in the form of n-rings around the local point. To ensure accurate curvedness computation without over smoothing, the optimal number of n-rings is determined to be 5. Then, for each segment, we determine the *Regional Curvedness* index, as the mean of the *Local Curvedness* indices in the segment.

The values of *Regional Curvedness* are determined for normal and MI patients, at end-diastolic and end-systolic instants. For normal patients, the *Regional Curvedness* index changes significantly from diastole to systole, as given by the *Diastole-to-Systole Change in Curvedness* (%ΔC) [2]

$$\%\Delta C = \frac{C_{ED} - C_{ES}}{C_{ED}} \times 100 \tag{1}$$

wherein C_{ED} and C_{ES} are end-diastolic and end-systolic curvedness, defined later on. The mean alteration in regional curvedness index value of %ΔC is determined for all the segments, for normal subjects and MI patients (and logged in Table 3). It can be seen that for MI patients, the regional curvedness index or (%ΔC) does not change significantly from end-diastole to end-systole, in comparison with normal subjects.

4.1.2 Clinical application methodology

Our study involved 10 normal subjects and 11 patients after myocardial infarction (MI). The hemodynamic and volumetric parameters of the subjects are summarized in Table 2.

Human Subjects and MRI Scans: The study to characterize regional curvedness index or (%ΔC) involved ten normal subjects and 11 patients after myocardial infarction [2]. All subjects underwent diagnostic MRI scans. For each subject, short-axis MRI images were taken along the plane which passes through the mitral and aortic valves of the heart at an interval of 8mm thickness. Each image has a spatial resolution of 1.5mm, acquired in a single breath hold, with 25 temporal phases per heart cycle. Of these images, the set of images corresponding to the cardiac cycle at end-diastole and end-systole are then used for the study.

LV Endocardial Surface Reconstruction and Segmentation: The MRI images were processed, by using a semi-automatic technique that is included in the CMRtools suite (Cardiovascular Solution, UK). The contours demarcating the myocardium and the LV chamber were defined by means of B-spline curves. The endocardial surface of the LV was reconstructed by joining the series of contours to form a triangle mesh. In order to facilitate quantification of the LV segmental regional curvedness, the endocardial surface was partitioned into 16 segments; the method of segmentation of the LV endocardial surface is provided in our paper [2].

Left Ventricular Shape Analysis: To quantify LV remodeling, we first define a measure known as the *Local Curvedness* index [2]. This is essentially a shape descriptor used to quantify how curved the surface is in the vicinity of a vertex on the LV endocardial surface. This is done by using the 3-d mesh of the LV endocardial surface as an input. Each vertex of the mesh is processed by fitting a quadric surface over a local region around the vertex as described in our paper [2]. The extent of this local region is determined by the n-ring parameter. Next, the *Local Curvedness* index of each vertex can be calculated from the coefficients of the fitted quadric surface by [2]:

$$C = \sqrt{\frac{\kappa_1^2 + \kappa_2^2}{2}} = \frac{1}{A^2}\sqrt{\frac{2B^2 + A^2\left(4ac - b^2\right)}{A}} \tag{2}$$

such that

$$A = \sqrt{d^2 + e^2 + 1}$$

$$B = a + ae^2 + c + cd^2 + bde$$

where a, b, c, d and e are the coefficients of the fitted quadric surface at the vertex.
In order to derive the *Regional Curvedness*, the endocardial surface is partitioned into 16
segments (Fig 3). The *Regional Curvedness* for the each segment is the mean of the *Local
Curvedness* indices in the segment. The flowchart of the overall workflow for the regional LV
shape analysis is shown in Fig. 4.

4.1.3 Clinical studies results
In our clinical studies, it was found that (1) MI patients exhibit decreased curvedness and
%ΔC, (2) MI patients exhibit increased variation of curvedness and variation of %ΔC, and (3)
LV ejection fraction is positively correlated with curvedness and %ΔC, and inversely
correlated with variation of %ΔC.
The *Diastole-to-Systole Change in Curvedness* (%ΔC), as defined by equation (1), is a measure
of regional deformity due to contraction. Positive values of %ΔC indicate regions of
increasing inward concavity of the LV wall during systole, while negative values of %ΔC
indicate wall regions of decreasing inward concavity. The %ΔC measure can be employed to
relate the regional differences in hypokinesis due to myocardial infarction.
Variation of curvedness: The extent of LV surface inhomogeneity is characterized by a
coefficient of variation of curvedness at end-diastole (CV_C_{ED}) and end-systole (CV_C_{ES}):

$$CV\_C = \frac{\sigma(C)}{\mu(C)} \tag{3}$$

where $\sigma(C)$ is the standard deviation of the regional curvedness and $\mu(C)$ is the mean of the
regional curvedness of the segments of the LV mesh.
To evaluate the extent of functional non-uniformity of LV regions, the index CV_DC was
determined as [2]:

$$CV\_\Delta C = \frac{CV\_C_{ED} - CV\_C_{ES}}{CV\_C_{ED}} \times 100 \tag{4}$$

In general, the larger the values of CV_C_{ED} and CV_C_{ES}, the more inhomogeneous the LV
endocardial surface appears. Hence, the larger the value of index CV_ΔC, the more
functionally non-uniform are the LV shape changes due to LV contraction.
*Curvedness, Variation of curvedness and Diastole-to-systole change %ΔC in MI patients compared
to Normal subjects:* The hemodynamic and volumetric parameters of the subjects are
summarized in Table 2. For patients after MI, the LV ejection fraction was significantly
lower than that in the control subjects. In addition, their LV end-diastolic and end-systolic
indexed volumes were greater than those in the control subjects.
The values of regional curvedness from apex to base in MI patients and normal subjects are
given in Table 3 and Fig. 5, to highlight the regional variations of the LV curvature. In the
normal group, there was a significant increase in the curvedness at the apex from diastole to
systole. However, in MI patients, there was no significant difference in curvedness in all

segments. Significant differences in end-diastolic curvedness C_{ED} and end-systolic curvedness C_{ES} were noted between MI and normal groups. Among the 16 segments of the LV, the variation coefficient of C_{ES} (CV_C_{ES}) was significantly lower in MI patients than in the normal group (18±4% in MI vs 31±8% in normal, p<0.0001), indicating fair homogeneity of LV shape in MI at end-systole. Correspondingly, the diastole-to-systole change in curvedness (%ΔC) was significantly lower, and the variation of %ΔC was higher in MI patients compared to normal group, indicating ventricular functional non-uniformity due to the pathologic state.

	Control (n=10)	MI (n=11)	p value
Age (years)	41 ± 16	60 ± 6	0.003
Weight (kg)	67 ± 15	65 ± 14	0.30
Height (cm)	169 ± 8	165 ± 10	0.86
Diastolic pressure (mmHg)	73 ± 12	74 ± 18	0.79
Systolic pressure (mmHg)	122 ± 17	116 ± 20	0.50
HR (beats/min)	70 ± 9	84 ± 13	0.012
CI (L/min/m²)	3.3 ± 0.4	2.2 ± 0.5	<0.001
EDVI (ml/m²)	73 ± 10	148 ± 40	<0.001
ESVI (ml/m²)	26 ± 6	122 ± 38	<0.001
EF (%)	65 ± 5	18 ± 5	<0.001
Sphericity index	0.52 ± 0.06	0.62 ± 0.08	0.01
LV mass index	56 ± 12	83 ± 13	0.004

Table 2. Characteristics of Normal Control and Patients after MI, involved in the study.

Segment	Controls (n=10)			MI (n=11)		
	C_{ED}(x10⁻² mm⁻¹)	C_{ES}(x10⁻² mm⁻¹)	ΔC (%)	C_{ED}(x10⁻² mm⁻¹)	C_{ES}(x10⁻² mm⁻¹)	ΔC (%)
1. Basal anterior	4.1 ± 0.8	5.6 ± 0.7	38 ± 22	3.4 ± 0.5*	3.7 ± 0.5ξ	7 ± 17#
2. Basal anterior septal	3.4 ± 0.6	5.3 ± 1.1	57 ± 30	3.7 ± 1.0	3.7 ± 0.9*	4 ± 26ξ
3. Basal inferior septal	3.1 ± 0.5	4.8 ± 0.6	61 ± 25	3.6 ± 0.9	4.0 ± 1.0*	12 ± 23ξ
4. Basal inferior	4.0 ± 0.5	5.8 ± 1.0	45 ± 27	3.7 ± 1.0	4.1 ± 1.0*	13 ± 21*
5. Basal inferior lateral	3.5 ± 0.5	5.3 ± 0.9	50 ± 22	3.0 ± 0.6*	3.3 ± 0.9ξ	13 ± 25*
6. Basal anterior lateral	3.6 ± 0.8	5.0 ± 0.7	46 ± 26	3.1 ± 0.8	3.2 ± 0.8ξ	8 ± 24*
7. Middle anterior	3.9 ± 0.5	6.0 ± 0.1	56 ± 20	3.4 ± 0.2*	3.6 ± 0.3ξ	6 ± 12ξ
8. Middle anterior septal	3.9 ± 0.4	0.6 ± 1.4	55 ± 26	3.3 ± 0.6*	3.4 ± 0.4ξ	5 ± 17ξ
9. Middle inferior septal	3.6 ± 0.6	5.2 ± 1.1	44 ± 18	3.5 ± 0.7	3.4 ± 0.5ξ	1 ± 16ξ
10. Middle inferior	4.1 ± 0.5	6.0 ± 1.2	51 ± 31	3.8 ± 0.7	4.0 ± 0.7ξ	7 ± 17*
11. Middle inferior lateral	3.6 ± 0.3	5.5 ± 0.8	52 ± 25	3.1 ± 0.6*	3.4 ± 0.6ξ	10 ± 14ξ
12. Middle anterior lateral	3.2 ± 0.3	4.7 ± 0.9	45 ± 19	2.9 ± 0.5	3.0 ± 0.3ξ	4 ± 14ξ
13. Apical anterior	4.8 ± 1.0	9.3 ± 2.0	96 ± 33	3.8 ± 0.7*	4.1 ± 1.0ξ	6 ± 15ξ
14. Apical septal	4.9 ± 0.6	9.0 ± 1.9	83 ± 34	4.4 ± 0.7	4.6 ± 0.9ξ	4 ± 14ξ
15. Apical inferior	5.7 ± 0.9	11 ± 2.8	90 ± 44	4.7 ± 1.0*	4.9 ± 1.0ξ	5 ± 14ξ
16. Apical lateral	4.4 ± 0.7	8.8 ± 2.2	103 ± 65	3.8 ± 0.4	4.0 ± 0.7ξ	2 ± 12ξ
Mean	4.0 ± 0.4	6.5 ± 1.0	61 ± 18	3.6 ± 0.5*	3.8 ± 0.5ξ	7 ± 9ξ
Coefficient of variation (%)	21 ± 5	31 ± 8	51 ± 14	19 ± 5	18 ± 4ξ	392 ± 501*

*, p<0.05; #, p<0.01; ξ, p<0.001

Table 3. Left ventricular Regional curvedness, Diastole-to-systole change in curvedness (%ΔC), Variation of curvedness and %ΔC in MI compared to normal state, as defined by equation (1). This table is related to our work in Ref [2].

Cardiac Myocardial Disease States Cause Left Ventricular Remodeling with Decreased Contractility and Lead to Heart Failure; Interventions by Coronary....

221

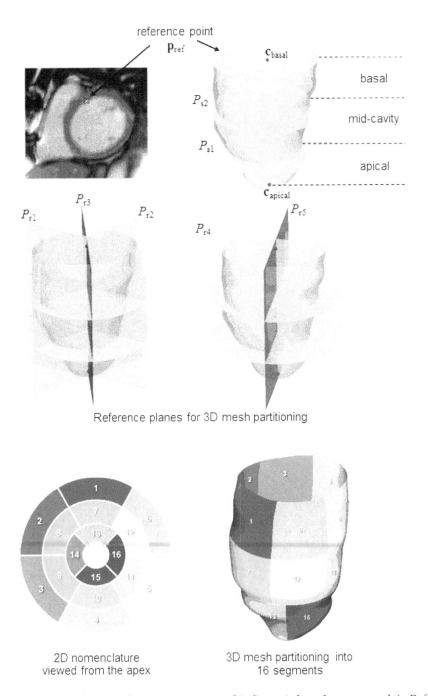

reference point

P_{ref}

c_{basal}

basal

P_{s2}

mid-cavity

P_{s1}

apical

c_{apical}

P_{r3}

P_{r1} P_{r2} P_{r5}

P_{r4}

Reference planes for 3D mesh partitioning

2D nomenclature
viewed from the apex

3D mesh partitioning into
16 segments

Fig. 3. Partitioning of LV mesh into 16 segments; this figure is based on our work in Ref. [2].

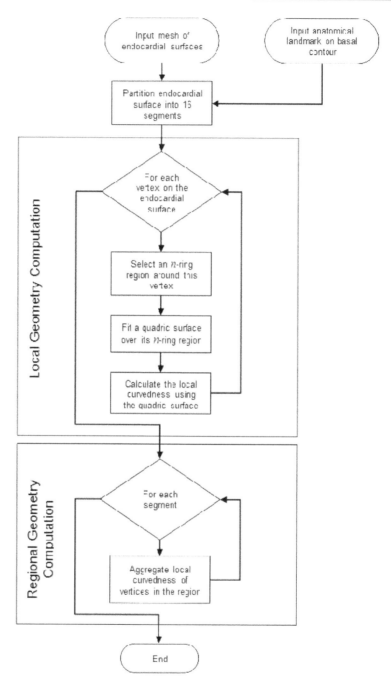

Fig. 4. Flowchart of the overall workflow for the regional LV shape analysis; this figure is based on our work in Ref. [2].

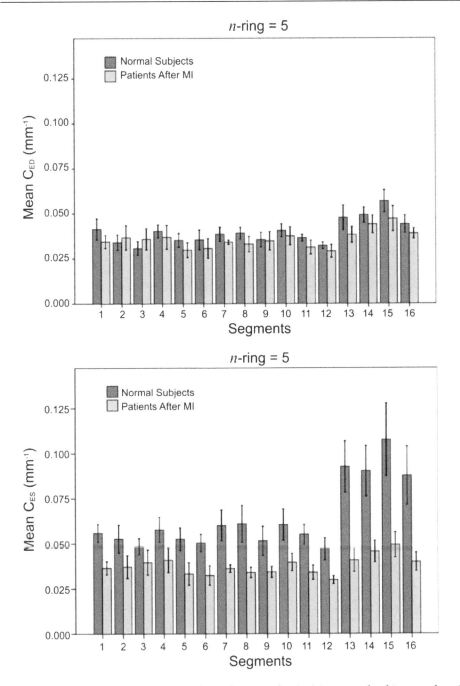

Fig. 5. Comparison of regional curvedness (5-ring selection) in normal subjects and patients
after myocardial infarction; this figure is based on our work in Ref. [2].

5. Employment of cardiac contractility Index dσ*/dt$_{max}$ to demonstrate (i) its excellent correlation with traditional contractility index dP/dt$_{max}$, in patients with varying ejection fractions, and (ii) its capacity to diagnose Heart failure with normal ejection fraction (HFNEF, or diastolic heart failure) and with reduced ejection fraction (HFREF, or systolic heart failure)

5.1 Cardiac contractility expressed in terms of maximum rate of change of pressure-normalized LV wall stress, dσ*/dt$_{max}$

Left Ventricular (LV) contractility can be termed as the capacity of the LV to develop intra-myocardial stress, and thereby intra-cavitary pressure, to eject blood volume as rapidly as possible. It is hence rational and appropriate to formulate a LV contractility index on the basis of LV wall stress. The traditional dP/dt_{max} is based on left ventricular intra-cavitary pressure which is generated by an active myocardial stress. Hence, analogous to dP/dt_{max}, which is based on LV intra-cavitary pressure, we have formulated a novel LV contractility index based on LV wall stress, namely the maximum rate of change during systole of LV wall stress normalized to LV intra-cavitary pressure, $d(\sigma/P)/dt_{max}$ or $d\sigma^*/dt_{max}$, where $\sigma^* = \sigma/P$ [3].

Model Analysis and Index Formulation: For mathematically simplicity, we have approximated the LV as a thick-wall spherical shell consisting of incompressible, homogeneous, isotropic, elastic material. The maximum circumferential wall stress (σ_θ) can be expressed at the endocardium, as:

$$\sigma_\theta(r_i) = P\left[\frac{r_i^3/r_e^3 + 1/2}{1 - r_i^3/r_e^3}\right] \tag{5}$$

where r_i and r_e are the inner and outer radii, P is LV intracavitary pressure.
By normalizing wall stress to LV intra-cavitary pressure (P), we obtain:

$$\sigma^*(r_i) = \frac{\sigma_\theta}{P} = \frac{r_i^3}{r_e^3 - r_i^3}\left(1 + \frac{r_e^3}{2r_i^3}\right) \tag{6}$$

Since the maximum wall stress occurs at the inner endocardial wall, we have:

$$\sigma^*(r = r_i) = \left(\frac{V/(V_m+V)+1/2}{1-V/(V_m+V)}\right) = \left(\frac{3V+V_m}{2V_m}\right) = \left(\frac{3V}{2V_m} + \frac{1}{2}\right) \tag{7}$$

where P is LV intra-cavitary pressure; σ_θ is the wall stress; V ($= 4\pi r_i^3/3$) denotes LV volume; V_m ($= 4\pi\left(r_e^3 - r_i^3\right)/3$) denotes LV myocardial volume; r_i and r_e are the inner and outer radii of the LV, respectively. Differentiating equation (7) with respect to time, we get:

$$d\sigma^*/dt_{max} = \left|\frac{d(\sigma_\theta/P)}{dt}\right|_{max} = \frac{3}{2V_m}\left|\left(\frac{dV}{dt}\right)\right|_{max} \tag{8}$$

It can be thus noted that in contrast to the indices of dP/dt_{max}, E_{es} and $E_{a,max}$, our $d\sigma^*/dt_{max}$ index can be determined solely from non-invasive assessment of LV geometry and flow. Normalizing LV wall stress to LV pressure obviates the need for invasive LV pressure measurement.

Our LV contractility formulation has been based on the premise that LV wall stress (due to LV myocardial sarcomere contraction) is responsible for the development of LV pressure. Hence, it is more rational to base LV contractile function on LV wall stress per pressure. Hence, analogus to dP/dt_{max}, this LV contractility index is formulated as the maximal rate of pressure-normalized wall stress (as given by the above equation 8), to represent the maximal flow rate out of the ventricle (dV/dt) normalized to myocardial volume (V_m). This is somewhat in keeping with cardiac output or maximal volume change having been used as a measure of myocardial contractility in rats or human, provided that the influence of afterload is taken into account.

5.2 Medical application to subjects with varying ejection fractions, to demonstrate excellent correlation of $d\sigma^*/dt_{max}$ with dP/dt_{max}

We have validated $d\sigma^*/dt_{max}$ against dP/dt_{max} and $E_{a,max}$ in 30 subjects with disparate ventricular function in Figure 6, and demonstrated the index's load independence, albeit under conditions of limited preload and after load manipulations [3].

For this study, thirty volunteers [mean 58.1 (range 48–77) yr of age, 13:2 male-to-female ratio] with diverse cardiac conditions were recruited. From their LV pressure-volume data, LVEF and dP/dt_{max} were computed directly from these traces. Active elastance Ea at various times was also computed from the pressure-volume loops, from the data in Table 4, based on our earlier work on Ea definition and determination [4]. $E_{a,max}$ was extrapolated from the peak of the Ea-time curve [4]. The single-beat estimation of end-systolic elastance Ees (SB) was determined, using bilinearly approximated time-varying elastance [5].

The patients were divided into three groups on the basis of tertiles of LVEF, with 10 individuals in each group, as shown in Table 5. Intergroup comparisons show significant differences between the mean values of dP/dt_{max}, $E_{a,max}$, Ees(SB), and $d\sigma^*/dt_{max}$ in those in the highest tertile compared with those in lowest and middle tertiles. There is agreement with regard to the index $d\sigma^*/dt_{max}$ with dP/dt_{max}, $E_{a,max}$, and Ees(SB) across the three tertiles of ascending LVEF values, with statistically significant differences in LV contractility indexes among the three groups. Values of dP/dt_{max}, Ees(SB), and $d\sigma^*/dt_{max}$ were statistically significantly lower in patients in the lowest and middle tertiles had than those in the highest tertile.

Figure 6 summarizes the correlation between $d\sigma^*/dt_{max}$, dP/dt_{max}, and $E_{a,max}$, as well as Ees(SB). Linear regression analysis revealed good correlation between $d\sigma^*/dt_{max}$ and dP/dt_{max}, $E_{a,max}$, and Ees(SB), with significant correlation coefficients in each case: $d\sigma^*/dt_{max}=0.0075dP/dt_{max}- 4.70$ ($r = 0.88$, $P < 0.01$), $d\sigma^*/dt_{max}=1.20E_{a,max} + 1.40$ ($r = 0.89$, $P<0.01$), and $d\sigma^*/dt_{max}= 1.60Ees$(SB) $+ 1.20$ ($r = 0.88$, $P < 0.01$). In contrast, the correlation between $d\sigma^*/dt_{max}$and LVEF is less strong ($r = 0.71$), as is the correlation between Ees (SB) and LVEF ($r = 0.78$), underscoring the lack of specificity of LVEF as an index of myocardial contractility.

Frame No. (i)	t, s	P, mmHg	V, ml	Ea, mmHg/ml
		Isovolumic contraction		
1	0	18	136.7	0
2	0.02	22	135.7	0.0295
3	0.04	32	134.6	0.1038
4	0.06	52	133.5	0.2536
5	0.08	80	132.5	0.4636
		Isovolumic relaxation		
18	0.34	74	85.0	0.0590
19	0.36	50	85.5	0.1778
20	0.38	30	86.4	0.3127
21	0.40	17	90.6	0.4636

Table 4. Active Elastance Ea *computed at discrete time points during isovolumic contraction and relaxation in a sample subject:* i, time instant in the cardiac cycle (frame number from end-diastole); t, time from start of isovolumic contraction; P, measured left ventricular intracavitary pressure; V, measured left ventricular intracavitary volume; $E_{a,i}$, calculated active elastance at instant i. This table is related to our work in Ref [4].

	Lowest tertile	Middle tertile	Highest tertile
Ejection fraction	0.38 ± 0.12*	0.49 ± 0.13*	0.63 ± 0.05
Age,yr	58.30 ± 8.86	56.10 ± 6.15	59.90 ± 6.17
Heart rate,beats/min	71.18 ± 10.72	71.77 ± 10.68	71.46 ± 9.09
dP/dt_{max},mmHg/s	960 ± 115*	1,121 ± 113*	1,360 ± 97
Ea,max, mmHg/ml	0.95 ± 0.32*	1.85 ± 0.59*	3.61 ± 0.62
Ees(SB),mmHg/ml	0.72 ± 0.26*	1.51 ± 0.20*	2.81 ± 0.51
$d\sigma^*/dt_{max}$,s^{-1}	2.30 ± 0.58*	3.60 ± 1.06*	5.64 ± 1.13

Table 5. *LV contractility indexes classified into tertiles of LVEF:* Left ventricular contractility indices classified into tertiles of left ventricular ejection fraction. Values are expressed as mean ± standard deviation. Asterisks denote statistically significant difference ($p<0.05$) when compared with corresponding values in the highest tertile of left ventricular ejection fraction. dP/dt_{max}, peak first time-derivative of the ventricular pressure; $E_{a,max}$, maximum left ventricular elastance; $Ees(SB)$, single-beat LV end-systolic elastance; $d\sigma^*/dt_{max}$,left ventricular contractility index. This table is related to our work in Ref [3].

Cardiac Myocardial Disease States Cause Left Ventricular Remodeling with Decreased Contractility and Lead to
Heart Failure; Interventions by Coronary....

227

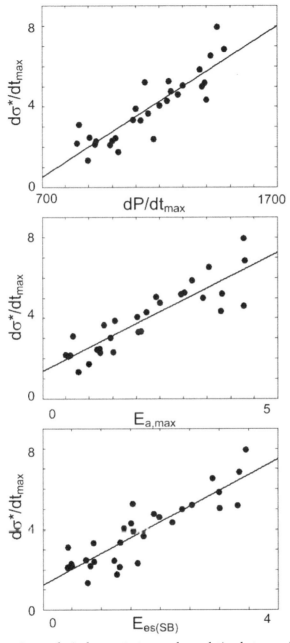

Fig. 6. Linear regression analysis demonstrates good correlation between $d\sigma^*/dt_{max}$ and
dP/dt_{max} : [$d\sigma^*/dt_{max}$ = 0.0075dP/dt_{max}-4.70, r=0.88; top figure], between $d\sigma^*/dt_{max}$ and $E_{a,max}$
[$d\sigma^*/dt_{max}$ = 1.20$E_{a,max}$+1.40, r=0.89; middle figure], and between $d\sigma^*/dt_{max}$ and $E_{es(SB)}$
[$d\sigma^*/dt_{max}$ = 1.60$E_{es(SB)}$+1.20, r=0.88; bottom figure]. This figure is based on our work in Ref [3].

5.3 Use of cardiac contractility index $d\sigma^*/dt_{max}$ to diagnose heart failure with normal ejection fraction (HFNEF) and with reduced ejection fraction (HFREF)

In another study [6], we assessed the capacity of $d\sigma^*/dt_{max}$ to diagnose heart failure in patients with normal ejection fraction (HFNEF) and with reduced ejection fraction (HFREF).

5.3.1 Introduction

Heart failure (HF) is a major health care burden: it is the leading cause of hospitalization in persons older than 65 years, and confers an annual mortality of 10%. HF can occur with either normal or reduced LV ejection fraction (EF), depending on different degrees of ventricular remodeling. Both heart failure with normal ejection fraction (HFNEF) and heart failure with reduced ejection fraction (HFREF), also commonly known as diastolic and systolic heart failure respectively, have equally poor prognosis [7]. Medical therapy targets to reduce load, by using vasodilators and/or to alter contractile strength using inotropic agents. Alternatively, some therapies target to affect cardiac remodeling, such as passive cardiac support devices, surgical restoration of LV shape (i.e. the Dor procedure), and stem cells therapies.

Assessment of left ventricular (LV) contractility is important for HF management and evaluation of the heart's response to medical and surgical therapies. Although approaches based on pressure-volume analysis, stress-strain analysis, and dP/dt_{max}-EDV relations [8]can provide assessments of contractile function, these relations generally require invasive data measured at several chamber loads and thus are difficult to apply in routine or long-term clinical studies. This is an important limitation, because heart failure often requires longitudinal evaluation. The ideal measure of contractility should have the following characteristics: sensitivity to inotropic changes, independence from loading conditions as well as heart size and mass, ease of application, and proven usefulness in the clinical setting. LV ejection fraction (EF) is the index overwhelmingly used to assess cardiac function in both clinical and experimental studies, despite the fact that it is highly dependent upon preload and afterload. Based on the National Heart Lung and Blood Institute's Framingham Heart Study, an LVEF50% as cut-off for the presence of normal LVEF has been used in the present study [9].

Usefulness of $d\sigma^*/dt_{max}$ **as Contractility Index:** During LV systole, LV wall stress is generated intrinsically by sarcomere contraction and results in the development of extrinsic LV pressure. We have shown earlier that our novel LV contractility index, $d\sigma^*/dt_{max}$ (maximal change rate of pressure-normalized wall stress) correlates well with LV dP/dt_{max} [3]. We have proposed and validated a new LV contractility index, $d\sigma^*/dt_{max}$,based on the maximal rate of development of LV wall stress with respect to LV pressure. From the right-hand side of equation (8), this index is also seen to represent the maximal flow rate from the ventricle (cardiac output) normalized to myocardial volume (or mass).

This index is easily measured non-invasively (i.e. from echocardiography or magnetic resonance imaging), is sensitive to LV inotropic changes, and has been demonstrated by us to be preload and afterload independent [3]. Importantly, it is measured at a single steady-state condition, as opposed to the multiple variably loaded cardiac cycles required for many of the other indices. Thus $d\sigma^*/dt_{max}$has several qualities that make it a useful LV contractility index. This study [6] has constituted an important step toward establishing the clinical utility of $d\sigma^*/dt_{max}$ as a tool for diagnosis of HF (both HFNEF and HFREF) as well as for

Cardiac Myocardial Disease States Cause Left Ventricular Remodeling with Decreased Contractility and Lead to
Heart Failure; Interventions by Coronary....

229

follow-up surveillance of LV function. Hence we see a great potential for application of this novel index to evaluate heart function in diverse heart conditions.

5.3.2 Clinical methodology

Patients referred to our echocardiography service with symptoms and signs of heart failure underwent echocardiography and electrocardiography (ECG). Patients with atrial fibrillation, more than mild mitral or aortic valvular regurgitation, and unsatisfactory echocardiographic images were excluded. Clinical signs of heart failure were defined as presence of at least one of the following: raised jugular venous pressure, peripheral pedema, hepatomegaly, basal inspiratory crepitation or gallop rhythm. Patients with LVEF $\geq$ 50% and LVEF < 50% on echocardiography were classified into HFNEF and HFREF, respectively.

Echocardiography Study: With the subject in the left lateral decubitus position, 2D examinations, M-mode measurements and Doppler recordings were performed from the standard left parasternal long- and short-axis as well as the apical four chamber views with simultaneous ECG. The LVEF was assessed by using a 2-dimensional method by an experienced observer; normal LVEF was defined as greater or equal to 50%. Mitral flow velocities were obtained from the apical 4-chamber view using pulsed wave Doppler technique with the sample volume at the tips of the corresponding valve leaflets. LV outflow tract velocity was obtained from apical 5-chamber view, using pulsed wave Doppler technique with the sample volume at the aortic valve level.

The measurements included peak E (peak early trans-mitral filling velocity during early diastole) and A (peak trans-mitral atrial filling velocity during late diastole); wave velocities (cm/s) were measured and E/A ratio was calculated. The E wave deceleration time (DT) was also calculated as the time elapsed between peak E velocity and the point where the extrapolation of the deceleration slope of E velocity crosses the zero baseline measured in milliseconds. LVOT maximal velocity V_{peak} was measured, and LV mass was calculated by using ASE methods [10, 11]. Myocardial tissue Doppler (TDI) velocities were also estimated at the atrioventricular ring, septal positions, in the apical 4 chamber view. All measurements were averaged over two or three cardiac cycle.

Calculation of $d\sigma^*/dt_{max}$ *from Echocardiography:* The contractility index was computed by the above equation (8). M-mode echocardiographic measurements of the LV were obtained, and LV mass calculated using standardized methodology [10, 11]. Myocardial volume was calculated by dividing LV mass with myocardial density (assumed to be 1.05 g/ml). Furthermore, two-dimensional apical four- and two-chamber views of the LV were acquired, and end-diastolic and –systolic endocardial contours were manually outlined. The corresponding LVEDV were then automatically determined using biplane Simpson's method.

From Pulse-wave echo-Doppler interrogation of the LV outflow tract (LVOT), we calculated (in the absence of significant mitral regurgitation or aortic valve dysfunction) the maximal LV volume rate (dV/dt_{max}) during ejection: $dV/dt_{max}=V_{peak}*AVA$, where V_{peak} is the peak velocity sampled at the LVOT and AVA is the aortic valve area (= $\pi D^2/4$, where D is the LVOT diameter measured in the two-dimensional parasternal long-axis image of the heart), as shown in Figure 7. Upon substituting values of myocardial volume and dV/dt_{max} into equation (8), we determined the value of $d\sigma^*/dt_{max}$.

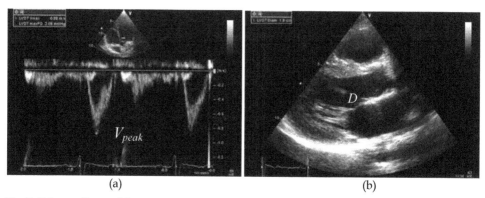

(a)   (b)

Fig. 7. Echocardiographic measurement on (a) peak velocity V_{peak} sampled at the LVOT and (b) LVOT diameter D measured in the two-dimensional parasternal long-axis image of the heart. This figure is adopted from Ref. [6].

Clinical Studies: The study involved 26 age- and sex-matched subjects in each of the groups of normal controls, HFNEF and HFREF. The characteristics of 78 subjects are shown in Table 6. It summarizes the subjects' age, BSA, LVEF, peak E, peak A, E/A ratio, DT, heart rate (HR), septal E/E', lateral E/E' and our index $d\sigma^*/dt_{max}$. Mean $d\sigma^*/dt_{max}$ was 3.91 s^{-1} (95%CI, 3.56-4.26 s^{-1}) in control subjects; it was reduced in heart failure, HFNEF, to 2.90 s^{-1} (95%CI, 2.56-3.24 s^{-1}); and in HFREF, to 1.84 s^{-1} (95%CI, 1.60-2.07 s^{-1}). There exists no substantially difference between the average values of LVEF, peak E, peak A, E/A ratio, DT, heart rate (HR), septal E/E', and lateral E/E' in HENEF compared to normal controls, except for $d\sigma^*/dt_{max}$ (2.90 ± 0.84 vs. 3.91 ± 0.87, p<0.001). However, there exists significant difference between the average values of LVEF, peak E, peak A, E/A ratio, DT, septal E/E', lateral E/E' and $d\sigma^*/dt_{max}$ in HEREF compared to HFNEF.

Discussion on the Usefulness of $d\sigma^*/dt_{max}$: During LV systole, LV wall stress is generated intrinsically by sarcomere contraction and results in the development of extrinsic LV pressure. LV wall stress is dependent on wall thickness, LV geometry and chamber pressure and sarcomere contraction. Hence, it is rational to quantify the LV wall stress as an intrinsic measure of myocardial contractility. We have proposed and validated a new LV contractility index, $d\sigma^*/dt_{max}$,based on the maximal rate of development of LV wall stress with respect to LV pressure. From the right-hand side of equation (8),this index is also seen to represent the maximal flow rate from the ventricle (cardiac output) normalized to myocardial volume (or mass).

Assessment of heart failure with normal ejection fraction (HFNEF) and reduced ejection fraction (HFREF): Heart failure may be viewed as a progressive disorder that is initiated after an "index event" with a resultant loss of functioning cardiac myocytes, thereby preventing the heart from contracting normally. HF can occur with either normal or reduced LV ejection fraction (LVEF), depending on different degree of ventricular remodeling. Perhaps 50% of patients with heart failure have a normal or minimally impaired LVEF (HFNEF) [12, 13].

Although mechanisms for HFNEF remain incompletely understood, diastolic dysfunction is said to play a dominant role: impaired relaxation, increased passive stiffness, raised end-diastolic pressure (EDP) [14]. The diagnostic standard for HFNEF is cardiac catheterization, which demonstrates increased EDP. However, a more practical noninvasive alternative is echocardiography. Our study has shown that E/A ratio (1.26 ± 0.90 vs. 0.96 ± 0.38, p<0.05) and DT (157 ± 41 ms vs. 214 ± 47 ms, p<0.05) are significantly different between HFREF and normal controls, and not so between HFNEF and normal controls (Table 6).

Our contractility index, of change rate of normalized wall stress index $d\sigma^*/dt_{max}$, is dependent on lumen and wall volume of LV chamber and represents an integrated assessment of LV systolic performance [3], based on our findings relating $d\sigma^*/dt_{max}$ with HFNEF and HFREF [6]. In this study, as shown in Table 7, we find that there exists significant difference in dV/dt_{max} between HFREF and HFNEF (233 ± 48 ml/s vs. 355 ± 65 ml/s, p<0.05), while there exists no difference between HFNEF and normal controls (355 ± 65 ml/s vs. 353 ± 80 ml/s, NS). Similarly, there exists significant difference in LV mass between normal controls and HFNEF (147 ± 41 g vs. 202 ± 47 g, p<0.05), while there is no difference between HFREF and HFNEF (213 ± 60 g vs. 202 ± 47 g, NS).

Our $d\sigma^*/dt_{max}$ index, using dV/dt_{max} normalized with LV mass, can clearly differentiate HFREF, HFNEF and normal controls (p<0.05) (Table 7). The average value of $d\sigma^*/dt_{max}$ decreases in HFNEF and HFREF, in relation to normal controls. The mean value of $d\sigma^*/dt_{max}$ was found to be 3.91 s^{-1} (95%CI, 3.56-4.26 s^{-1}) in control subjects; the index was reduced in heart failure patients: in HFNEF, to 2.90 s^{-1} (95%CI, 2.56-3.24 s^{-1}) and in HFREF, to 1.84 s^{-1} (95%CI, 1.60-2.07 s^{-1}). This suggests that poor systolic function of LV is associated with lower $d\sigma^*/dt_{max}$ values. Therefore, it can again be concluded that $d\sigma^*/dt_{max}$ is an appropriate index for representing assessment of LV contractile function in heart failure with/without preserved LV ejection fraction.

	Controls (n=26) (n=21)	HFNEF (n=26)	HFREF (n=26)
Age (years)	72 ± 8	70 ± 8	70 ± 8
Gender (male:female)	16:10	16:10	16:10
BSA (m^2)	1.69 ± 0.20	1.71 ± 0.20	1.61 ± 0.20
LVEF (%)	68.3 ± 5.1	66.5 ± 4.9	33 ± 13.7§*
E/A ratio	0.96 ± 0.38	0.78 ± 0.24	1.26 ± 0.90*
DT (ms)	214 ± 47	214 ± 67	157 ± 41§*
HR (beats/min)	66 ± 10	72 ± 16	80 ± 11§
Septal E/E′	8.48 ± 2.10	9.79 ± 3.29	13.68 ± 4.78§*
Lateral E/E′	6.64 ± 1.55	8.39 ± 2.76	10.28 ± 3.40§*
$d\sigma^*/dt_{max}$ (s^{-1})	3.91 ± 0.87	2.90 ± 0.84§*	1.84 ± 0.59§*

A, mitral atrial flow velocity on echo-Doppler; BSA, body surface area; DT, mitral E deceleration time; E, mitral early velocity; E′, septal mitral annular myocardial velocity on tissue Doppler imaging; HR, heart rate, LVEF, let ventricular ejection fraction.

The values are expressed as mean $\pm$ SD. § and * denote statistically significant difference of HF compared to controls, HFREF compared to HFNEF patients, respectively (Bonferroni pairwise test, p value <0.05)

Table 6. Patients characteristics and echocardiographic measurements in Group 1 (Controls), Group 2 (HFNEF) and Group 3 (HFREF). This table is related to our work in Ref [6].

	Controls (95% CI)	HFNEF (95% CI)	HFREF (95% CI)
dV/dt_{max} (ml/s)	353 (320, 385)	355 (329, 381)	233 (213, 252)[§*]
V_{peak}(cm/s)	106 (98, 115)	112 (104, 119)	73 (68, 78) [§*]
LV mass (g)	147 (131, 164)	202 (183, 221)[§]	213 (189, 297)[§]
$d\sigma^*/dt_{max}$ (s^{-1})	3.91 (3.56, 4.26)	2.90 (2.56, 3.24)[§*]	1.84 (1.60, 2.07)[§*]

[§] and * denote statistically significant difference of HF compared to controls, HFREF compared to HFNEF patients, respectively

Table 7. Comparison of the maximal flow rate dV/dt_{max}, V_{peak}, LV mass, and $d\sigma^*/dt_{max}$ in Group 1 (Controls), Group 2 (HFNEF) and Group 3 (HFREF). This table is related to our work in Ref [6].

6. Coronary Arterial Bypass Grafting (CABG) to salvage ischemic myocardial segments

As is well known, coronary artery bypass graft (CABG) surgery has been the standard treatment for serious blockages in the coronary arteries and for re-perfusing myocardial ischemic segments to restore them to normal contractile state. During the surgery, one end of the graft is sewn to the aorta (or its subsidiary branches) to create proximal anastomosis, while the other end is attached to coronary artery below the area of blockage to create distal anastomosis. In this way, the oxygen-rich blood is taken directly from the aorta, bypasses the obstruction, and flows through the graft to perfuse and nourish the heart muscle. The most commonly used graft is the saphenous vein. Besides this vein graft, some arterial conduits (such as internal mammary artery, gastroepiploic artery and radial artery etc), and synthetic veins (such as Dacron, Teflon and Polytetrafluoroethylene-PTFE veins) are also suitable for CABG.

Although the number of bypass operations keeps increasing, the CABG has not been without complications. Approximately 15% to 20% vein grafts occlude in the first year, and 22.5% to 30% occlude within the first 2 years. At 10 years, approximately 60% of vein grafts are patent; only 50% of these vein grafts remain free of significant stenosis. In order to intensively investigate the coronary arterial stenosis symptom, numerous research works have been carried out. One direction of these studies is to investigate the pathogenic mechanism of bypass graft failure. In this regard, vascular injury and biomechanical factors (such as wall shear stress related factors, compliance mismatch, etc.) are believed to stimulate cellular responses for pathological changes. In particular, hemodynamic flow patterns of CABG have considerable relevance to the causes and sites of pathogenesis. Hence, we have carried out simulation of hemodynamic flow patterns in CABG models, to look into the hemodynamic causes and mechanisms of lesions in coronary bypass grafts.

The flow characteristics and hemodynamic parameters distributions in a complete CABG model (as shown in Figure 8) have been investigated computationally by us [15]. It is found that disturbed flow (flow separation and reattachment, vortical and secondary flows) patterns occur at both proximal and distal anastomoses, especially at the distal anastomosis. In addition, regions of high-OSI & low-WSS and low-OSI & high-WSS are found in the proximal and distal anastomoses, especially at the toe and heel regions of distal anastomosis. These regions are suspected to initiate the atherosclerotic lesions and are further worsened by the increasing permeability of low-density lipoprotein as indicated by high WSSG. The comparisons of segmental average of HPs (in the Table of Figure 8) further imply that intimal hyperplasia is more prone to form in the distal anastomosis than the proximal anastomosis, especially along the suture line at the toe and heel of distal anastomosis, which was in line with the in-vivo observations.

We then investigated the fluid dynamics of blood flow in two complete models of CABG for the right and left coronary artery separately, as shown in Figure 9 [16]. The results reveal that blood flow through the coronary artery bypass graft primarily occurs only during the diastolic phase of the cardiac cycle, which is in agreement with the physiological observation. However, at the onset of ejection, some backflow from the coronary artery into the bypass graft is found for the CABG to left coronary artery, which is absent for the right coronary artery. This reversal of flow during systole can be explained by the predominant intra-cardiac course of the left coronary artery system. As the same time, this study also found a low WSS region near the heel and a high WSS in the toe region of the anastomosis domain.

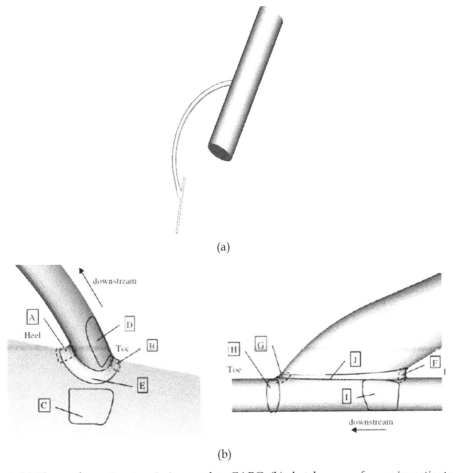

(a)

(b)

Fig. 8. (a) The configuration to mimic complete CABG; (b) sketch maps of areas investigated for segmental averages of hemodynamic parameters (HPs) in proximal and distal anastomoses;

Name		Labels in maps	Area (mm²)	<WSS> (Pa)	<WSSG>	<OSI>
Proximal anastomosis	heel	A	4.23	4.48	11.88	0.41
	toe	B	1.59	8.58	20.24	0.36
	part3	C	9.46	3.17	3.24	0.49
	part4	D	25.50	6.56	14.25	0.07
	suture_line	E	12.00	5.06	9.01	0.29
Distal anastomosis	heel	F	0.55	2.92	12.75	0.24
	toe	G	0.55	**28.04**	**147.17**	**0.02**
	part3	H	9.70	**14.63**	**64.24**	**0.07**
	part4	I	6.81	0.85	3.59	0.22
	suture_line	J	3.18	**9.20**	**41.43**	**0.11**

(c)

Fig. 8. (c) the segmental averages of HPs at these locations. This figure is adopted from our work in Ref 15 (Zhang et al., 2008).

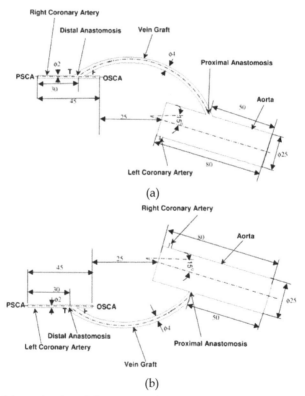

(a)

(b)

Fig. 9. Geometry (plane view) and dimensions (in mm) of the bypass models of: (a) The aorta-right coronary artery bypass model; (b) The aorta-left coronary artery bypass model (PSCA-Perfused Segment of the Coronary Artery; OSCA-Occluded Segment of the Coronary Artery; T-Toe; H-Heel). This figure is adopted from our work in Ref 16: (Sankaranarayanan et al., 2005).

Cardiac Myocardial Disease States Cause Left Ventricular Remodeling with Decreased Contractility and Lead to
Heart Failure; Interventions by Coronary....

235

CABG performance is based on flow characteristics at both the proximal and distal anastomoses. So then, let us summarise the optimal geometrical parameters for proximal and distal anastomoses. For proximal anastomosis, a detailed study, on the effect of three grafting angles (viz. 45° forward facing, 45° backward facing, and 90°), has been carried out by Chua et al. [17]. The results show a flow separation region along the graft inner wall immediately after the heel at peak flow phase, which decreases in size with the grafting angle shifting from 45° forward facing to 45° backward facing. The existence of nearly fixed stagnating location, flow separation, vortex, high-WSS-low-OSI, low-WSS-high-OSI, and high WSSG is suspected to lead to graft stenosis. Among these three models, the 45° backward-facing graft is found to have the lowest variation range of time-averaged WSS and the lowest segmental average of WSSG, as shown in Figure 10; these parameters are then recommended for obtaining higher expected patency rates in bypass operations.

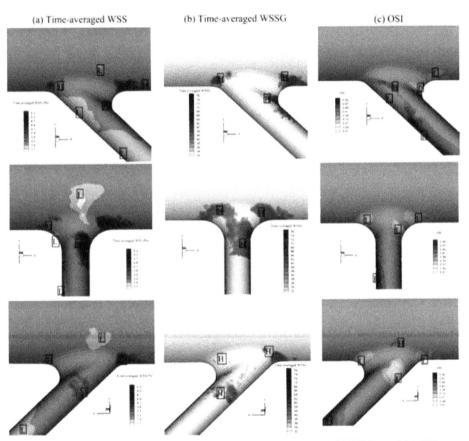

Fig. 10. Contours of HPs: (a) time-averaged WSS; (b) time-averaged WSSG; and (c) OSI, on the surfaces of 45° forward-facing, 90°, and 45° backward-facing models ('H' means the region has high value, 'L' means the region has low values. This figure is adopted from our work in Ref 17 (Chua et al., 2005b).

At the distal anastomosis junction of CABG, the important geometrical parameters for smooth flow are the angle (α) and the diameter ratio (ϕ) between the graft and host artery. The hemodynamics associated with these parametric were investigated by Xiong and Chong [18], over a range of ϕ (1:1, 1.5:1 and 2:1) and α (15°, 30°, 45° and 60°) in physiological coronary flow conditions. It is found that increasing ϕ from 1:1 to 1.5:1 almost eliminates both low WSS and high OSI at the toe. However, further increasing ϕ to 2:1 causes elevated OSI for most of the host artery segment in the anastomotic region and in almost the entire graft. On the other hand, varying α is also found to change certain aspects of hemodynamics, although less than those changes related with different values of ϕ. A smaller value of α is found to be associated with a higher OSI in the anastomotic region, whereas a larger α causes higher WSSG on the artery floor. Therefore, it is suggested that for distal coronary anastomosis (with a 20:80 proximal to distal flow division ratio maintained in the host artery), the geometry associated with $\phi = 1.5$ and $\alpha = 30$–45° is favorable for enhancing long-term performance.

7. Surgical Ventricular Restoration (SVR), combined with CABG, restores LV shape and improves cardiac contractility

7.1 Introduction

Ischemic heart disease is one of the most widely spread, progressive and prognostically unfavorable diseases of the cardiovascular system. In ischemic dilated cardiomyopathy (IDC) patients, the remodeling process involves a lesser systolic LV curved shape, increase of peak wall stress, and decrease of contractile functional index, compared with normal subjects. Surgical ventricular restoration (SVR) is performed in chronic ischemic heart disease patients with large non-aneurysmal or aneurysmal post-myocardial infarction zones. It involves operative methods, that reduce LV volume and 'restore' ventricular ellipsoidal shape, by exclusion of anteroseptal, apical, and anterolateral LV scarred segments by means of intra-cardiac patch or direct closure.

For patients in heart failure (HF) resulting from serious myocardial diseases of ischemic dilated cardiomyopathy and myocardial infarction (MI), surgical ventricular restoration (SVR), designed to restore the LV to its normal shape (reversal of LV remodeling), is performed usually in conjunction with coronary artery bypass grafting (CABG). In our study [19], in 40 ischemic dilated cardiomyopathy (IDC) patients who underwent SVR and CABG, there was found to be: (i) decrease in end-diastolic volume from 318 ± 63 ml to 206 ± 59 ml ($p<0.01$) and in end-systolic volume from 228 ± 58 ml to 133 ± 61 ml ($p<0.01$), (ii) increase in LV ejection fraction from $26 \pm 7\%$ to $31 \pm 8\%$ ($p<0.01$), (iii) decrease in LV mass (from 204 ± 49 g to 187 ± 53 g, $p<0.01$), (iv) decrease in peak normalized wall stress (PNWS) (from 4.30 ± 0.95 to 3.31 ± 0.75, $p<0.01$) , (v) increase in end-systolic sphericity index SI (from 0.57 ± 0.094 to 0.67 ± 0.13, $p<0.01$), (vi) increased value of shape (S) index (from 0.44 ± 0.085 to 0.54 ± 0.089, $p<0.01$) during end-systole indicating that LV became more spherical after SVR, and most importantly (vii) improvement in LV contractility index $d\sigma^*/dt_{max}$ (from 2.69 ± 0.74 s^{-1} to 3.23 ± 0.73 s^{-1}, $p<0.01$).

Thus, in IDC patients, surgical ventricular restoration (in combination with CABG) aiming to reverse LV remodeling, has shown to (i) improve ventricular function and decrease wall stress, along with making a more curved apex, and (ii) improve cardiac contractility. It is not the LV shape alone that defines LV contractility. Rather, a complex interaction of the rate of change of shape factor (dS/dt_{max}) along with LV maximal flow rate and LV mass may explain the improvement in LV contractility.

Cardiac Myocardial Disease States Cause Left Ventricular Remodeling with Decreased Contractility and Lead to
Heart Failure; Interventions by Coronary....

237

7.2 Clinical study

The study was carried to retrospectively evaluate (with cardiac MRI) the changes on systolic function and LV wall stress, the relationships between LV geometry (shape) and dimensions and systolic function after SVR performed in chronic ischemic heart disease patients with aneurismal postmyocardial infarction zones. The study consisted of 40 patients with ischemic dilated cardiomyopathy who had SVR; the age of the patients averaged 69 years (range, 52-84 years). MRI scans were performed 2 weeks before surgery (pre-surgery) and 1 week after the surgery; the details of the MRI procedure are reported in our earlier work [19].

Cardiac magnetic resonance imaging (MRI) provides the means to study heart structure and function: the ventricular systolic and diastolic volumes (and hence ejection fraction) are easily assessed reproducibly and accurately; the regional wall motion of the asynergy area and the remote myocardium can be measured by several quantitative means, including with myocardial tagging; the presence or absence of nonviable, irreversible scar can be detected with gadolinium-based interstitial contrast agents.

Data Analysis, 3-dimensional modeling of LV: For analysis, the images were displayed on a computer monitor in a cine-loop mode using CMRtools, to reconstruct the 3-dimensional model of the left ventricle (LV). The LV epicardial and endocardial borders were outlined, and all the frames were delineated to produce a volume curve from end-diastolic and end-systolic phases. These measurements were used to determine the end-diastolic volume (EDV), end-systolic volume (ESV), stroke volume (SV), ejection fraction (EF), and LV mass.

Ellipsoidal Shape factor, Eccentricity (E) and sphericity index and normalized wall stress: The LV is modeled as a prolate spheroid, truncated 50% of the distance from equator to base, as shown in figure 11 [20, 21]. Then, the left ventricular cavity wall volume is calculated, from the endocardial anterior-posterior (AP) and base-apex (BA) lengths [20], as:

$$V_m = \frac{9}{8}\left[\left(\frac{BA}{1.5}+h\right)\left(\frac{AP}{2}+h\right)^2 - \left(\frac{BA}{1.5}\right)\left(\frac{AP}{2}\right)^2\right] \qquad (9)$$

wherein the BA and AP dimensions are identified in figure 12. The mean wall thickness (h) is calculated at each cavity volume, from the above equation, by assuming that myocardial wall volume (V_m) remains constant throughout the cardiac cycle. The endocardial minor axis dimension (SA) and major axis dimension (LA), shape factor (S), eccentricity (E) and sphericity index (SI) were then calculated as follows (refer figures 11 and 12):

$$SA = AP / 2; \ LA = BA / 1.5, \ S = SA / LA, \ E = \left(\frac{BA^2 - AP^2}{BA^2}\right)^{0.5}, \ SI = AP / BP \qquad (10)$$

wherein BA (the LV long axis) is defined as the longest distance from the apex to the base of the LV (defined as the mitral annular plane), as measured on the four-chamber cine MRI view of the heart; AP is defined as the widest LV minor axis (Figure 12). A small value of SI implies an ellipsoid LV, whereas values approaching "1" suggest a more spherical LV. The SI at end-diastole (SI_{ed}) and end-systole (SI_{es}), the % shortening of the long and minor axes, as well as the difference between end-diastolic and end-systolic values of SI, (SI_{ed} - SI_{es}) were calculated and are tabulated in Table 10 below.

The time-varying circumferential normalized wall stress, NWS(t), is calculated from the instantaneous measurements of LV dimensions and wall thickness, by treating the LV as a prolate spheroid model truncated 50% of the distance from equator to base [20, 21]

$$NWS(t) = \frac{AP(t)}{2h(t)}\left[1 - \frac{\frac{9AP(t)}{32h(t)}(SI)^2}{\frac{AP(t)}{h(t)}+1}\right] \tag{11}$$

The LV wall thickness, h(t), is calculated from the following formula (based on the above equation 9), by assuming that the myocardial wall volume (V_m) remains constant throughout the cardiac cycle:

$$\frac{9}{8}\left[\left(\frac{BA(t)}{1.5}+h(t)\right)\left(\frac{AP(t)}{2}+h(t)\right)^2\right] = V_m(t)+\left(\frac{BA(t)}{1.5}\right)\left(\frac{AP(t)}{2}\right)^2 \tag{12}$$

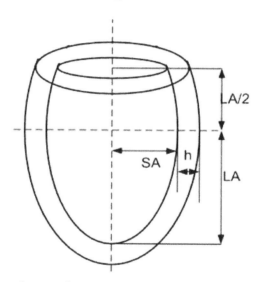

Fig. 11. LV model geometry, showing the major and minor radii of the inner surface of the LV (LA & SA) and the wall-thickness (h).

Cardicac Contractility $d\sigma^*/dt_{max}$: In order to compute the contractility index by employing the above equation (8), a 6-order polynomial function to curve-fit the volumes-time data to calculate the volume rate (dV/dt) by differentiating it. Then the contractility index $d\sigma^*/dt_{max}$ is calculated as:

$$d\sigma^*/dt_{max} = \left|\frac{d(\sigma_\theta/P)}{dt}\right|_{max} = \frac{3}{2V_m}\left|\left(\frac{dV}{dt}\right)\right|_{max}$$

where V_m is myocardial volume at the end-diastolic phase.

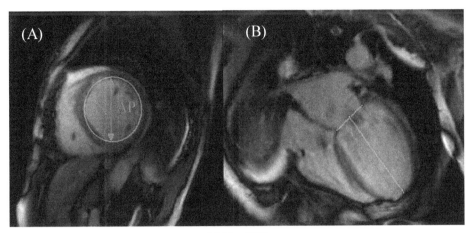

(a) Pre-SVR

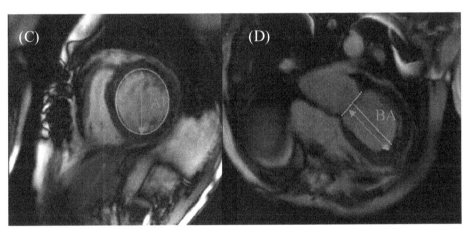

(b) Post-SVR

Fig. 12. Short-axis (panels A and C) and long-axis (panels B and D) magnetic resonance
images of patients, before (panels A and B) and after (panels C and D) surgical ventricular
restoration (SVR). Multiple short-axis cines from the apex to the base of the heart
(or orientated axial) and long-axis cines are used to quantify LV function. Anterior-posterior
(AP) (panels A and C) and base-apex (BA) (panels B and D) were measured from 2-D CMR
imaging before (panels A and B) and after (panels C and D) SVR during cardiac cycle. The
shape factor S, eccentricity index and sphericity index (SI) were calculated from Equation
(6). It can be noted that the long-axis decreased more dramatically compared with the short-
axis dimension, thereby producing a more spherical ventricle. This figure is adopted from
our work presented in Ref 19.

7.3 Clinical results

All 40 patients were treated with CABG and SVR (endoventricular circular patch plasty). The age of the patients averaged 69 years (range, 52-84 years). Among them, 19 patients had severe mitral regurgitation and received additional MVS. 28 patients had CHF. The baseline patient characteristics are summarized in Table 8.

Variables	Value
Male : female	36:4
Age (years)	69 ± 9
Body surface area (m²)	1.98 ± 0.18
Coronary artery disease	39 (98%)
Hypertension	19 (48%)
Diabetes mellitus	13 (33%)
Tobacco	23 (58%)
Congestive heart failure	28 (70%)
Peripheral arterial disease	3 (8%)
Stroke	2 (5%)
Creatinine (mg/dL)	1.17 ± 0.29
Prior cardiac surgery	19 (48%)
New York Heart Association class	
I-II	26 (65%)
III-IV	14 (35%)
Surgery	
Surgical ventricular restoration + coronary artery bypass grafting	21 (52%)
Surgical ventricular restoration + coronary artery bypass grafting + mitral valve surgery	19 (48%)
Values are mean ± SD or numbers of patients (percentages).	

Table 8. Patients' characteristics and clinical data (n=40). This table is related to our work in Ref [19].

7.4 Left ventricular functional indexes changes pre and post-surgery (MRI parameters)

Figure 12 shows typical short-axis and long-axis magnetic resonance images of patient pre- and post-SVR and how to calculate shape, eccentricity and sphericity index. It is noted that the long-axis decreased more dramatically compared with short-axis dimension, thereby producing a more spherical LV. Figure 13 shows typical 3-dimensional modeling of LV from CMR images using LVtools pre and post SVR.

The intraobserver and interobserver data for EDV, ESV and mass for pre- and post-surgery groups are shown in Table 9. Table 10 summarizes the mean LV functional indexes pre and post-SVR. Following SVR, there was a significant decrease in the dimensions of both the long- and short-axes of the LV. However, the long-axis dimension of the LV decreased more than the short-axis dimension, resulting in a more spherical ventricle post-SVR. There was a significantly reduction in end-diastolic volume index (EDVI), end-systolic volume index (ESVI), LV stroke volume index (SVI), LV mass index, and peak normalized wall stress after

Cardiac Myocardial Disease States Cause Left Ventricular Remodeling with Decreased Contractility and Lead to
Heart Failure; Interventions by Coronary....

241

SVR (Table 3). The values of LV EDVI, ESVI, LVEF and the contractility index $d\sigma^*/dt_{max}$ pre- and post-SVR are also shown in the scatter plots of Figure 14.

Table 10 provides the sphericity index (SI) values in end-diastole and end-systole and its diastolic-systolic change, as well as the % shortening of the long- and short-axes. During a cardiac cycle, LV shape becomes less spherical in systole (SI smaller) than in diastole (SI closer to '1'). The diastolic-systolic change in SI (SI_{ed}-SI_{es}) is significantly augmented by the operation, despite the LV chamber becoming more spherical. The % shortening of long-axis is not significantly altered, but the % shortening of the short-axis is significantly increased by the operation. Despite the seemingly unfavorable spherical LV shape post-SVR, the LV contractile function is significantly improved, as indicated by the increased value of $d\sigma^*/dt_{max}$.

The scatter plots of figure 14 graphically illustrate pre- and post-SVR values of ventricular end-diastolic volume, end-systolic volume, LVEF and contractility index $d\sigma^*/dt_{max}$. From Tables 9 and 10, we can note a significant reduction in end-diastolic volume (318 ± 63 ml vs. 206 ± 59 ml, p<0.01), end-systolic volume (228 ± 58 ml vs. 133 ± 61 ml, p<0.01), LV mass (204 ± 49 g vs. 187 ± 53 g, p<0.01), and peak normalized wall stress (PNWS) (4.64 ± 0.98 vs. 3.72 ± 0.87, p<0.01). Increased sphericity index SI (0.57 ± 0.094 vs. 0.67 ± 0.13, p<0.01) and increased shape factor (S) (0.44 ± 0.085 vs. 0.54 ± 0.089, p<0.01) during end-systole indicates that the LV became more spherical after SVR.

	Pre-SVR			Post-SVR		
	End diastolic volume (ml)	End systolic volume (ml)	LV mass (g)	End diastolic volume (ml)	End systolic volume (ml)	LV mass (g)
Intraobserver						
Mean	318 ± 63	228 ± 58	204 ± 49	206 ± 59	133 ± 61	187 ± 53
Mean difference	1.1 ± 8.60	-1.4 ± 8.38	-5.1 ± 7.96	0.3 ± 6.94	-1.5 ± 3.20	0.6 ± 8.51
Correlation coefficient	0.99	0.99	0.99	0.99	0.99	0.99
t-test p	N.S.	N.S.	N.S.	N.S.	N.S.	N.S.
% variability	1.50 ± 1.99	2.13 ± 2.45	3.74 ± 3.71	2.42 ± 2.32	1.79 ± 1.77	4.09 ± 3.02
Interobserver						
Mean	318 + 65	231 + 61	206 ± 52	207 ± 60	135 ± 62	189 ± 51
Mean difference	0.3 ± 10	7.9 ± 10	7.6 ± 14	2.0 ± 8.9	5.5 ± 10	3.1 ± 9.8
Correlation coefficient	0.99	0.99	0.97	0.99	0.99	0.98
t-test p	N.S.	N.S.	N.S.	N.S.	N.S.	N.S.
% variability	2.73 ± 1.58	4.11 + 2.93	5.90 ± 3.13	3.44 ± 2.94	5.87 ± 6.23	4.81 ± 2.49
N.S., not significant. Data are mean ± SD.						

Table 9. Reproducibility data in patients pre- and post-SVR. This table is related to our work in Ref [19].

The prime effect of SVR may be viewed as: (i) effecting a decrease in myocardial oxygen consumption by reduction of LV peak normalized wall stress, resulting in improved functioning of LV, and (ii) augmentation of value of the contractility index $d\sigma^*/dt_{max}$ (2.69 ± 0.74 s^{-1} vs. 3.23 ± 0.73 s^{-1}, p<0.01). This improvement may be attributed to (i) increased maximal flow dV/dt_{max} with reduced LV mass, and (ii) improved regional contraction and contractility of the remote myocardium. The improvement in remote myocardial performance is likely due to reduced myocardial stress, along with effective and complete revascularization. This is because the SVR procedure reduces the volume by more dramatically reducing long-axis dimension compared with the short-axis dimension, and producing a more spherical ventricle.

Based on Table 10, increased LV contractile function $d\sigma^*/dt_{max}$ can be not only associated with increased maximal flow dV/dt_{max}, reduced LV mass, and also increased maximal change rate of shape factor dS/dt_{max} (r=0.414, p<0.001). There was also good correlation between $d\sigma^*/dt_{max}$ and LVEF (r=0.69, p<0.001, pre-SVR; r=0.77, p<0.001, post-SVR) (Figure 15).

Variables	Pre SVR (n=40)	Post SVR (n=40)
Cardiac index (L/min/m^2)	2.84 ± 0.74	2.59 ± 0.74
Mean arterial pressure (mmHg)	85 ± 14	84 ± 8
Systolic blood pressure (mmHg)	115 ± 20	113 ± 10
Diastolic blood pressure (mmHg)	71 ± 12	70 ± 8
End diastolic volume index (ml/m^2)	156 ± 39	110 ± 33*
End systolic volume index (ml/ m^2)	117 ± 39	77 ± 31*
Stroke volume index (ml/ m^2)	39 ± 9	33 ± 8*
Left ventricular ejection fraction (%)	26 ± 7	31 ± 10*
LV mass index (g/m^2)	112 ± 25	101 ± 23*
End-diastolic long axis, BA$_{ed}$(cm)	10.89 ± 1.16	8.31 ± 1.00*
End-diastolic short axis, AP$_{ed}$(cm)	7.00 ± 0.80	6.64 ± 0.78*
End-systolic long axis, BA$_{es}$(cm)	10.37 ± 1.20	7.87 ± 1.05*
End-systolic short axis, AP$_{es}$ (cm)	5.86 ± 0.98	5.23 ± 1.06*
End-diastolic sphericity Index, SI$_{ed}$	0.65 ± 0.087	0.81 ± 0.11*
End-systolic sphericity index, SI$_{es}$	0.57 ± 0.094	0.67 ± 0.13*
Difference between end-diastolic and end-systolic sphericity index, SI$_{ed}$ - SI$_{es}$	0.077 ± 0.043	0.14 ± 0.059*
Long axis shortening (%)	4.8 ± 3.6	5.4 ± 4.4
Short axis shortening (%)	16.4 ± 6.8	22 ± 9.7*
dV/dt_{max} (ml/s)	364 ± 83	401 ± 81*
Pressure normalized wall stress	4.30 ± 0.95	3.31 ± 0.75*
Stroke work (mmHg·L)	6.61 ± 1.96	5.46 ± 1.64*
$d\sigma^*/dt_{max}$ (s^{-1})	2.69 ± 0.74	3.23 ± 0.73*
*p < 0.05. Values are mean ± SD.		

Table 10. Patients' data pre- and post-SVR. This table is related to our work in Ref [19].

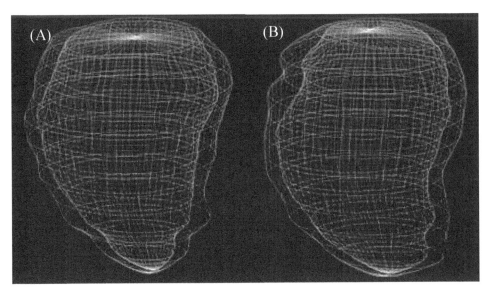

(a)Pre-SVR ED & ES

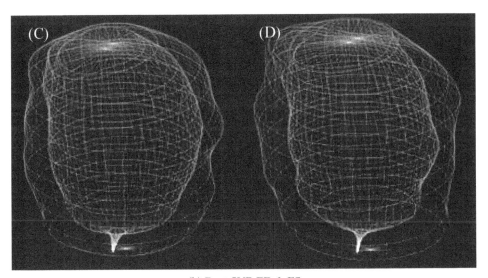

(b) Post-SVR ED & ES

Fig. 13. 3-dimensional reconstructions during end-diastole (panels A and C) and end-systole (panels B and D) phases before (panels A and B) and after (panels C and D) SVR using LVtools. It is created from the endocardial and epicardial contours, which were drawn for calculations of ventricular volumes and function from the multiple short-axis cines (Figure 12). This figure is based on our work presented in Ref. 19.

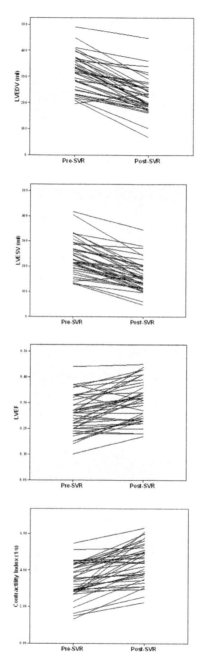

Fig. 14. Changes in end-diastolic volume (EDV), end-systolic volume (ESV), LVEF and contractility index $d\sigma^*/dtmax$ after SVR. This figure is adopted from our work in Ref. 19.

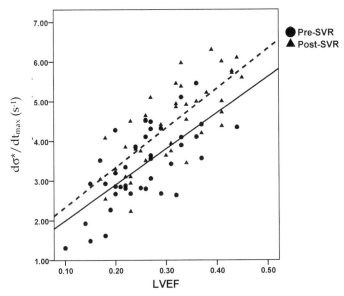

Fig. 15. Association between $d\sigma^*/dt_{max}$ and left ventricular (LV) ejection fraction (EF) pre- and
post-SVR. (Solid line: $d\sigma^*/dt_{max}=9.045\times EF+1.091$, r=0.69, p<0.001 for pre-SVR; dash line:
$d\sigma^*/dt_{max}=9.969\times EF+1.337$, r=0.77, p<0.001 for post-SVR). This figure is adopted from our
work in Ref 19.

8. Conclusion

Myocardial infarct patients' hearts undergo substantial remodeling and loss of contractility,
and can progress to heart failure. Echocardiographic texture analysis enables us to
determine the percentage volume of the infarcted segments in the ventricular volume.

The ventricular remodelling is defined in terms of reduced alteration in curvedness index
(%ΔC) from end-diastole to end-systole. The cardiac contractility is assessed in terms of the
value of the index $d\sigma^*/dt_{max}$.

With the help of these three indices of percentage volume of MI segments, curvedness index
and contractility index, we can track and assess LV progression to heart failure, and its
recovery following CABG and SVR.

9. References

[1] Detection of myocardial scars in neonatal infants form computerised echocardiographic
    texture analysis, by M.V Kamath, R.C Way, D.N. Ghista, T.M. Srinivasan, C. Wu, S
    Smeenk, C. Manning, J. Cannon, *Engineering in Medicine*, Vol. 15, Vo.3, 1986

[2] A Geometrical Approach for Evaluating Left Ventricular Remodeling in Myocardial
    Infarct Patients,Yi Su, Liang Zhong, Chi-Wan Lim, Dhanjoo Ghista, Terrance Chua,
    Ru-San Tan, *Computer Methods and Programs in Biomedicine*, In Press, 2011

[3] Zhong L, Tan RS, Ghista DN, Ng EYK, Chua LP, Kassab GS (2007), Validation of a novel
    noninvasive cardiac index of left ventricular contractility in patients. *Am J Physiol
    Heart CircPhysiol*, 292(6):H2764-H2772.

[4] Zhong L, Ghista DN, Ng EYK, Lim ST.Passive and active ventricular elastances of the left ventricle. Biomed Eng Online4: 10, 2005.

[5] Shishido T, Hayashi K, Shigemi K, Sato T, Sugimachi M, Sunagawa K. Single-beat estimation of end-systolic elastance using bilinearly approximated time-varying elastance. *Circulation* 102: 1983–1989, 2000.

[6] Liang Zhong, Kian-Keong Poh, Li-Ching Lee, Thu-Thao Le, Ru-San Tan, Attenuation of Stress-based Ventricular Contractility in Patients with Heart Failure and Normal Ejection Fraction,Annals, Academy of Medicine, Singapore, April 2011, Vol 40, No 4

[7] Hogg K, Swedberg K, McMurray J. Heart failure with preserved left ventricular systolic function: epidemiology, clinical characteristics, and prognosis. *J Am Coll Cardiol.*2004;43:317-327.

[8] Little WC. The left ventricular (dP/dt)max-end-diastolic volume relation in closed-chest dogs. *Circ Res.* 1985;56:808-815.

[9] Vasan RS, Levy D. Defining diastolic heart failure: a call for standardized diagnostic criteria. *Circulation.* 2000;101:2118-2121.

[10] Devereux RB, Alonso DR, Lutas EM, Gottlieb GJ, Campo E, Sachs I, Reichek N. Echocardiographic assessment of left ventricular hypertrophy: comparison to necropsy findings. *Am J Cardiol.* 1986;57:450-458.

[11] Lang RM, Bierig M, Devereux RB, Flachskampf FA, Foster E. Recommendations for chamber quantification. *J Am SocEchocardiogr.* 2005;18:1440-1463.

[12] Kitzman DW, Gardin JM, Gottdiener JS, et al. Importance of heart failure with preserved systolic function in patients > or =65 years of age. CHS Research Group Cardiovascular Health Study. *Am J Cardiol.* 2001;87:413-419.

[13] Banerjee P, Banerjee T, Khand A, et al. Diastolic heart failure: neglected or misdiagnosed? *J Am CollCardiol.* 2002;39:138-141.

[14] Burkhoff D, Maurer MS, Packer M. Heart failure with a normal ejection fraction: is it really a disorder of diastolic function? *Circulation.* 2003;107:656-658

[15] Zhang, J.M., Chua, L.P., Ghista, D.N., Yu, S.C.M. and Tan, Y.S., Numerical investigation and identification of susceptible site of atherosclerotic lesion formation in a complete coronary artery bypass model, *J. of Medical and Biological Engineering and Computing,* 2008, Vol. 46 (7), pp. 689-699.

[16] Sankaranarayanan, M., Chua L. P., Ghista D. N. and Tan Y. S., Computational model of blood flow in the aorta-coronary bypass graft, *BioMedical Engineering Online,*2005,4:14.

[17] Chua, L.P., Zhang, J.M., Yu, S.C.M., Ghista, D.N. and Tan, Y.S., Numerical study on the pulsatile flow characteristics of proximal anastomotic model, *J. of Engg. in Medicine, Proceedings of the Institution of Mechanical Engineers Part H,* 2005, Vol. 219, pp 361-379.

[18] Xiong, F.L., and Chong, C.K., A parametric numerical investigation on haemodynamics in distal coronary anastomoses, *Medical Engineering and Physics,* 2008, Vol. 30(3), pp. 311-320.

[19] Zhong L, Sola S, Tan RS, Ghista DN, Kurra V, Navia JL, Kassab GS. Effects of surgical ventricular restoration on left ventricular contractility assessed by a novel contractility index in patients with ischemic cardiomyopathy. *Am J Cardiol* 103: 674-679, 2009.

[20] Streeter Jr DD, Hanna WT. Engineering mechanics for successive states in canine left ventricular myocardium. I Cavity and wall geometry.*Circ Res.* 1973;33:639-655.

[21] Dhanjoo N. Ghista, *Applied Biomedical Engineering,* CRC Press, 2009

11

Lung Ventilation Modeling for Assessment of Lung Status: Detection of Lung Disease and Indication for Extubation of Mechanically-Ventilated COPD Patients

Dhanjoo N. Ghista[1], Kah Meng Koh[2], Rohit Pasam[3] and Yi Su[4]
[1]Department of Graduate and Continuing Education, Framingham State University,
Framingham, Massachusetts,
[2]VicWell BioMedical,
[3]Quodient, Inc,
[4]Institute of High Performance Computing, Agency for Science, Technology and Research,
[1,3]USA
[2,4]Singapore

1. Introduction

In pulmonary medicine, it is important to detect lung diseases, such as chronic obstructive pulmonary disease (COPD), emphysema, lung fibrosis and asthma. These diseases are characterized in terms of lung compliance and resistance-to-airflow. Another important endeavour of pulmonary medicine is mechanical ventilation of COPD patients and determining when to wean off these patients from the mechanical ventilator. In both these medical domains, lung ventilation dynamics plays a key role.

So in this chapter, we develop the lung ventilation dynamics model in terms of monitored lung volume (V) and driving pressure (P_L), in the form of a differential equation with parameters of lung compliance (C) and resistance-to-airflow (R). We obtain the solution of this equation in the forms of lung volume (V) function of P_L, C and R. For the monitored lung volume V and pressure P_L data, we can evaluate C and R by matching the model solution expression with the monitored lung volume V and driving pressure P_N data. So what we have done here is to develop the method for determining lung compliance (C) and resistance-to-airflow (R) as average values of C (= C_a) and R (= R_a) during the ventilation cycle.

Now in some lung diseases such as in emphysema, the lung compliance (C) is high. In other lung diseases such as in asthma, the airway resistance (R) is high. So we need to determine the ranges of C and R for normal lung status as well as for lung disease states. Then, we can develop a 2-parameter R-C diagnostic coordinate plane, on which we can demarcate the zones for different diseases. Then, for any patient's (R, C) value, we can plot the (R, C) point in the R-C diagnostic coordinate plane, and based on its location designate the lung disease state of the patient. A more convenient way for detecting lung disease is to combine R and C along with some ventilator data (such as tidal volume and breathing rate) into a non-dimensional lung ventilator index (LVI). Then, we can determine the ranges of LVI for normal and disease states, and thereby employ the patient's computed values of LVI to designate a specific lung disease for the patient. The LVI concept for detecting lung disease is more convenient to adopt in clinical practice, because it enables detection of lung disease states in the form of just one lung-ventilation number.

Now, in this methodology, we need to monitor (i) lung volume, by means of a spirometer, and (ii) lung pressure (P_L) equal to P_{mo} (pressure at mouth) minus pleural pressure (P_p). The pleural pressure measurement involves placing a balloon catheter transducer through the nose into the esophagus, whereby the esophageal tube pressure is assumed to be equal to the pressure in the pleural space surrounding it. Now this procedure cannot be carried out non-traumatically and routinely in patients. Hence, for routineand noninvasive assessment of lung ventilation for detection of lung disease states, it is necessary to have a method for determining R and C from only lung volume data. So, then, we have shown how we can compute R, C and lung pressure values non-invasively from just lung volume measurement.

Finally, we have presented how the lung ventilation modeling can be applied to study the lung ventilation dynamics of COPD patients on mechanical ventilation. We have shown how a COPD patient's lung C and R can be evaluated in terms of the monitored lung volume and applied ventilatory pressure. We have also formulated another lung ventilator index to study and assess the lung status improvement of COPD patients on mechanical ventilation, and to decide when they can be weaned off mechanical ventilation.

2. Lung ventilation model

2.1 Scope
In this section, we have developed a lung ventilation model by modeling the lung volume response to mouth minus pleural driving pressure (by means of a first order differential equation) in terms of resistance-to-airflow (R) and the lung compliance (C). The lung volume solution of the differential equation is matched with the clinical volume data, to evaluate the parameters, R and C. These parameters' values can help us to distinguish lung disease states, such as obstructive lung and lung with stiffened parenchyma, asthma and emphysema.

2.2 Role of lung ventilation
Lung ventilation constitutes inhalation of appropriate air volume under driving pressure (= mouth pressure – pleuralpressure), so as to: (i) provide adequate alveolar O_2 amount at

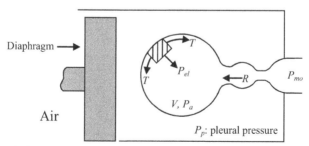

Fig. 1. Lumped Lobule Lung Model. In the figure, P_a is the alveolar pressure; P_{mo} is the pressure at the mouth; P_p is the pleural pressure; $P_{el} = P_a - P_p = 2h\sigma/r = 2T/r$, the lung elastic recoil pressure; r is the radius of the alveolar chamber and h is its wall thickness; T is the wall tension in the alveolar chamber; V is the lung volume; R is the resistance to airflow; and C is the lung compliance. This figure is adopted from our work in Ref [1].

appropriate partial pressure, (ii) oxygenate the pulmonary blood, and (iii) thereby provide adequate metabolic oxygen to the cells. Hence, ventilatory function and performance assessment entails determining how much air volume is provided to the alveoli, to make available adequate alveolar oxygen for blood oxygenation and cellular respiration.

In this lumped lobule lung model [1], we have (i) a lumped alveolar chamber of volume V and pressure P_a, and (ii) lumped airway having airflow resistance R. In this airway, the pressure varies from P_{mo} at the mouth to P_a in the alveolar chamber. The pleural pressure is P_p.

2.3 Lung ventilation analysis (using a linear first-order differential equation model)

We first analyze Lung Ventilation function by means of a model represented by a first-order differential equation (De_q) in lung-volume (V) dynamics in response to the driving pressure P_L (= mouth pressure − pleural pressure). In this model [2], the lung lobes and the alveoli are lumped into one lung lobule, as depicted in Figure 1. Figure 2 displays typical data of lung volume and flow, alveolar and pleural pressure.

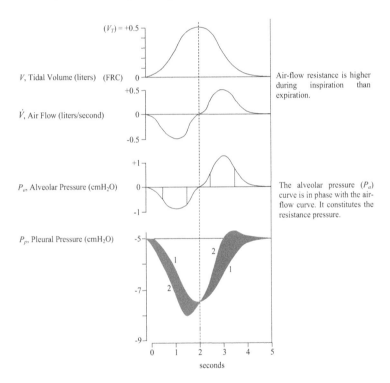

Fig. 2. Lung ventilatory model and lung-volume and pleural-pressure data. In the bottom figure, Curve 1 represents the negative of P_{el}, the pressure required to overcome lung elastance (=1/C) plus elastic recoil pressure at the end of expiration. Curve 2 represents $P_p = -P_{el} + P_a$. Now, as can be noted from Figure 1, $P_a - P_p = P_{el}$. The lung driving pressure $P_L = P_{mo} - P_p$, and the net driving pressure $P_N(t)$ in Equation (1-b) equals P_L minus P_{el} at end-expiration. We define resistance-to-airflow (R) as $(P_{mo} - P_a) / \dot{V}$. We define lung compliance $C = V/(P_{el} - P_{el0}) = V/(P_a - P_p) - P_{el0}$. This figure is adopted from our work in Ref [1].

Based on Figures 1 and 2, we can put:

(i) $(P_a - P_p) - P_{el} = 0$

(ii) $P_{el} = (2\sigma h) / R = 2T / R = V/C + P_{el0}$ (at end – expiration)

(iii) $(P_{mo} - P_a) = R(dV / dt)$

(iv) $P_L = P_{mo} - P_p = (P_{mo} - P_a) + (P_a - P_p)$

(v) $R(dV / dt) + V/C = P_L - P_{el0}$ (lung elastic recoil pressure at end – expiration)

$$\left. \phantom{\begin{matrix}1\\1\\1\\1\\1\end{matrix}} \right\} \text{(1-a)}$$

The lung ventilation model governing equation is hence formulated as:

$$R\dot{V} + \frac{V}{C} = P_L(t) - P_{el0} = P_N(t)$$

$$\text{(1-b)}$$

wherein:

i.   the values of pressure $P_N(t)$ are obtained from the P_L $(= P_{mo} - P_p)$ data relative to P_{el0}

ii. the parameters of this Governing De_q are lung compliance (C) and airflow-resistance (R); in the equation both R and C are instantaneous values

iii. $V = V(t) - V_0$ (the lung air volume at the end-expiration = lung air volume inspired and expired during a single breath)

iv. P_{el0} is the lung elastic-recoil pressure at the end of expiration, and

$$P_{el0} = P_{el} - \frac{V}{C}$$

$$\text{(1-c)}$$

v.   At end-expiration when $\omega t = \omega T,$ $P_L = P_{el0}$

Now, in order to evaluate the lung model parameters C and R, we need to simulate this governing equation to lung volume (V) and pressure (P_N) data. This clinical data is shown in Figure 3. The lung volume is measured by integrating the airflow velocity-time curve, where the airflow velocity can be measured by means of a ventilator pneumatograph; the lung volume can also be measured by means of a spirometer. Inhalation and exhalation pressures are measured by means of a pressure transducer connected to the ventilatory tubing; likewise, a pressure transducer can also be similarly connected to the spirometer tubing. The pleural pressure is measured by placing a balloon catheter transducer through the nose into the esophagus; it is assumed that the esophageal tube pressure equals the pressure in the pleural space surrounding it.

In Equation (1-a), we have put $P_N(t) = \sum_{i=1}^{3} P_i \sin(\omega_i t + c_i)$, expressed as a Fourier series. The governing equation (1-b) now becomes:

$$R\dot{V} + \frac{V}{C} = P_N(t) = \sum_{i=1}^{3} P_i \sin(\omega_i t + c_i)$$

$$\text{(2-a)}$$

where the right-hand side represents the net driving pressure minus pleural pressure: $P_N = (P_{mo} - P_p) - P_{el0}$. This P_N is in fact the driving pressure $(P_{mo} - P_p)$ normalized with respect to its value at end-expiration. Equation (2-a) can be rewritten as follows:

$$\dot{V} + \frac{V}{RC} = \frac{1}{R}\sum_{i=1}^{3} P_i \sin(\omega_i t + c_i) \tag{2-b}$$

wherein the $P(t)$ clinical data (displayed in Figure 3) is represented by:

$$P(t) = \sum_{i=1}^{3} P_i \sin(\omega_i t + c_i) \tag{3}$$

P_1= 1.581 cmH$_2$O	P_2 = -5.534 cmH$_2$O	P_3 = 0.5523 cmH$_2$O
ω_1 = 1.214 rad/s	ω_2 = 0.001414 rad/s	ω_3 = 2.401 rad/s
c_1 = -0.3132 rad	c_2 = 3.297 rad	c_3 = -2.381 rad

The pressure curve (in Figure 3) represented by the above Equation (3) closely matches the pressure data of Figure 3. If, in Equation (1), we designate R_a and C_a as the average values (R and C) for the ventilatory cycle, then the solution of Equation (2) is given by:

$$V(t) = \sum_{i=1}^{3} \frac{P_i C_a[\sin(\omega_i t + c_i) - \omega_i R_a C_a \cos(\omega_i t + c_i)]}{1 + \omega_i^2 (R_a C_a)^2} - He^{-\frac{t}{R_a C_a}} \tag{4}$$

wherein the term $(R_a C_a)$ is denoted by τ_a. We need to have $V = 0$ at $t = 0$. Hence, putting V (at $t = 0$) = 0, gives us:

$$H = \sum_{i=1}^{3} \frac{P_i C_a[\sin(c_i) - \omega_i R_a C_a \cos(c_i)]}{1 + \omega_i^2 (R_a C_a)^2} \tag{5}$$

Then from Equations (4) and (5), the overall expressions for $V(t)$ becomes

$$V(t) = \sum_{i=1}^{3} \frac{P_i C_a[\sin(\omega_i t + c_i) - \omega_i \tau_a \cos(\omega_i t + c_i)]}{1 + \omega_i^2 \tau_a^2} - \sum_{i=1}^{3} \frac{P_i C_a[\sin(c_i) - \omega_i \tau_a \cos(c_i)]}{1 + \omega_i^2 \tau_a^2} e^{-\frac{t}{\tau_a}} \tag{6}$$

We also want that $dV/dt = 0$ at $t = 0$, implying no air-flow at the start of inspiration. So then by differentiating Equation (6), we get the expression for air-flow ($\dot{V}$), as:

$$\dot{V} = \sum_{l=1}^{3} \frac{P_i C_a[\omega_i \cos(\omega_i t + c_i) + \omega_i^2 \tau_a \sin(\omega_i t + c_i)]}{1 + \omega_i^2 \tau_u^2} + \sum_{i-1}^{3} \frac{P_i C_a[\sin(c_i) - \omega_i \tau_a \cos(c_i)]}{(1 + \omega_i^2 \tau_u^2)\tau_u} e^{-\frac{t}{\tau_a}} \tag{7}$$

For the above values of τ_a= 0.485 s and for ω_i and $c_i$given by Equation (3), we get $\dot{V}(t = 0) = \sum_{i=1}^{3} (P_i / R_a)\sin(c_i) \approx 0$, to satisfy the initial condition.

Now by matching the above $V(t)$ expression in Equation (6) with the $V(t)$ data in Figure 3, and carrying out parameter-identification, we can determine the in vivo values of C_a, R_a and τ_a to be:

$$C_a = 0.218\, L(\text{cmH}_2\text{O})^{-1}, R_a = 2.275(\text{cmH}_2\text{O})\text{sL}^{-1}, \tau_a = 0.485\, s \tag{8}$$

The computed $V(t)$ curve, represented by Equation (6) for the above values of C_a and R_a, is shown in Figure 3.

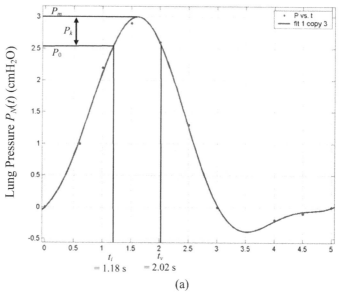

(a)

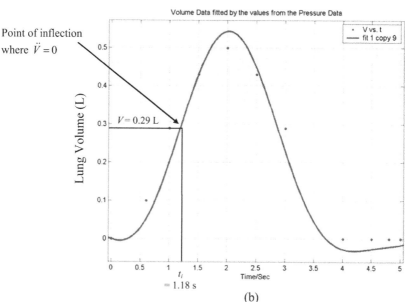

(b)

Fig. 3. (a)The pressure curve represented by Equation (3) matched against the pressure data (represented by dots). (b) The volume curve represented by Equation (6), for $C_a = 0.2132$ L(cmH$_2$O)$^{-1}$ and $R_a = 2.275$ (cmH$_2$O)sL^{-1}, matched against the volume data represented by dots. In Figure 3(a), the terms P_0, P_m and P_k refer to Equation (11). At $t = t_v$, V is maximum and $\dot{V}$ is zero. This figure is adopted from our work in Ref [1].

Let us have some validation of the average values of C and R obtained by parameter-identification scheme, by determining the values of C and R at some specific time instants. For that purpose, we can put down from Equation (1-b),

$$R\ddot{V} + \frac{\dot{V}}{C} = \dot{P}_N(t) \tag{9}$$

Now the volume (V) curve in Figure 3(b) has an inflection point at $t = t_i = 1.18$ s, at which $\ddot{V} = 0$. At $t = 1.18$ s, $V = 0.29$ L, $\dot{V} = 0.48$ Ls^{-1}, $P_N = 2.53$ cmH$_2$O, and $\dot{P}_N = 1.66$ (cmH$_2$O)s^{-1}. Upon substituting these values into Equation (9), we get $C = 0.289$ L(cmH$_2$O)$^{-1}$. Then substituting this value of C along with the values of V, $\dot{V}$ and P_N into Equation (1-a), we get $R = 3.18$ (cmH$_2$O)sL^{-1}. These values of C and R at $t = t_i = 1.18$ s are of the same order of magnitude as the average values of C and R given by Equation (8). This provides us a measure of confidence to our parameter-identification scheme for obtaining the average values C_a and R_a.

Now since Lung disease will influence the values of R and C, these parameters can be employed to diagnose lung diseases. For instance in the case of emphysema, the destruction of lung tissue between the alveoli produces a more compliant lung, and hence results in a larger value of C. In asthma, there is increased airway resistance (R) due to contraction of the smooth muscle around the airways. In fibrosis of the lung, the membranes between the alveoli thicken and hence lung compliance (C) decreases. Thus by determining the normal and diseased ranges of the parameters R and C, we can employ this simple Lung-ventilation model for differential diagnosis.

3. Non-dimensional ventilatory index

For disease detection, it is more convenient to formulate and employ a non-dimensional number to serve as a ventilatory performance index LVI (to characterize ventilatory function), as:

$$LVI_1 = \left[(R_a C_a)\left(\text{Ventilatory rate in s}^{-1} \right)60 \right]^2 = \tau_a^2 (BR)^2 60^2 \tag{10}$$

where BR is the breathing rate.

Now, let us obtain its order-of-magnitude by adopting representative values of R_a and C_a in normal and disease states. Let us take the above computed values of $R_a = 2.275$ (cmH$_2$O)sL^{-1} and $C_a = 0.2132$ L(cmH$_2$O)$^{-1}$ and $BR = 12$ m^{-1} or 0.2 s^{-1}, computed by matching Equation (6) to the data of Figure 3.

Then, in a supposed normal situation, the value of LVI_1 is of the order of 33.88. In the case of obstructive lung disease (with increased R_a), let us take $R_a = 5$ (cmH$_2$O)sL^{-1}, $C_a = 0.12$ L(cmH$_2$O)$^{-1}$ and $BR = 0.3$ s^{-1}; then we get $LVI_1 = 118.6$. For the case of emphysema (with enhanced C_a), let us take $R_a = 2.0$ (cmH$_2$O)sL^{-1}, $C_a = 0.5$ L(cmH$_2$O)$^{-1}$ and $BR = 0.2$ s^{-1}; then we obtain $LVI_1 = 144$. In the case of lung fibrosis (with decreased C_a), we take $R_a = 2.0$ (cmH$_2$O)sL^{-1}, $C_a = 0.08$ L(cmH$_2$O)$^{-1}$ and $BR = 0.2$ s^{-1}; then we obtain $LVI_1 = 3.7$.

We can, hence summarize that LVI_1 would be in the range of 2-5 in the case of fibrotic lung disease, 5-50 in normal persons, 50-150 in the case of obstructive lung disease and 150-200 for the case of emphysema,. This would of course be needed to be verified by analyzing a big patient population.

Now, all of this analysis requires pleural pressure data, for which the patient has to be intubated. If now we evaluate the patient in an outpatient clinic, in which we can only monitor lung volume and not the pleural pressure, then let us develop a non-invasive method for determining lung compliance (C), resistance-to-airflow (R) and ventilatory index.

4. For non-invasive assessment of lung status and determination of lung compliance and resistance-to-airflow

Our primary need is to be able to determine lung pressure $P_N(t)$ non-invasively. If we observe the $P_N(t)$ curve in Figure 3(a), we can note that, during the period of time from $t = t_i = 1.18$ s to $t = t_v = 2.02$ s, we can represent it as:

$$P = P_N = P_k \sin \omega_p (t - t_i) + P_0$$

$$(11\text{-}a)$$

$$\cong P_k \sin \omega_p (t - 1.18) + 2.5$$

$$(11\text{-}b)$$

where (i) $P_k = P_m - P_0$, and $P_k = 0.5$ cmH$_2$O in Figure 3(a), (ii) $t = t_i$, the inflection point on the lung volume curve (= 1.18 s in Figure 3(b)), and (iii) $P_0 = 2.5$ cmH$_2$O.

We can determine the value of ω_p, by invoking the condition that the pressure P becomes maximum (= P_m) at $t = t_m = (t_v + t_i)/2$. Hence, at $t = t_m$,

$$P_m = P_k \sin \omega_p (t_m - t_i) + P_0$$

But since $P_k = P_m - P_0$, we get

$$P_m - P_0 = (P_m - P_0) \sin \omega_p (t_m - t_i)$$

or

$$\sin \omega_p (t_m - t_i) = 1$$

wherein

$$t_m = (t_v + t_i)/2 = (2.02 + 1.18)/2 = 1.6$$

Therefore, we have

$$\omega_p (t_m - t_i) = \omega_p (1.6 - 1.18) = \pi/2 = 1.57$$

$$(12)$$

In Equation (12), $t_m = (t_v + t_i)/2$ and both t_v (= 2.02 s) and t_i (= 1.18 s) can be known from the lung volume curve. Hence, ω_p can be determined. For Figure 3 data, we get $\omega_p = 3.73$ rad/sec.

Hence, in our case of Figure 3 data, we can represent lung pressure P_N (= P) between t_i and t_v as:

$$P = P_k \sin \omega_p (t - t_i) + P_0$$

$$(13\text{-}a)$$

$$= 0.5 \sin 3.73 (t - 1.18) + 2.5$$

$$(13\text{-}b)$$

and

$$\dot{P} = 1.87\cos 3.73(t - 1.18) \tag{13-c}$$

So, in general, the parameters of the lung pressure curve are P_k and P_0, since t_i and ω_p can be determined in terms of t_v and t_i, as per Equation (12) and Figure 3. Likewise, we can also represent lung volume between t_i and t_v as:

$$V = V_T \sin \omega_v (t - t_i) + V_0 \tag{14-a}$$

where V_T is the tidal volume and V_0 is the lung volume at $t = t_i$. Based on Figure 3(b) data, we can rewrite Equation (14-a) as follows:

$$V = 0.25\sin \omega_v (t - 1.18) + 0.3$$

At $t = t_v = 2.02$ s, $V = V_T = 0.55$ L. Hence, $\omega_v(2.02 - 1.18) = 1.57$ (or $\pi/2$), so that $\omega_v = 1.87$ rad. So then Equation (14-a) can be written as:

$$V = 0.25\sin 1.87(t - 1.18) + 0.3 \tag{14-b}$$

$$\dot{V} = 0.47\cos 1.87(t - 1.18) \tag{14-c}$$

$$\ddot{V} = -0.88\sin 1.87(t - 1.18) \tag{14-d}$$

Now based on Equation (13), we can represent the governing lung ventilation model Equation (1-b) as

$$R\dot{V} + \frac{V}{C} = P_K \sin 3.73(t - 1.18) + P_0 \tag{15-a}$$

or as,

$$R\dot{V} + \frac{V}{C} = 0.5\sin 3.73(t - 1.18) + 2.5 \tag{15-b}$$

Let us employ this equation to determine the values of C and R at some specific points in the ventilation cycle. At $t = t_i$ (the inflection point) = 1.18 s, $\ddot{V} = 0$ Ls^{-2}, $\dot{V} = 0.48$ Ls^{-1} and $V = 0.3$ L. Now, we can differentiate Equation (15) as:

$$R\ddot{V} + \frac{\dot{V}}{C} = (P_K \omega_p)\cos 3.73(t - 1.18) \tag{16-a}$$

or as,

$$R\ddot{V} + \frac{\dot{V}}{C} = (0.5)(3.73)\cos 3.73(t - 1.18)$$

$$= 1.86\cos 3.73(t - 1.18) \tag{16-b}$$

Hence, from this Equation (16), we get:

$$\frac{\dot{V}(=0.48)}{C} = 1.86 \text{ , or } C = 0.258 \text{ L(cmH}_2\text{O)}^{-1}.$$

Then, upon substituting this value of C into Equation (15), we get:

$$R(0.48) + \frac{0.3}{0.258} = 2.5 \text{ , or } R = 2.79 \text{ (cmH}_2\text{O)sL}^{-1}.$$

Hence at $t = t_i = 1.18$ s, $C = 0.258$ L(cmH$_2$O)$^{-1}$ and $R = 2.79$ (cmH$_2$O)sL^{-1}.   (17)

Let us now evaluate C and R at $t = t_k = 1.6$ s, the time associated with the peak lung pressure. From Equations (15) and (16), we can put down:

$$R\dot{V} + \frac{V}{C} = 0.5\sin 3.73(1.6 - 1.18) + 2.5 = 3 \tag{18-a}$$

$$R\ddot{V} + \frac{\dot{V}}{C} = 1.86\cos 3.73(1.6 - 1.18) = 0 \tag{18-b}$$

At $t = t_k = 1.6$ s, we get from Equations (13) and (14), $V = 0.48$ L, $\dot{V} = 0.33$ Ls^{-1}, $\ddot{V} = -0.622$ Ls^{-2}, $P = 3$ cmH$_2$O and $\dot{P} = 0$ (cmH$_2$O)s^{-1}.

Substituting these values into Equations (18-a) and (18-b), we get:

$$0.33R + \frac{0.48}{C} = 3 \tag{19-a}$$

$$-0.622R + \frac{0.33}{C} = 0 \tag{19-b}$$

from which we obtain for $t = t_k = 1.6$ s,

$$C = 0.22\left(\text{cmH}_2\text{O}\right)\text{sL}^{-1}, R = 2.51\left(\text{cmH}_2\text{O}\right)\text{sL}^{-1} \tag{20}$$

Finally, let us evaluate C and R at $t = t_v = 2.02$ s. From Equation (15), we get:

$$R\dot{V} + \frac{V}{C} = 0.5\sin 3.73(2.02 - 1.18) + 2.5 \tag{21}$$

Now at $t = t_v = 2.02$ s, $V = 0.55$ L, $\dot{V} = 0$ Ls^{-1}. So then, from Equation (21), we obtain:

$$\frac{0.55}{C} = 2.5 \text{ , or } C = 0.22 \text{ (cmH}_2\text{O)sL}^{-1} \tag{22}$$

It can be noted that the values of C and R given by Equations (17), (20) and (22) are similar to their average values $C_a = 0.218$ L(cmH2O)$^{-1}$ and $R_a = 2.275$ (cmH2O)sL^{-1}. This lends a measure of confidence to our Equation (15), for which V and $\dot{V}$ are given by Equations (14-b) and (14-c).

Let us now proceed to how we can determine the values of lung pressure function parameters P_k and P_0 along with R and C from the monitored values of lung volume.

At $t = t_i = 1.18$ s, we have from Equation (15-a)

$$R\dot{V} + \frac{V}{C} = P_K \sin 3.73(t - 1.18) + P_0 \tag{23}$$

so that by substituting $V = 0.3$ L, $\dot{V} = 0.48$ Ls^{-1}, we get:

$$R(0.48) + \frac{0.3}{C} = P_0 \tag{24}$$

Also, from Equation (16), we get by substituting the values of $\dot{V} = 0.48$ Ls^{-1} and $\ddot{V} = 0$ Ls^{-2}

$$\frac{0.48}{C} = 3.73 P_k \tag{25}$$

At $t = t_m = 1.6$ s, we have from Equations (15) and (16) as well as by substituting the values of $V = 0.48$ L, $\dot{V} = 0.33$ Ls^{-1} and $\ddot{V} = -0.622$ Ls^{-2}, we get:

$$0.33R + \frac{0.48}{C} = P_k + P_0 = P_m \tag{26}$$

$$-0.622R + \frac{0.33}{C} = 0, \text{ or } R = 0.53/C \tag{27}$$

Then at $t = t_v = 2.02$ s, we get from Equation (15), along with $V = 0.55$ L, $\dot{V} = 0$ Ls^{-1},

$$\frac{0.55}{C} = P_0 \tag{28}$$

We hence have Equations (24), (25), (26), (27) and (28) to solve and determine the best values of the four unknowns R, C, P_k and P_0. For this purpose, we define the ranges of these four terms, as:
R: 2.1, 2.2, 2.3 (cmH$_2$O)sL^{-1}; C: 0.20, 0.21, 0.22 L(cmH$_2$O)$^{-1}$; P_k = 0.4, 0.5, 0.6 (cmH$_2$O); P_0 = 2.4, 2.5, 2.6 (cmH$_2$O)
Then, in order to satisfy these equations, we obtain for the best values of R, C, P_k and P_0, based on the solution in Appendix, as

$$R = 2.29(cmH_2O)sL^{-1}, \ C = 0.22L(cmH_2O)^{-1}, \ P_k = 0.58(cmH_2O), \ P_0 = 2.47(cmH_2O) \tag{29}$$

As can be noted, these values of C and R correspond to the average values of C and R given by Equation (8). This then lends credibility to our procedure for non-invasive determination of C and R, for lung disease detection. This procedure enables us to in fact determine lung pressure toward evaluation of C and R.
Now since this procedure enables us to determine maximum lung driving pressure $P_m = P_k + P_0$, we can also formulate the non-dimensional lung ventilatory index as:

$$LVI_2 = \frac{R}{C} \frac{(TV)^2}{(P_m)^2} (BR)(60)^2 \tag{30}$$

wherein BR is in s^{-1}. For our case, $R = 2.275$ (cmH$_2$O)sL^{-1}, $C = 0.2132$ L(cmH$_2$O)$^{-1}$, $P_m = 3$ cmH$_2$O, and $TV = 0.55$ L. This gives $LVI_2 = 25.8$. By using this LVI_2 index, we can expect its value to be of the order of 30 for normal subjects, 300 for COPD patients, 5 for emphysema patients, and 100 in the case of lung fibrosis.

Here again, we need to determine *LVI* for normal lung states as well as for different lung disease states. We can then compare which of the formulas (10) or (30) enable better separation of lung disease states from the normal state.

Comments related to values of the ranges of the parameters: Before proceeding to the next section, let us address the basis of providing the above indicated ranges of parameters. The lung ventilation volume and driving pressure curves in Figure 3 are for a normal case. By carrying out this procedure for other normal subjects, we can define and confirm the above mentioned normal ranges for these parameters, for obtaining their best values.

Now then how do we distinguish subjects with disorders, such as obstructive lung disease (with increased value of R_a), emphysema (with enhanced value of C_a), lung fibrosis (with decreased value of C_a)? This can be made out from the shape and values of the lung ventilation volume curve. So then by repeating this procedure for subjects with these disorders, we will be able to characterize the shapes of the lung ventilation curves for normal subjects and for prescribing appropriate ranges of the parameters, for obtaining the best values of these parameters.

5. Lung-status evaluation and indicators for extubation of mechanically-ventilated COPD patients

5.1 Introduction

In mechanically ventilated patients with chronic-obstructive-pulmonary-disease (COPD), elevated airway resistance and decreased lung compliance (i.e., stiffer lung) are observed with rapid breathing. The need for accurate predictive indicators of lung-status improvement is essential for ventilator discontinuation through stepwise reduction in mechanical support, as and when patients are increasingly able to support their own breathing, followed by trials of unassisted breathing preceding extubation, and ending with extubation.

For this reason, we have developed an easy-to-employ lung ventilatory index (LVI), involving the intrinsic parameters of a lung ventilatory model, represented by a first-order differential equation in lung-volume response to ventilator driving pressure. The LVI is then employed for evaluating lung-status of chronic-obstructive-pulmonary-disease (COPD) patients requiring mechanical ventilation because of acute respiratory failure.

5.2 Scope and methodology

We recruited 13 mechanically ventilated patients with chronic obstructive pulmonary disease (COPD) in acute respiratory failure [3]. All patients met the diagnostic criterion of COPD. The first attempt of discontinuation (or weaning off the ventilator) for every patient was made within a short duration (not exceeding 88 hours). The patients in the study were between the ages of 54-83 years. All the patients were on synchronized intermittent mandatory ventilation (SIMV) mode with mandatory ventilation at initial intubation. Based on the physician's judgment, the modes were changed for eventual discontinuation of mechanical ventilation. The time period for recording observations was one hour. For all purposes in this study, a successful ventilator discontinuation is defined as the toleration to extubation for 24 hours or longer and a failed ventilator discontinuation is defined as either a distress when ventilator support is withdrawn or the need for reintubation. Our LVI was then employed to distinguish patients who could be successfully weaned off the mechanical ventilator.

Hence, the scope of this section is that we have developed a lung-ventilatory index (LVI), based on a lung-model represented by a first-order differential equation in lung-volume dynamics to assess lung function and efficiency in the case of chronic-obstructive-pulmonary-disease (COPD) patients requiring mechanical ventilation because of acute respiratory failure. Herein, we have attempted to evaluate the efficacy of the LVI in identifying improving or deteriorating lung condition in such mechanically ventilated chronic-obstructive-pulmonary-disease (COPD) patients, and consequently if LVI can be used as a potential indicator to predict ventilator discontinuation. In our bioengineering study of 13 COPD patients who were mechanically ventilated because of acute respiratory failure, when their LVI was evaluated, it provided clear separation between patients with improving and deteriorating lung condition. Finally, we formulated a lung improvement index (LII) representative of the overall lung response to treatment and medication, and a parameter m that corresponds to the rate of lung improvement and reflects the stability of lung-status with time. This chapter is based on our previous chapter [3] in the book on Human Respiration (edited by V. Kulish and published by WIT Press) and other works on this subject [4-9].

5.3 Lung ventilation model

From a ventilatory mechanics viewpoint, the lungs can be considered analogous to a balloon, which can be inflated and deflated (passively). The gradient between the mouth-pressure (P_{mo}) and the alveolar pressure (P_{al}) causes respiration to occur. During inspiration, $P_m > P_{al}$, which causes air to enter the lungs. During expiration P_{al} increases, and is greater than P_{mo}; this causes the air to be expelled out of the lungs passively. These pressure differentials provide a force driving the gas flow. The pressure difference between the alveolar pressure (P_{al}) and pleural pressure (P_p) counter balances the elastic recoil. Thus the assessment of respiratory mechanics involves the measurement of flows, volumes (flow integrated over time) and pressure-gradients. The lung ventilation model (shown in Figure 1) is based on the same dynamic-equilibrium differential equation (Equation 1-b), expressing lung volume response to pressure across the lung, as:

$$R\dot{V} + \frac{V}{C} = P_L(t) - P_{el0} = B\sin(\omega t) \tag{31}$$

wherein:
i.   the total positive pressure across the lungs, $P_L = P_{mo} - P_p$, wherein P_p is determined by intubating the patient, and assuming that the pressure in the relaxed esophageal tube equals the pressure in the pleural space surrounding it.
ii   the parameters of the governing Equation (31) are lung compliance (C) and airflow resistance (R), with both R and C being instantaneous values
iii. $V = V(t) - V_e$ (wherein V_e is the end-expiratory lung volume)
iv. P_{el0} is the end-expiratory pressure
v.   the net driving pressure $P_N = P_L - P_{el0}$
Let B be the amplitude of the net pressure wave form applied by the ventilator, C_a be the averaged dynamic lung compliance, R_a the averaged dynamic resistance to airflow, the driving pressure P_L be given as $P_L = P_{el0} + B\sin(\omega t)$, and the net pressure P_N be given by $P_N = B\sin(\omega t)$, as depicted in Figure 4. The governing equation (31) then becomes:

$$R_a\dot{V} + \frac{V}{C_a} = P_N = B\sin(\omega t) \tag{32}$$

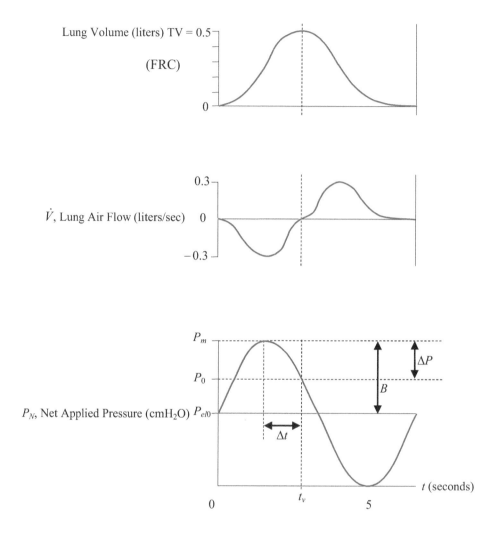

Fig. 4. Lung ventilatory model data shows air-flow ($\dot{V}$) and volume (V) and net pressure (P_N). Pause pressure (P_0) occurs at t_v, at which the volume is maximum (TV = tidal volume). Δt is the phase difference between the time of maximum volume and peak pressure (P_m). It is also the time lag between the peak and pause pressures. B is the amplitude of the net pressure waveform P_N applied by the ventilator. This P_N oscillates about P_{el0} with amplitude of B. The difference between peak pressure P_m and pause pressure P_0 is Δp. This figure is adopted from our work in Ref [3].

Lung Ventilation Modeling for Assessment of Lung Status: Detection of Lung Disease and Indication for Extubation of
Mechanically-Ventilated COPD Patients

261

The volume response to P_N (the solution to Equation (32)) is given by:

$$V(t) = \frac{BC_a[\sin(\omega t) - \omega k_a \cos(\omega t)]}{1 + \omega^2 k_a^2} + He^{\left(\frac{-t}{k_a}\right)}$$
(33)

wherein:

i. k_a (= $R_a C_a$) is the averaged time constant,

ii.   the integration constant H is determined from the initial conditions,

iii.   the model parameters are C_a and k_a (i.e., C_a and R_a), and

iv. ω is the frequency of the oscillating pressure profile applied by the ventilator

An essential condition is that the flow rate is zero at the beginning of inspiration and end of expiration. Hence, the flow rate $dV/dt = 0$ at $t = 0$. Applying this initial condition to our differential Equation (33), the constant H is obtained as:

$$H = \frac{BC_a \omega k_a}{1 + \omega^2 k_a^2}$$
(34)

Then, from Equations (33) and (34), we obtain:

$$V(t) = \frac{BC_a[\sin(\omega t) - \omega k_a \cos(\omega t)]}{1 + \omega^2 k_a^2} + \frac{BC_a \omega k_a}{1 + \omega^2 k_a^2} e^{\left(\frac{-t}{k_a}\right)}$$
(35)

For $t = t_v$, $V(t)$ is maximum and equal to the tidal volume (TV). Now in a normal person, k_a is of the order of 0.1 and 0.5 in ventilated patients with respiratory disorders, which is relevant to our study of COPD patients. At $t = t_v$ at which the lung volume is maximum, we note from Figure 4 that t_v is of the order of 2 s. Hence t_v/k_a is of the order 20-4, so that e^{-t/k_a} is of the order of e^{-20} to e^{-4}, which is very small and hence negligible. Hence, in Equation (35), we can neglect the exponential term so that,

$$V(t) = \frac{BC_a[\sin(\omega t) - \omega k_a \cos(\omega t)]}{1 + \omega^2 k_a^2}$$
(36)

Figure 4 illustrates a typical data of V, $\dot{V}$ and P_N. For evaluating the parameter k_a, we will determine the time at which $V(t)$ is maximum and equal to the tidal volume (TV), Hence, putting $dV/dt = 0$ in Equation (36), we obtain:

$$\cos(\omega t) + \omega k_a \sin(\omega t) = e^{\left(\frac{-t}{k_a}\right)}, \text{ for } t = t_v$$
(37)

Hence from Equation (37), we obtain the following expression for k_a:

$$\tan(\omega t) = -1 / \omega k_a, \text{ for } t = t_v$$
(38-a)

or,

$$k_a = -(1 / \omega)\tan(\omega t_v) \tag{38-b}$$

Since both ω and t_v are known, we can evaluate k_a from Equation (38-b).
Now from Equation (38-b), we can put down:

$$TV = \frac{BC_a[\sin(\omega t_v) - \omega k_a \cos(\omega t_v)]}{1 + \omega^2 k_a^2} \tag{39}$$

Since ω, t_v and k_a are known, we can now determine C_a in terms of TV and applied pressure amplitude B.
Then knowing k_a and C_a, we can determine

$$R_a = k_a / C_a \tag{40}$$

For our COPD patients, the ranges of the computed values of these parameters are:

$$R_a = 9 - 43(cmH_2 0)sL^{-1}; C_a = 0.020 - 0.080 \ L(cmH_2 0)^{-1} \tag{41}$$

Now that we have determined the expressions for the parameters R_a and C_a, the next step is to develop an integrated index lung ventilatory incorporating these parameters.

5.4 Formulating a Lung Ventilatory Index (*LVI*) incorporating R_a and C_a

We believe that the correlations between average airflow-resistance (R_a), average lung-compliance (C_a), tidal volume (TV), respiratory rate (RF), and maximum inspiratory pressure or peak pressure (P_m) can be used as a possible indicator for determining lung-status in a mechanically ventilated COPD patient with acute respiratory failure. We hence propose that a composite index (*LVI*), incorporating these isolated parameters, can have a higher predictive power for assessing lung status and determining when a patient on a mechanical ventilator.

For this purpose, we note that COPD patients have higher R_a, lower C_a, lower TV, higher P_m and higher respiratory rate (or breathing frequency) RF. If we want the non-dimensional lung-ventilatory index (*LVI*) to have a high value for a COPD patient, further increasing *LVI* for deteriorating lung-status and decreasing *LVI* for improving lung-status in a mechanically ventilated COPD patient in acute respiratory failure, then the non-dimensional lung-ventilatory index (*LVI*) can be expressed, as given by Equation (30):

$$LVI_2 = \left[\frac{R_a(TV)^2(RF)}{C_a(P_m)^2} \right] \times (60)^2 \tag{42}$$

where RF is the respiratory-rate frequency.
Let us obtain the order-of-magnitude values of this LVI_2 index for a mechanically ventilated COPD patient in acute respiratory failure (by using representative computed values of the parameters R_a, C_a, RF, TV, and P_m), in order to verify that the formula for LVI_2 (given by Equation (42)) can enable distinct separation of COPD patients in acute respiratory failure from patients ready to be weaned off the respirator. For an intubated COPD patient, we have

$$LVI_2 \text{ (Intubated COPD)} = \frac{[15(\text{cmH}_2\text{O})\text{sL}^{-1}][0.5\text{ L}]^2[0.33\text{ s}^{-1}]}{[0.035\text{ L}(\text{cmH}_2\text{O})^{-1}][20\text{ cmH}_2\text{O}]^2} \times (60)^2 = 318 \qquad (43)$$

wherein $R_a = 15\ (\text{cmH}_2\text{O})\text{sL}^{-1}$, $C_a = 0.035\ \text{L}(\text{cmH}_2\text{O})^{-1}$, $RF = 0.33\ \text{s}^{-1}$, $TV = 0.5\ \text{L}$ and $P_m = 20\ \text{cmH}_2\text{O}$.

Now, let us obtain the order-of-magnitude of LVI (by using representative computed values of R_a, C_a, RF, TV, and P_m) for a COPD patient with improving lung-status just before successful discontinuation. For a successfully weaned COPD patient (examined in an outpatient clinic), we have

$$LVI_2 \text{ (Outpatient COPD)} = \frac{[5(\text{cmH}_2\text{O})\text{sL}^{-1}][0.35\text{ L}]^2[0.33\text{ s}^{-1}]}{[0.10\text{ L}(\text{cmH}_2\text{O})^{-1}][12\text{ cmH}_2\text{O}]^2} \times (60)^2 = 50.5 \qquad (44)$$

wherein $R_a = 5\ (\text{cmH}_2\text{O})\text{sL}^{-1}$, $C_a = 0.10\ \text{L}(\text{cmH}_2\text{O})^{-1}$, $RF = 0.33\ \text{s}^{-1}$, $TV = 0.35\ \text{L}$ and $P_m = 12\ \text{cmH}_2\text{O}$.

Hence for LVI_2 to reflect lung status improvement in a mechanically ventilated COPD patient in acute respiratory failure, there should be a pronounced decrease in the value of LVI_2. This shows that the LVI given by Equation (42) can enable effective decision making to wean off a COPD patient from mechanical ventilator.

Appendix:

Solution procedure to obtain the best values of R, C, P_k and P_0 provided in Equation (29).
For the four unknowns R, C, P_k and P_0, where

R: 2.1, 2.2, 2.3 $(\text{cmH}_2\text{O})\text{sL}^{-1}$;
C: 0.20, 0.21, 0.22 $\text{L}(\text{cmH}_2\text{O})^{-1}$;
P_k: 0.4, 0.5, 0.6 (cmH_2O);
P_0: 2.4, 2.5, 2.6 (cmH_2O);

we want to find the best values of R, C, P_k and P_0 such that they satisfy the following equations:

$$R(0.48) + \frac{0.3}{C} = P_0 \qquad (24)$$

$$\frac{0.48}{C} = 3.73 P_k \qquad (25)$$

$$0.33R + \frac{0.48}{C} = P_k + P_0 \qquad (26)$$

$$-0.622R + \frac{0.33}{C} = 0 \qquad (27)$$

$$\frac{0.55}{C} = P_0 \qquad (28)$$

Solution:

We can rewrite Equations (26) to (30) so that the terms are collected at the LHS, i.e.,

$$R(0.48) + \frac{0.3}{C} - P_0 = 0 \tag{A-1}$$

$$\frac{0.48}{C} - 3.73 P_k = 0 \tag{A-2}$$

$$0.33R + \frac{0.48}{C} - P_k - P_0 = 0 \tag{A-3}$$

$$-0.622R + \frac{0.33}{C} = 0 \tag{A-4}$$

$$\frac{0.55}{C} - P_0 = 0 \tag{A-5}$$

Since we are trying to find the best values of R, C, P_k and P_0, the RHS of Equations (A-1) to (A-5) can be replaced by an error term so that

$$R(0.48) + \frac{0.3}{C} - P_0 = e_1 \tag{A-6}$$

$$\frac{0.48}{C} - 3.73 P_k = e_2 \tag{A-7}$$

$$0.33R + \frac{0.48}{C} - P_k - P_0 = e_3 \tag{A-8}$$

$$-0.622R + \frac{0.33}{C} = e_4 \tag{A-9}$$

$$\frac{0.55}{C} - P_0 = e_5 \tag{A-10}$$

In general, we can represent the overall error incurred at any values of R, C, P_k and P_0 by an objective function F so that

$$F(R,C,P_k,P_0) = \sum_{i=1}^{5} \|e_i\| \tag{A-11}$$

We can then find the best values of R, C, P_k and P_0 but solving Equation (A-11) as an optimization problem with the aim to minimize F, subjected to the bounded constraints:

$2.1 \leq R \leq 2.3$

$0.20 \leq C \leq 0.22$

$2.4 \leq P_0 \leq 2.6$

$0.4 \leq P_k \leq 0.6$

The optimal solution is obtained at $F = 0.239539$, and the associated best values of R, C, P_k
and P_0 are

$R = 2.29425$ cmH$_2$O)sL^{-1}

$C = 0.219758$ L(cmH$_2$O)$^{-1}$

$P_k = 0.57926$ (cmH$_2$O)

$P_0 = 2.46702$ (cmH$_2$O)   (A-12)

6. References

[1] Ghista D N, Loh K M, Lung Ventilation Modelling and assessment, in *Human Respiration: Anatomy and Physiology, Mathematical Modelling and Applications*, ed by Vladimir Kulish, WIT Press, 2006.

[2] Ghista D N, Sankaranarayanan M, Lung Ventilation Modeling for Lung Disease Diagnosis, in *Applied Biomedical Engineering Mechanics*, ed by Dhanjoo N. Ghista, CRC Press (Taylor & Francis Group), Boca Raton, Florida (US), 2009.

[3] Ghista D N, Pasam R, Vasudev S B, Bandi P, Kumar R V, Indicator for Lung-status in mechanically-ventilated COPD patients using Lung Ventilation Modeling and Assessment, in *Human Respiration: Anatomy and Physiology, Mathematical Modeling, Numerical Simulation and Application*, ed byVladimir Kulish, WIT Press (Southampton, UK), 2005.

[4] MacIntyre N R., Evidence-based Guidelines for Weaning and Discontinuing Ventilatory support by A Collective task Force of the American College of Chest Physicians, *Respiratory Care*, 47(1), 69-90, 2002.

[5] Menzies R, Gibbons W, Goldberg P., Determinants of weaning and survival among patients with COPD who require mechanical ventilation for acute respiratory failure, *Chest*, 95(2), 398-405, 1989.

[6] Yang KL, Tobin MJ., N Engl, A prospective study of indexes predicting the outcome of trials of weaning from mechanical ventilation, *J Med.*, 324(21), 1445-50, 1991.

[7] Alvisi R, Volta CA., Righini ER., Capuzzo M, Ragazzi R, Verri M, Candini G, Gritti G, Milic-Emili J, Predictors of weaning outcome in chronic obstructive pulmonary disease patients, *Eur Respir J.*, 15(4), 656-62, 2000.

[8] Pauwels RA., Buist AS., Calverley PM., Jenkins CR., Hurd SS., Global strategy for the diagnosis, management, and prevention of chronic obstructive pulmonary disease, NHLBI/WHO Global Initiative for Chronic Obstructive Lung Disease (GOLD) Workshop summary, *Am J Respir Crit Care Med.*, 163(5), 1256-76, 2001.

[9] Meade M, Guyatt G, Cook D, Griffith L, Sinuff T, Kergl C, Mancebo J, Esteban A, Epstein S, Predicting success in weaning from mechanical ventilation, *Chest.*, 120(6) Suppl, 400S-24S, 2001.

[10] Ubran A, Van de Graaff WB, Tobin MJ, Variability of patient-ventilator interaction with pressure support ventilation in patients with chronic obstructive pulmonary disease, *Am J Respiratory Crit Care Med.*, 152(1), 129-136, 1995.

Physiological Nondimensional Indices in Medical Assessment: For Quantifying Physiological Systems and Analysing Medical Tests' Data

Dhanjoo N. Ghista

Department of Graduate and Continuing Education,
Framingham State University, Framingham, Massachusetts,
USA

1. Introduction

1.1 Concept of Non-Dimensional Physiological Indices (NDPIs) or Physiological Numbers (PHYNs)

In this chapter, we are providing a new concept in physiological systems analysis (or organ systems analysis), in terms of nondimensional physiological indices (or physiological system numbers), for qualifying patient health and disease status as well as patient improvement.

The concept of a Nondimensional Physiological Index (or NDPI) is quite new, and has been adopted from Engineering, wherein nondimensional numbers (made up of several terms) are employed to characterize disturbance phenomena. For example, in a cardiovascular fluid-flow regime, the Reynold's number

$$N_{re} = \rho VD \Big/ \mu \qquad (1)$$

is employed to characterize the conditions when N_{re} exceeds a certain critical value, at which laminar blood flow changes to turbulent flow. This can occur in the ascending aorta when either the aortic valve is stenotic (giving rise to murmurs) or in the case of anaemia (decreased blood viscosity).

In physiological medicine, the use of nondimensional indices or numbers can provide a generalized approach by which unification or integration of a number of isolated but related events into one nondimensional physiological index (NDPI)can help to characterize an abnormal state associated with a particular organ or physiological system or an anatomical structure. The evaluation of the distribution of the values of such NDPI(s), in a big patient-population, can then enable us to designate normal and disordered ranges of NDPI, with a critical value of NDPI separating these two ranges. In this way, NDPI(s) can help us to formulate patient-health indices (PHIs), not only to facilitate differential diagnosis of patients but to assess the severity of the disease or disorder as well[1,2].

This chapter is based on the author's paper: Nondimensional Indices for Medical Assessment, in Mechanics in Medicine and Biology, Vol 9, No 4, 2009, World Scientific Publishers.

In medicine, assessment tests are carried out to (i) determine the functional performance of an organ (such as the heart) or a physiological system (such as the glucose-insulin regulatory system), and (ii) diagnose an anatomical structure's pathological condition, such as a calcified mitral valve or osteoporosis. In many cases, the medical tests do not quantifiably assess the concerned oral or physiological system and do not quantifiably diagnose the pathological condition of the anatomical structures.

So for some conventional tests (such as the Treadmill test to assess heart function, and Oral glucose tolerance test to diagnose diabetes), we have developed NDPI (s) made up of parameters that (i) are associated with the methodology of the tests, and (ii) characterise the function and disease states of organs (such as the heart) and physiological systems (such as the glucose-insulin regulatory system). We have also developed some new medical tests to detect anatomical structures' pathology (such as arteriosclerosis and mitral valve calcification) in terms of NDPI (s) to characterize their pathological state. In this chapter, we have formulated nine medical tests and their associated NDPI s).

We would like that the NDPI (s) developed in this chapter, for both conventional tests as well as newly formulated tests, can be applied clinically to set the stage for more accurate medical assessment. These medical tests and their associated NDPI (s) need to be applied to large patient population, to determine the normal and abnormal (or pathological) ranges of these NDPI (s). This will enable incorporation of our newly formulated NDPI (s) into medical practice.

2. Cardiac contractility index

Let us provide an example of one of our NDPI(s) which has been clinically employed. In cardiology, the index $(dP/dt)_{max}$ (of maximum rate-of-increase of left ventricular chamber pressure) has been traditionally employed as a measure of cardiac contractility. Diminished cardiac contractility affects cardiac output and can lead to heart failure. Hence, this is an important index of left ventricle (LV) functional capability. However, this index requires the invasive measurement of LV chamber pressure by catheterization. So, we developed an alternative cardiac contractibility index in terms of the normalised wall stress of the LV with respect to LV chamber pressure, $\sigma^* = \sigma/P$. Now, corresponding to $(dP/dt)_{max}$, we have formulated the cardiac contractility index (CCI) of $d\sigma^*/dt_{max}$, which does not require the measurement of LV chamber pressure. This contractility index can be conveniently expressed in terms of LV chamber cavity volume (V) and myocardial volume (V_m), as indicated in Ref 4. as well as in this section 5 of chapter 34 on how cardiac disease states cause decreased contractility and how surgical ventricular restoration improves contractility. By employing a thick-walled spherical LV model with inner and outer radii r_i and r_e, we can express LV pressure-normalised wall stress (σ^*) as [3,4]:

$$\sigma^*(r = r_i) = \left(\frac{3V}{2V_m} + \frac{1}{2} \right) \tag{2}$$

where $V(=4\pi r_i^3/3)$ denotes LV volume, and $V_m(=4\pi(r_e^3-r_i^3)/3)$ denotes LV myocardial volume. Then by differentiating σ^* with respect to time, we get the expression for CCI as:

$$CCI, d\sigma^*/dt_{max} = \left| \frac{d(\sigma_\theta/p)}{dt} \right|_{max} = \frac{3}{2V_m} \left| \left(\frac{dV}{dt} \right) \right|_{max} \tag{3a}$$

Physiological Nondimensional Indices in Medical Assessment: For Quantifying Physiological Systems and
Analysing Medical Tests' Data

269

At the National Heart Center (in Singapore Gerneral Hospital), we have validated do^*/dt_{max} against dP/dt_{max} in subjects with disparate ventricular function (as illustrated by figure 6 in chapter 34) and demonstrated the index's load independence. For normal subjects, this index value range is 4 - 4.5 s^{-1}. We have also successfully employed this $(do^*/dt)_{max}$ contractility index to assess (i) reduced LV contractility in ischemic cardiomyopathy patients in the range of 2. $64 \pm 0.74 s^{-1}$ and (ii) improved contractility in the range of 3.32 ± 0.73 s^{-1} following surgical ventricular restoration [5].

We could even divide this CCI index by the heart rate, and make it nondimensional, as:

$$CCI - 2 = (100) \, [do^*/dt_{max}] \, / \, HR \left(ins^{-1} \right) \tag{3b}$$

For normal hearts (with HR in the range of 60-80 per min), this index would give values in the range of 300-500, while for ischemic cardiomyopathic hearts (and HR greater than 100 per min) this index would be in the range of 200 and below. This CCI-2 index would even more reliably represent cardiac contractility, by more distinctly differentiating ischemic and infracted hearts from normal hearts.

3. Assessing cardiac fitness and heart function

In this section, we show how the conventional Treadmill test can be formulated in biomedical engineering terms, to derive a cardiac fitness index to detect a malfunctioning heart due to, for instance, an infarcted heart caused by coronary occlusion.

The Treadmill test's cardiac fitness model consists of a first-order differential-equation system models, describing the heart rate (HR) response (y) to exertion (exercise, jogging, etc.) monitored in terms of a constant work-load or power (W), where y is defined as follows [6,2]

$$y = \frac{HR(t) - HR(rest)}{HR(rest)} \tag{4}$$

In a Treadmill test, the subject is asked to exercise on the treadmill for a period of time, t_e(min). During this period, the $HR(t)$ (and hence y) is monitored. Now we develop a model to simulate (i) the $HR(t)$ response to during exercise, i.e. $t < t_e$ and (ii) thereafter for $HR(t)$ decay after the termination of exercise (as illustrated in figure 1).

The DEq for y response to exercise on the treadmill at a constant work-load or power exerted (W) is given by

$$\frac{dy}{dt} + k_1 y = C_0 W , \text{ for } t \le t_e$$

$$\frac{dy}{dt} + k_2 y = 0 , \text{ for } t \ge t_e \tag{5}$$

where (1) k_1 and k_2 are the model parameters, and (2) C_0 is a conversion factor to express W in the same units as the other terms of the equation. The y solutions to equations (5) are represented by:

$$y = \frac{C_0 W}{k_1} \left(1 - e^{-k_1 t} \right), \text{ for } t \le t_e \text{ during the exercise} = \frac{y_e \left(1 - e^{-k_1 t} \right)}{\left(1 - e^{-k_1 t_e} \right)} \tag{6}$$

$$y = y_e e^{-k_2(t-t_e)}, \text{ for } t \geq t_e \text{ during the recovery period when W=0} \qquad (7)$$

where $y_e = y(t=t_e)$, and k_1 and k_2 are the model parameters which can serve as cardiac fitness parameters (in units of min^{-1}).

We now carry out parametric simulation to the monitored HR data on treadmill, by making equations (6 and 7) fit the HR data. The parameters k_1 and k_2 can be combined into a non-dimensional fitness index CFI given by:

$$CFI = k_1 k_2 t_e^2 \qquad (8)$$

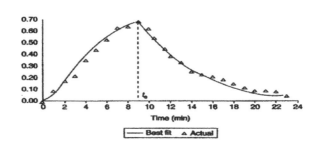

Fig. 1. Sample subject's monitored y versus t data (Adopted from Ref 6: Lim GeokHian, Dhanjoo N. Ghista, Koo TseYoong, John Tan Cher Chat, Philip EngTiew& Loo Chian Min; *International Journal of Computer Application in Technology: Biomedical Engineering & Computing Special Issue*, Vol 21, No 1/2, 2004.).

According to this formulation of *CFI*, for subjects exercised at identical workloads, a healthier subjects would have (1) greater k_1 (i.e., slower rate of increase of HR during exercise), (2) greater k_2 (i.e., faster rate of decrease of HR following exercise), (3) greater t_e(i.e., exercise endurance); and hence (4) higher value of *CFI*.

We need to evaluate *CFI* for a big spectrum of patients, and then compute its distribution curve, to determine the efficacy of this index, in order to yield distinct separation of *CFI* ranges for healthy subjects and cardiac patients. This *CFI* can then also be employed to assess improvement in cardiac fitness following cardiac rehabilitation regime.

4. Lung ventilation Index to detect lung disorders

Herein, we are formulating a new test involving: (i) monitoring of lung volume by means of a spirometer; (ii) the biomedical engineering model of the lung volume response to lung inflation driving pressure, which is equal to mouth pressure minus pleural pressure monitored by placing a balloon catheter transducer through the nose into the esophagus; (iii) derivation of the lung ventilation index made up of the parameters of the lung volume response model, and its employment to detect lung disorders.

The differential equation of lung ventilation volume (V) response to lung inflation-pressure (P_L) is given (as illustrated in figure 2) by [1, 2]:

$$R\frac{dV}{dt} + \frac{V}{C} = P_D = \left(P_0 - P_p\right) - \left|\left(P_p @ end - expiration\right)\right|$$
$$= P_1 - P_1 \cos \omega t + P_2 \cos \omega t \qquad (9)$$

Physiological Nondimensional Indices in Medical Assessment: For Quantifying Physiological Systems and
Analysing Medical Tests' Data

271

wherein V is the lung volume in litres (L), P_0 is the presence at the mouth(in cm H_2O) and P_p is the pleural pressure (in cm H_2O). The right-hand side terms constitute the fourier series representation of the lung driving pressure (P_D) in cm H_2O, R is the resistance to airflow (in cm $H_2O \cdot S \cdot L^{-1}$), C is the lung compliance (in L/cm H_2O), and P_1 and P_2 are the magnitudes of fourier series terms of the lung driving (oscillatory) pressure P_D [= (mouth-pressure minus pleural-pressure), with respect to the absolute value of end-expiratory pleural pressure]. For a typical P_D cyclic pressure profile (Fig.2), represented by

$$P_1 = 1.84\text{cm } H_2O, \qquad P_2 = 3.16\text{cm } H_2O, \qquad \omega = 0.5\pi s^{-1},$$

the solution to the above Eq. (9) is given by

$$V = P_1 C\left(1 - e^{-t/\tau}\right) - P_1 C\frac{\left(\cos\omega t + \omega\tau\sin\omega t\right)}{1 + \omega^2\tau^2} + P_2 C\frac{\left(\sin\omega t + \omega\tau\cos\omega t\right)}{1 + \omega^2\tau^2}$$
$$+ \frac{e^{-t/\tau}}{\left(1 + \omega^2\tau^2\right)}\left[P_1 C\left(1 + 2\omega^2\tau^2\right) + P_2 C\left(\omega\tau\right)\right] \tag{10}$$

wherein $\tau = RC$. By fitting this lung volume solution to the clinically monitored lung-volume parameters, we get: $R = 1.24$(cm H_2O) sL^{-1}, $C = 0.21L$(cm $H_2O)^{-1}$.
Now, then, let us formulate the nondimensional lung ventilator performance index (LVPI-1) given by:

$$\text{LVPI-1} = RC \text{ (BR per min)}, \tag{11}$$

wherein BR, the breathing rate $=30\omega/\pi$ per min $= 15/\text{min}$ or $0.25/\text{sec}$ for the data provided in Fig.2. For our case study, the value of $LVPI$ is 3.9.

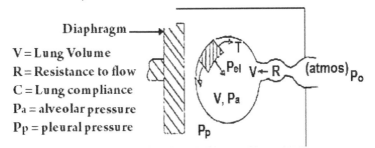

$$P_L - P_o - P_p$$
$$= \left(P_o - P_a\right) + \left(P_a - P_p\right) = \left(P_o - P_a\right) + P_{el}$$
$$= \left(R\frac{dV}{dt}\right) + \left(\frac{V}{C} + P_{el}@t = T\right)$$
$$= \left(R\frac{dV}{dt}\right) + \left(\frac{V}{C} - P_p@t = T\right)$$

Driving Pressure, $P_D = P_L - \left|P_p@t = T\right|$

Fig. 2. Lung Ventilation Lumped parameter model.

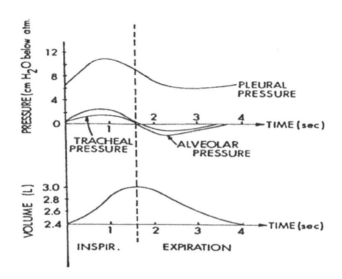

Fig. 3. Lung pressure and volume as functions for normal breathing; note that the pressure extremes occur before the volume extremes.

Let us now see how lung disease will influence R, C and hence $LVPI$. For instance in emphysema, the destruction of lung tissue will produce a more compliant lung and hence a larger value of C = 0.5L (cm $H_2O)^{-1}$, say, yielding a value of $LVPI$ of about 10. In asthma, there is increased airway resistance (due to contraction of the smooth muscles around the airways) to say R=5(cm H_2O) sL^{-1}. The breathing rate can go also go up to BR = 20/min, say. Hence, the value of $LVPI$ can go up to 20. In the case of lung-congestion, due to mitral-valve disease, it would be important to determine $LVPI$, so as to serve as an indicator for determining cardiac condition (in end-stage heart-disease). By determining the distribution of a big patient population, we can determine the $LVPI$ ranges for normal and disease states.

Now this procedure refers to monitoring of pleural pressure (P_P) by placing a balloon catheter transducer through the nose into the esophagus (assuming that the esophageal tube pressure equals the pressure in the pleural space surrounding it). So, let us now develop a procedure whereby we do not need to monitor pleural pressure and only need to monitor lung volume by means of a spirometer. For this purpose, we now identify three model parameters (P_1C), (P_2C) and τ. These parameters can be determined by having equation (10) match or stimulate the lung volume data in figure 3. We now formulate a non-invasively determinable nondimensional ventilator index ($LVPI$-2), as

$$LVPI-2 = \frac{(BR)\tau(TV)^2}{(P_1C)(P_2C)}, \quad TV = \text{tidal volume} \tag{12}$$

Upon evaluating $LVPI$-2 for a number of patients, we can determine its ranges for normal and disease states, to employ it diagnostically. We can expect that subjects with chronic obstructive lung disease (COPD) and asthma subjects (with a high value of R) will have a high value of $LVPI$ -2, while emphysema subjects (with high value of C) will have a low value of $LVPI$ -2. So in the distribution curve of $LVPI$ -2, emphysema subjects will be at the

Physiological Nondimensional Indices in Medical Assessment: For Quantifying Physiological Systems and
Analysing Medical Tests' Data

273

low end of distribution curve, COPD and asthma subjects will be at the high end of the distribution curve, while normal subjects will be in the middle of the distribution curve.

Now for noninvasive assessment of lung disease state in terms of lung compliance (C) and resistance-to-flow (R), we need to be able to determine lung pressure (P_D) function noninvasively. In section 4 of chapter 36 on lung ventilation modeling for assessment of lung station we have shown how we can determine the lung driving pressure functional parameters along with C and R in termo of the monitored values of lung volume.

We have also formulated a non-dimensional lung ventilatory index (equation 30) in terms of R, C, tidal volume (TV), lung pressure value at TV and breating rate. After this index is evaluated for different disease states, it will enable reliable noninvasure assessment of lung status.

5. Diabetes diagnosis from oral-glucose-tolerance test by means of a diabetes index (OGTT)

In this section, we have developed a biomedical engineering model for OGTT and demonstrated how it can be applied to OGTT data in (figure 4) to formulate and evaluate a diabetic NDPI, to more reliably diagnose diabetes.

For oral-glucose tolerance test simulation (entailing digestive and blood-pool chambers), the differential equation, and governing blood-glucose response (y) to oral ingestion of glucose bolus (G, gm L^{-1}hr^{-1}), given by[7]:

$$y'' + 2Ay' + \omega_n^2 y = G\delta(t) \ ; \ y \ in \ gL^{-1}, \ G \ in \ gL^{-1}hr^{-1}$$

$$y'' + \lambda T_d y' + \lambda y = G\delta(t) \tag{13}$$

where $\omega_n (= \lambda^{1/2})$ is the natural oscillation-frequency of the system, A is the attenuation or damping constant of the system, $\omega = (\omega_n^2 - A^2)^{1/2}$ is the (angular) frequency of damped oscillation of the system, $\lambda (2A/T_d = \omega_n^2)$ is the parameter representing regulation proportional to rate-of–change of glucose concentration (y), and λT_d is the parameter representing regulation proportional to rate-of-change of glucose concentration (y').

Figure 4 illustrates the OGTT data for typical normal and diabetic subjects. For an impulse glucose ingestion input, we can simulate a normal patient's blood-glucose concentration databy means of the solution of the Oral glucoseregulatory (second-order system) model, as an under-dampedglucose-concentration response curve, given by:

$$y(t) = \left(\frac{G}{\omega}\right) e^{-At} \sin \omega t, \tag{14}$$

When this solution is made to simulate the normal subjects OGTT data, we get $A = 1.4$ hr^{-1}, $\omega = 0.775$ rad/hr, $G = 1.04$ gL^{-1} hr^{-1}, $\lambda = 2.6$ hr^{-2}, $T_d = 1.08$ hr. The simulated curve is also depicted in figure 4.

For a potential diabetic subject, we adopt the solution of the above Differential equation model, as an over-damped response function:

$$y(t) = \left(\frac{B}{\omega}\right) e^{-At} \sinh \omega t. \tag{15}$$

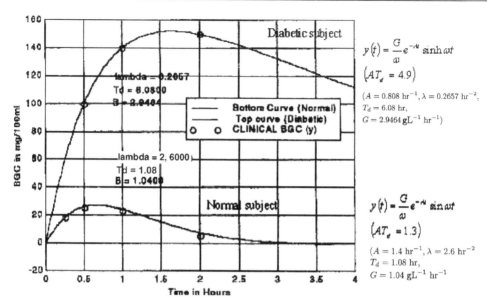

Fig. 4. OGTT Response Curves: $A = 1.4 \mathrm{hr}^{-1}$ (i.e. higher damping-coefficient value) for the normal subject for the diabetic patient, $A = 0.808 \mathrm{hr}^{-1}$. Also, for the normal subject, the regulation parameters λ and λT_d are 2.6 hr^{-2} and 2.8 hr^{-1} respectively, which are greater than their values of 0.26 hr^{-2} and 1.62 hr^{-1} for the diabetic subject. Further, the non-dimensional number for the normal subject is 1.3, compared to 4.9 for the diabetic subject.

For this OGTT data simulated function (figure 4), the parameters values are: $A = 0.808 \mathrm{hr}^{-1}$, $\omega = 0.622$ rad hr^{-1}, $G = 2.95 \mathrm{gL}^{-1} \mathrm{hr}^{-1}$, $\lambda = 0.266 \mathrm{hr}^{-2}$, $T_d = 6.08$ hr.

Now, we come to the interesting part of this model, by formulating the nondimensional Diabetes index (DBI) as:

$$DBI = AT_d = \frac{2A^2}{\lambda} = \frac{2A^2}{\omega_n^2}. \qquad (16)$$

The value of DBI for the normal subject is found to be 1.5, whereas that for the diabetic subject is 4.9. It is further seen (in our initial clinical tests) that DBI for normal subjects is < 1.6, while the DBI for diabetic patients is > 4.5. This is a testimony of the efficacy of the model, and especially for the nondimensionalDBI.

Now between these two cases of under-damped and over-damped responses, we have the case of a critically-damped response, for which the solution of the OGTT model differential equation (13) is given by

$$y(t) = Gte^{-At}, \text{ for which } \omega = 0, \text{ and } A^2 = \lambda = \omega_n^2$$

This critically-damped response corresponds to cases of subjects who are not distinctly normal or diabetic but are at the risk of becoming diabetic. It can be seen that DBI for the critically-damped response is 2. So, in the distribution curve of DBI (to be obtained by applying this method to a large patient population), the DBI range of less than 1.6 would

Physiological Nondimensional Indices in Medical Assessment: For Quantifying Physiological Systems and
Analysing Medical Tests' Data

275

refer to normal subjects, the range of greater than 4.5 would refer to diabetic subjects, and range of 2-4 would refer to subjects at risk of becoming diabetic. This would make the use of the model and the *DBI* to be so convenient for the physician.

6. Characterization of arterial stiffness or arteriosclerosis by means of NDI

In this section, we are formulating a new test to noninvasively determine the arterial constitutive property so as to be able to diagnose arteriosclerosis.

For a circular cylindrical arterial tube of radius a and wall-thickness h, we can express the stress σ and elastic-modulus E, as follows:

$$\sigma = \frac{Pa}{h} = \frac{130 Pa}{h} N / m^2; \quad E = \frac{2(PWV)^2 a\rho}{h}; \quad E = E_0 + m\sigma \tag{17}$$

in terms of (i) the arterial dimensions a and h, the auscultatory (or automatedly) measurable diastolic pressure (P) and pulse-wave velocity (PWV) determined by ultrasound [8]. The table below then depicts the computed values of σ and E at four independent times.

P (mmHg)	PMV (m/s)	A (mm)	h (mm)	$E\left[\dfrac{N}{m^2}\right]$	$\sigma\left[\dfrac{N}{m^2}\right]$
80	5.3	4.1	1.10	2.13×10^5	3.38×10^4
85	5.4	4.5	1	2.6×10^5	4.97×10^4
90	5.42	5.0	0.90	3.01×10^5	5.97×10^4
95	5.5	5.0	0.90	3.38×10^5	6.68×10^4

Table 1.

$$\text{Result}: \quad E(N/m^2) = 4.2\sigma + 0.5 \times 10^5 (N/m^2) = mo + E_0. \tag{18}$$

We will now define the arteriosclerotic non-dimensional index

$$ART - NDI = mE_0 / (\text{mean diastolic pressure}) \tag{19}$$

For the above patient, the value of the $ART - NDI$ is

$$ART - NDI = \frac{(4.2)(0.5 \times 10^5 \, N/m^2)}{(87 \times 137 \, N/m^2)} = 17.6 \tag{20}$$

and will be much higher for arteriosclerotic patients, which we will determine by conducting clinical tests-applications of this analysis. This *ART-NDI* to detect arteriosclerosis requires echocardiographic determination of arterial dimensions and PWV[8], and auscultatory diastolic pressure.

7. To non-invasively determine aortic elasticity (m), peripheral resistance (R), aortic NDI, and aortic pressure profile

Herein, we have developed the analysis to noninvasively determine the aortic pressure profile, which can have significant diagnostic applications. This analysis is also employed to determine (i) aortic volume elasticity parameter m $(=dP/dV)$, (ii) periphal resistance parameters $R=P$(pressure)$/Q$(flow rate), and (iii) the aortic property NDPI, given by aortic number. Based on the aorta fluid mechanics model (figure 5), we obtain:

$$\frac{dV}{dt} = I(t) - Q(t) = I(t) - P(t)/R \tag{21-a}$$

$$\frac{dP}{dt} = \frac{dP}{dV}\frac{dV}{dt} = m\frac{dV}{dt} \tag{21-b}$$

We can then put down the aortic pressure (P) response to aortic inflow-rate or LV outflow-rate $I(t)$ as follows [9,1] :

$$\frac{dP}{dt} + \lambda P = mI(t); \tag{22}$$

wherein, $m =$ Volume elasticity of aorta (in Pa/m³), and $\lambda = (m/R)$ in s⁻¹.
The LV outflow-rate is represented as:

$$I(t) = (A)\sin(\pi / t_s)t + (A / 2)\sin(2\pi / t_s)t, \quad 0 < t < t_s \ (systole)$$
$$= 0; \quad t > t_s \ (diastole) \tag{23}$$

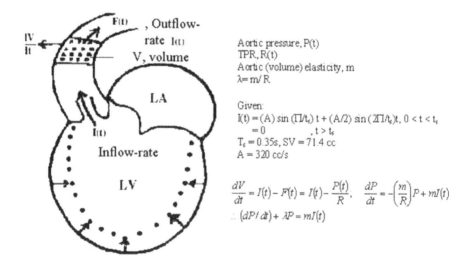

Fig. 5. To derive the equation for Aortic-pressure response to the stroke-volume or LV output rate $I(t)$.

If t_s= 0.35s, and Stroke vol(SV) is known (from, say, echocardiography), then we have

$$\int_0^{t_s}\left[(A)\sin(\pi/t_s)t + (A/2)\sin(2\pi/t_s)t\right] = SV \tag{24}$$

wherein $A = \pi\left(SV\right)/2t_s$

$$\text{So if } SV = 71.4cc, \text{ then } A = 320\,cc/s. \tag{25}$$

The solutions of Eq. (22) for the aortic diastolic and systolic periods are obtained as follows [9,1]:
Aortic Diastolic Pressure expression (fig 6):

$$P_d(t) = P_z e^{-\lambda(t-t_s)}; \quad P_2 = \text{ pressure at start of diastole}$$
$$= P_1\left(at\ t = T\right) = \text{ pressure at end of systolic phase} \tag{26}$$
$$\therefore P_d(t) = P_1 e^{\lambda(T-t)}, \quad \text{wherein } T = 0.8s.$$

This $P_d(t)$ function is equal to P_2 at $t = t_s$ (at the end of systolic phase) and P_1 at $t=T$ (end of diastotic phase).
Aortic Systolic Pressure expression (fig 6):

$$P_s(t) = \left(P_1 + \frac{Am\omega}{\lambda^2 + \omega^2} + \frac{2Am\omega}{\lambda^2 + 4\omega^2}\right)e^{-\lambda t} + mA\left(\frac{\lambda\sin\omega t - \omega\cos\omega t}{\lambda^2 + \omega^2}\right)$$
$$+ \frac{mA}{2}\left(\frac{\lambda\sin 2\omega t - 2\omega\cos 2\omega t}{\lambda^2 + 4\omega^2}\right); \quad \omega = \frac{\pi}{t_s}. \tag{27}$$

This $P_s(t)$ function is maximum at $t=t_m$, and equal to the monitored systolic auscultatory pressure $P_3 = 118\,mmHg$.
We now (i) incorporate into Eqs. (26) and (27) the auscultatory data of P_1(80mmHg) and P_3 (118mmHg) with $T = 0.8$ s, and us=0.35s, as well as (ii) invoke continuityin diastolic and systolic pressure expressions, to (iii) put down and solve the following three equations (in three unknowns: m, λ and t_m which $P_s = P_2$):

$$P_d\left(\text{at}_s = 0.35s\right) = P_s\left(\text{at}_s = 0.35s\right) \tag{28}$$

$$\frac{dP_s\left(\text{att}_m\right)}{dt} = 0 \tag{29}$$

$$P_s\left(t = t_m\right) = P_3\left(= 118mmHg\right) \tag{30}$$

to obtain: λ=0.66 s^{-1}, $m = 0.78$ mmHg cm^{-3}, $R = 1.18$mmHg cm^{-3} s, t_m= 0.25 s, for $T - 0.8$ s.
We now formulate the Aortic number (or index):

$$\text{Aortic number} = \lambda T = mT/R \tag{31}$$

whereinλ= m/R in the governing differential equation (22), m =103 x 10^6 Pa m^{-3}, R=157x 10^6 Pa m^{-3} s, and λ =0.66s^{-1}

We thereby obtain the $Aortic\ Number = \lambda T = \left(0.66s^{-1}\right)\left(0.8s\right) = 0.52.$ \tag{32}

In order to have a more convenient order-of-magnitude value of the Aortic number index (equation 31), we could employ the $Aortic\ number$ = 100 (λT). In the distribution of $Aortic$

number (obtained by applying this methodology to a large patient population), the low range of Aortic Number will correspond to patients with vasoconstriction, the high range of Aortic Number will associated with arteriosclerotic patients, and patients with normal healthy aorta will be in the middle of the distribution.

Finally with the help of the evaluated parameters m and R, we can now construct the aortic pressure profile based on equations (26 and 27), as illustrated in figure 6. This aortic pressure profile can have significant diagnostic implications. As we know, in Ayurvedic medicine and Chinese Traditional medicine, the physician feels the pulsation of the patients brachial artery (just proximal to the wrist), and based on it provides diagnosis of a wide spectrum of diseases. Essentially, the physician is feeling the magnitude and shape of the arterial pressure pulse.

Now, we have shown that we can in fact noninvasively determine the aortic pressure profile, which is more diagnostically indicative than the pressure pulse in the more distally located brachial artery. Hence, we can bring to bear this medical inferential and experiential knowledge to firstly characterise the magnitude and shape of the aortic pressure profile (by Fourier analysis), and then correlate the Fourier series parameters to the information about the associated disease states available from Ayurvedic and Chinese medicine systems.

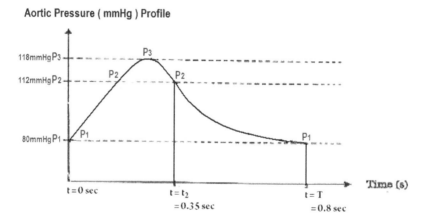

Fig. 6. Illustration of the Aortic pressure profile, based on the analysis. The systolic phase is from t = 0 to t = ts = 0.35s. The diastolic phase is from t = ts = 0.35s to t = T= 0.8s.

8. Mitral-Valve (MV) property determination for its pathology characterization (to provide interventional guidelines)

Determining the in-vivo constitutive property of the mitral valve (for a quantifiable estimate of its calcification and degeneration) constitutes another example combining "clinical-data monitoring and processing" with "modelling-for-clinical diagnosis".

The mitral valve opens at the start of the diastolic phase when the blood from the left atrium fills the left ventricle (LV). At the end of the diastolic phase and at the initiation of LV contraction phase, the rising LV pressure and the blood flow pattern in the LV chamber brings the valve cusps together to close the MV, and set its cusps into vibratory motion, which is monitored as the First Heart sound (FHS).

Physiological Nondimensional Indices in Medical Assessment: For Quantifying Physiological Systems and
Analysing Medical Tests' Data

279

From a biomedical engineering consideration, the mitral valve in its closed position (at the end of the diastolic phase) can be modelled as a semi-circular membrane, which is fixed along its circular edge to the heart chamber wall and supported along its straight edge by the chordae tendineae (as depicted in figure 7), so that its deflection is zero along its edges [10,11]. In this configuration, the MV vibrates after its cusps come together to close the valve. The frequency of MV vibration (f_{mv}) (as obtained from the FHS frequency spectrum) can be expressed in terms of the MV constitutive parameter property (E vso), which conveys information about its health state and pathology. This methodology provide a more reliable and quantitative approach for detecting a pathological MV (such as owing to its leaflets calcification) than by merely listening to the First Heart sound (aspractised clinically).

To this end, we provide the expressions for determining MV stress (σ) and its elastic modulus (E) from the physiological data of the MV vibration its closed configuration. We then develop expressions for mitral valves modulus-based property E^* and stress-based property σ *, and propose that the E^* vso * relationship be employed to characterise mitral valve pathology. Alternatively, we can also track mitral valve pathological deterioration, by monitoring the changes in valve of m (= E/σ) in terms of the changes in f_{mv} as the valve pathology progresses, and determine the time for intervention of replacing the pathological MV by means of a prosthetic MV.

We make use of:

- echocardiography (to determine the mitra-valve geometry) and spectral phonocardiography (of the first-heart sound associated with MV vibration), to determine the second-peak frequency (f_2) of the first heart-sound spectrum
- along with static and dynamic (vibration) analyses of the semi-circular mitral valve leaflet model (held along its circular boundary), as illustrated in Fig. 7, to obtain the following expressions [10 & 11], for

$$\text{Stress (}\sigma\text{) in the leaflet membrane} \qquad = \frac{\pi_2 f_2^2 a^2 \rho}{\left(K_{nm}/2\right)^2}, \qquad (33)$$

wherein a is the radius of the semi-circular leaflet; ρ is the leaflet membrane density per unit area; K_{nm} is the mth zero of the nth order Bessel function $J_n(k)$, m (number of nodal circles = 1, and (number of nodal diameters)=1, and k_{11} = 3.832.

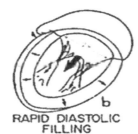
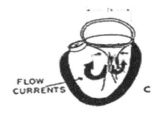

RAPID DIASTOLIC FILLING

Fig. 7. The mechanism of MV closure and subsequent vibration, in the genesis of the first-heart sound. Functional mechanics of the Mitral valve: (a) mitral valve opening at start of left centricular diastole; (b) as the filling left ventricle distends, traction is applied through the chordae tendineae to the valve cusps, pulling them together; (c) at the start of LV systole, the valve cusps are sealed together by the high internal pressure and the flow pattern in the ventricular chamber. It is at this point in time that the mitral valve starts vibrating.

Modulus (E) of the leaflet membrane $= \dfrac{\pi^8 f_2^2 \rho^3 t^2 a^4 (1-v)}{\left(K_{11}/2\right)^6 q_0^2 S_n}$, (34)

Wherein t = leaflet thickness, v is the Poisson's ratio, q_o is the pressure difference across the leaflet at time of occurrence of the closed Mitral-valve (MV) Vibration, and S_n (the summation of a series) = 0.105.

Based on eqs. (33) and (34), the nondimensional constitutive parameter (m) of the MV, is given by

$$m = \frac{E}{\sigma} = \frac{3\pi^6 f_2^4 \rho^2 t^2 a^2 (1-v)}{q_0^2 S_n \left(K_{11}/2\right)^4}$$ (35)

As a matter of interest, for the data: $f_2 = f_{mv} = 100\mathrm{Hz}$, $q_0 = 2$ mm Hg, $\rho = 1.02$ gm/cm^3, $a = 1$ cm, $t = 0.5$ mm, $v = 0.5$, and evaluating S_n (= 0 · 0234, Eq 2 · 16, Ref.2), we get $\sigma = 2.75 \times 10^3$ N/m^2 and E=1.6x10^5 N/m^2.

Now, changes in MV pathology will affect its density(ρ) and thickness (t), as well as its modulus (E) vs stress (σ) property which we want to determine by combining FHS power-spectrum analysis (to determine f_{mv}) and 2-d echocardiographic analysis (to determine the size parameter a).

We now designate a new stress-based property (σ^*) of MV (from equation 33), as

$$\sigma^* = \frac{\sigma}{\rho} = \frac{\pi^2 f_{mv}^2 a^2}{(1.916)^2}$$ (36)

as well as a new modulus –based property (E^*) of MV (from equation 34), as

$$E^* = \frac{E q_0^2}{\rho^3 t^2} == \frac{\pi^8 f_{mv}^6 a^4 (1-v)}{(1.916)^2 S_n}$$, (37)

We can now employ this E^* vs. σ^* relationship as a constitutive property of MV, to characterize and track its degeneration for timely intervention purpose.

This technology and methodology can provide the basis for timely surgical and/ or replacement intervention for a diseased MV. In order to apply this analysis, we can determine the valvular leaflet size parameter (a) from 2-D echocardiograms. The valvular leaflet vibrational frequency (f_{mv}) can be obtained from the frequency spectra of the FHS phonocardiographic signal associated with MV movement.

We can study a number of patients and determine the in vivo (E^*,σ^*) values of their valves, at a regular intervals during their degeneration process. We can also simultaneously and regularly monitor cardiac symptoms and chamber sizes and correlate them with the valcular constitutive E^*- σ^* property. By means of these correlations, we can determine the critical (E^*- σ^*) boundary at which intervention will have to be made to replace the degenerated natural valve by means of a prosthetic flexible leaflet MV.

In an alternative somewhat simpler approach, the mitral valve constitutive property parameter m (equation 35) can be employed diagnostically to track the deterioration due to calcification of the MV, in terms of Δm according to the relationship:

$$\Delta m = \left(\partial m / \partial f_{mn}\right) \Delta f_{mn} + \left(\partial m / \partial q_0\right) \Delta q_0; \text{ where in } f_{mv} = f_2$$ (38)

Physiological Nondimensional Indices in Medical Assessment: For Quantifying Physiological Systems and
Analysing Medical Tests' Data

281

$$\text{so that: } \frac{\Delta m}{m} = 4\left(\frac{\Delta f_{mn}}{f_{mn}}\right) - 2\left(\frac{\Delta q_0}{q_0}\right), \tag{39}$$

$$\text{or, } \frac{m'(=m+\Delta m)}{m} = 1 + 4\left(\frac{\Delta f_{mn}}{f_{mn}}\right) - 2\left(\frac{\Delta q_0}{q_0}\right) \tag{40}$$

Now at the time of occurrence of the first heart sound (FHS), the differential pressure or loading (q_0) across the mitral valve is very small. Hence the change in pressure loading valve (Δq_0), over the period of time of patient-tracking, will also be small compared to Δf_{mv}, and hence can be neglected in equation (40).
Hence from Eq. (39), we can compute

$$\frac{\Delta m}{m} = \left(\frac{4\Delta f_m}{f_m}\right) \tag{41}$$

to represent the change (Δm) in the parameter (m), by merely monitoring the change in frequency (Δf_{mv}) with respect to its earlier value (f_{mv}).

9. Noninvasive determination of bone osteoporosis index (in terms of bone flexural stiffness) for osteoporosis detection

Osteoporosis is a metabolic bone disease that is characterised by decreased bone mineral content and associated decreased in its mechanical strength. Thus, the osteoporotic bone is more prone to fracture.
Noninvasive measurement methods for osteoporosis detection include single and dual beam photon absorptiometry and a comparatively low cost low-frequency mechanical vibration (resonance and impedance) method [1, 12]. The low-frequency impedance response curve of (the first bending mode of) ulna yields the resonant frequency (f_r) value, which can be formulated in terms of the mechanical properties of the ulna bone, namely its bending stiffness (EI) and mass (M). It has been found that the difference between normal and osteoporotic bone is 20% in resonant frequency (f_r) and 80% in bending stiffness EI [13]. This is because f_r is the ratio of bone stiffness (EI) to mass (M), and in the pathologic osteoporotic condition both stiffness and mass decrease. Also, it has been shown that in fresh canine bone, the bending moment causing fracture has a correlation with EI of r =0.96 and with bone mineral content of r=0.90. Thus based on these results it appears that the ulna bending stiffness EI is a good indicator of bone fracture strength, which is diminished in osteoporosis. Now, both EI and M are contained in the expression for the natural frequency (f_r) of ulna vibrations, which in turn can be obtain from its resonance frequency [14]. In order to determine the resonance frequency of the ulna bone, it can be simply supported at its extremities and a vibrating probe pressed against the skin at the center of the forearm (as carried out by Steel and Gordon [14] and schematised in Fig. 8). The resonance frequency f_r (= natural frequency f_n of vibration of ulna) is obtained from the recording of the acceleration response as a function of the frequency.
If the bone is vibrating at an angular frequency p, the weight of the ulna bone per unit length is w radius of the ulna is R, and its length is ℓ, then the natural frequency f_n of the vibrating ulna beam, with its mass concentrated in the middle is given by :

$$f_n = p / 2\pi = \sqrt{(g / \delta st)} / 2\pi \qquad (42)$$

where δ_{st}, the maximum central deflection of the simply supported ulna bone, is given by

$$\delta_{st} = w\ell^4 / (77EI) \qquad (43)$$

wherein w is the ulna weight per unit length.
Hence, from (42) and (43), we get

$$f_n = \frac{1}{2\pi}\left(\frac{77gEI}{wl^4}\right)^{1/2} \qquad (44)$$

By putting $w = \rho A g$ ρ: bone density, A: cross-sectional area, we get

$$f_n = \frac{1}{2\pi}\left(\frac{77EI}{\rho Al^4}\right)^{1/2} = 1.4\left(\frac{EI}{Ml^3}\right)^{1/2}, \qquad (45)$$

where M is the mass of the ulna bone.
By altering the frequency of the vibrating probe, we set the ulna into resonance, and the resonance frequency will be equal to the natural frequency. For f_r resonance frequency $= f_n =$ 400Hz, $A = 50 \times 10^{-4} m^2$, $I = 3 \times 10^{-8} m^4$, length $(\ell) = 0.17m$, $\rho = 1.8 \times 10^3$ kg/m^3, we get $E = 20 \times 10^9$ N/m^2 and $EI = 30$Nm^2.
Thus, from equation (45), by modeling the ulna bone as a simply-supported vibrating beam, and determining its natural transverse-vibrational frequency (equal to its measured resonance frequency f_r), we can measure its flexural stiffness EI, to detect osteoporosis.

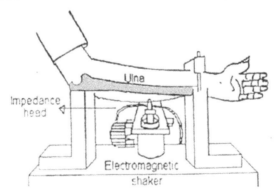

Fig. 8. Set-up used by Steel and Gordon [14] to determine the impedance of ulna. In this set up, the impedance head is attached to the moving element in the shaker. The probe, which contracts the skin, is attached to the impedance head.

10. Cardiac assessment based on Myocardial infarct detection and Intra-ventricular flow and pressure determination

In cardiology, a primary disorder is that of a heart with infarcted myocardium. This infarcted myocardial wall mitigates adequate contraction of the wall. So, the end-result of an

infarcted left ventricle (LV) is poor intra-LV velocity distribution and pressure-gradient
distribution, causing impaired outflow from the LV into the aorta.

In the infarcted myocardial segments, the myocardial infrastructure of actin and myosin
filaments (and their cross – bridges) is disrupted, and hence there is no contraction within
these infarcted myocardial segments. Figure 9 [15] illustrates a myocardial sarcomere
segment's bioengineering model, composed of two symmetrical myocardial structural units
(MSUs). In these MSU(s), the contractile elements represent the actin-myosin contractile
components of the sarcomere segment.

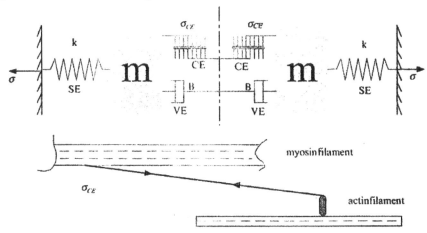

Fig. 9. Based on the conventional Hill three-element model and Huxley cross bridge theory, we
have developed a myocardial model involving the LV myocardial mass, series-elastic element
(CE). In this figure we have linked the anatomical associations of these myocardial model
elements with microscopic structure of the heart muscle. This figure illustrates the sarcomere
element contractile model, involving: the effective mass (m) of the muscle tissue that is
accelerated; elastic parameter k of the series element stress σ_{SE} (k = elastic modulus of the
sarcomere) viscous damping parameter B of the stress σ_{VE} in the parallel viscous element VE,
the generated contractile stress σ_{CE} between myosin (thick) and actin (thin) filaments.

The disruptions of these contractile elements impairs the contractile capability of that
sarcomere segment. Hence, a LV with infarcted myocardial segments will have diminished
contractility, inadequate and improper intra-LV flow, and poor ejection.

Detection of myocardial infarcted segments: Now, infarcted myocardial segments can be
detected as highly reflectile echo zones (HREZs) in 2- dimensional B-scan echocardiograms.
In this context, we have shown earlier [16] how infarcted myocardial segments can be
detected (in shape and size), by echo-texture analysis, as highly reflectile echo zones or
HREZs. Now, each tissue component of the heart generates a grey scale pattern or texture
related to the tissue density and fibrous content, and hence tissue stiffness.

In diseased states (such as myocardial ischemia, myocardial fibrosis, and infiltrative
diseases), changes in myocardial tissue stiffness have been recognised by employing echo
intensity and mean grey level of pixel as the basis for recognition of such myocardial
disorders. It was found that hyper- reflectile echoes (HREs) correlated well with diseased
cardiac muscle, and that myocardial tissue containing HREs corresponded with foci of sub
endorcardial necrosis and even calcification.

In our earlier study [15], in order to determine highly reflectile echo zones (HREZs), echocardiograms were recorded; each image was made up of 256 x 256 pixels, with each pixel having a resolution of 0-256 grey scales. The echocardiographic images were digitised into 256 grey scales. Then, echo intensity levels from normal infants were used to delineate the range of echo intensities for normal tissues. The upper bound of the echo intensity was set to 100 per cent in each normal infant, and the intensities from the rest of the image was referenced to this level. Normally, pericardium had the highest intensity level. It was found that the upper- bound of the echo intensity value for healthy tissue (expressed as a percentage of the pericardial echo intensity value) was 54.2.

Patient (Sex)	Region A	Region B	Region C	Region D	HRE and its location
B (M)	M: 167.44	54.76	51.02	82.20	105.74
	SD: 25.00	28.2	17.71	24.68	30.88
	N: 65	84	75	31	65
	P: 100	32.7	30.5	49.1	63.1
					Septum
P (F)	148.76	61.73	79.81	61.7	108.18
	26.78	23.02	22.05	24.2	13.03
	50	75	47	49	40
	100	41.5	53.8	41.5	72.6
					Septum
Br (M)	141.65	68.3	69.3	33.93	89.412
	29.56	26.8	24.8	24.4	28.0
	40	40	49	44	79
	100	41.5	53.8	41.5	73.1
					Septum
F (F)	157.34	50.1	60.8	53.8	112.1
	30.0	29.5	18.8	22.7	10.3
	35	45	49	44	31
	100	31.8	38.6	34.2	71.2
					R. ventricle
HI (M)	168.1	54.7	58.2	62.4	96.4
	21.35	21.8	16.9	20.0	14.7
	47	36	35	37	49
	100	32.5	34.6	37.1	57.3
					L. ventricle
G (M)	117.7	46.9	45.5	42.7	85.3
	20.6	19.0	20.6	19.1	22.6
	45	44	40	49	37
	100	39.8	38.7	36.2	72.5
					R. ventricle

A = Posterior Pericardium, B = Anterior Myocardium, C = Posterior Myocardium, D = Septum

Table 1. Echo intensity values for various anatomic regions of diseased pediatric hearts (based on long axis view). The numbers in the four rows represent Mean (M), Standard Deviation (SD), Number of Pixels (N), Percentage of Posterior Pericardial Intensity (P).

For patients whose echo-texture analysis showed presence of HREs, it was found that the echocardiographic intensities of the HREs from these patients intensities), were distinctly higher than the echo intensity range of normal tissue (as depicted in Table 1).

Figure 10(a) depicts an echo image of an infant with visible scars regions 1 and 2, while figures 10(b) depict printouts of the echo intensities from these two regions, wherein the infarcted segments are depicted in dark colour.

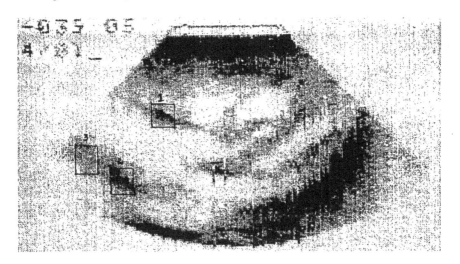

Fig. 10. (a) Long axis view of a pediatric patient's heart showing HRE regions 1 and 2 and a healthy region 3.(Adopted from the author's paper Ref 16).

Y/X	99	100	101	102	103	104	105	106	107	108	109	110	111	112	113	114
98	79	78	88	90	99	96	102	108	91	77	92	86	135	122	73	55
99	114	115	101	114	126	128	114	116	119	126	82	68	84	103	78	57
100	151	137	125	128	136	135	133	134	149	137	91	75	74	73	82	83
101	175	177	171	151	144	143	154	147	138	142	139	139	126	64	76	71
102	202	196	174	125	192	193	183	164	131	131	125	132	92	89	81	116
103	139	143	183	193	206	217	233	248	209	146	116	102	111	113	117	116
104	147	136	143	178	203	251	250	265	229	203	75	71	92	82	88	95
105	208	110	132	151	230	223	227	249	255	255	230	210	104	87	81	112
106	84	104	88	121	147	184	227	239	255	255	252	247	220	125	76	70
107	83	110	108	122	135	175	194	183	206	228	211	255	255	184	141	131
108	68	92	122	131	145	147	149	151	237	181	189	222	241	178	190	167
109	56	76	91	122	132	127	145	142	154	150	156	156	195	190	206	190
110	76	63	96	96	82	83	109	120	142	128	133	141	153	181	192	191
111	59	57	63	66	70	103	106	118	96	94	86	110	129	150	96	66
112	58	60	59	57	58	61	71	77	106	89	91	92	200	147	97	85
113	74	71	78	60	56	58	57	62	71	70	79	83	78	92	67	76
114	57	57	65	63	57	56	63	56	51	56	58	80	85	78	67	55
115	51	60	63	63	58	67	56	57	54	59	57	58	59	76	68	81

Fig. 10. (b) Pixel values corresponding to highly reflectile echo region 1. The central region having echo-intensity values greater than 200 is infarcted. (Adopted from the author's paper, Ref 16).

Myocardial tissue pixels having echo-intensity values greater than 200 were designated to be infarcted. This infarcted sub-region is seen to be surrounded by an ischemic sub-region whose pixels have echo intensity values between 100 and 200. The surrounding healthy tissue has echo intensity less than 100.

In this way, in each highly reflectile echo zone (HREZ) made up of , say, N number of pixels, we can determine the number (I) of infarcted pixels. The ration I/N represents the infarcted potion of that HREZ myocardial segment. The total number of all the infarcted pixels in all the HREZs provides an indication of the amount of infarcted myocardium of the heart or of the LV.

Intra-LV Blood Flow velocity and pressure distribution: Now, let us come to the outcome of an infarcted heart and LV. Figure 11 illustrates this outcome in the form of intra-LV blood-flow velocity and pressure (or pressure-gradient) distributions [17]. During LV diastole, from the monitored entrance velocity of blood at the mitral valve and the wall motion of the expanding LV, we can compute the intra-LV blood-flow velocity and pressure distributions, by computational fluid dynamics (CFD). During systole, from the monitored exit velocity or the aortic valve and the wall motion of the contracting LV, we can compute the intra-LV blood-flow velocity and pressure distribution by CFD.

Figure 11 illustrates the computed intra-LV blood-flow velocity and pressure distribution of a patient with an infarcted myocardium, before and after administration of nitroglycerin to determine the viability of the myocardial wall following bypass surgery. Referring to Fig 11, for the patient (with a myocardial infarct), Figs. 11(a1) depict super-imposed LV outlines at known equal intervals during diastole and systole before nitroglycerin administration, and Figs 11 (a2) depict super-imposed LV outlines at known equal intervals during diastole and systole after nitroglycerin administration; nitroglycerin is a myocardial perfusing agent, and hence a quasi-simulator of coronary bypass surgery or coronary angioplasty. From these images, we can determine the instantaneous wall displacements and hence the wall velocities at these time instants.

This data, along with the monitored entrance and exit velocities of blood into and from the LV, constitutes the data for our CFD analysis. For computational purposes, the intra-LV flow is determined from the boundary condition of LV wall-motion velocity and inlet/outlet blood flow velocity to the standard potential-flow equation $\nabla^2 \Phi = 0$. The intra-LV pressure gradient can then be computed from the Bernoulli equation for unsteady potential flow.

Figures 11(b1) and 11 (c1) depict intra-LV blood-flow velocity distributions during diastole and systole, before nitroglycerin administrations.

Figures 11(b2) and 11 (c2) depict intra-LV blood-flow velocity distributions during diastole and systole, after nitroglycerin administrations

Figures 11(d1) and 11 (e1) depict intra-LV pressure distributions during diastole and systole, before nitroglycerin administrations

Figures 11(d2) and 11 (e2) depict intra-LV pressure distributions during diastole and systole, after nitroglycerin administrations

In this patient, the poor motion of the infarcted LV wall offers resistance to proper filling of the LV (Fig 11-b1). However, it can be noted that following the administration of nitroglycerin, there is improved filling of the LV (Fig. 11-b2). During systolic ejection phase, the infarcted LV wall segments do not contract, and this results in inadequate intra-LV flow velocity distribution, which mitigates adequate emptying of the LV (Fig 11-c1). Following nitroglycerin administration, there is improved outflow velocity distribution. Likewise, figures (11-d1 and 11-e1) demonstrate adverse intra-LV pressure gradients during filling and ejection phases, which are improved after administration of nitroglycerin (Eq 11-d2 and 11-e2). This has provided the basis for advocating coronary revascularization by coronary bypass surgery, for this patient.

Physiological Nondimensional Indices in Medical Assessment: For Quantifying Physiological Systems and
Analysing Medical Tests' Data

287

The computed intra-LV blood-flow velocity and pressure distributions provide illustrative and quantitative outcome of an infarcted LV to the physician, which enables more distinct assessment of LV dysfunction. The cause of this LV dysfunction is provided by the echo-texture analysis of 2-d B-scan echocardiograms of HREZ(s), in terms of the amount (or number of echocardiogram image pixels) of the infarcted myocardial wall. Together, these two methodologies provide reliable and quantitative assessment of (i) how much of the LV myocardium is infarcted and its effect on the intra-LV blood flow and pressure- gradient, and (ii) intra-LV distributions of blood-flow velocity and pressure distributions, to assess candidacy for coronary bypass surgery.

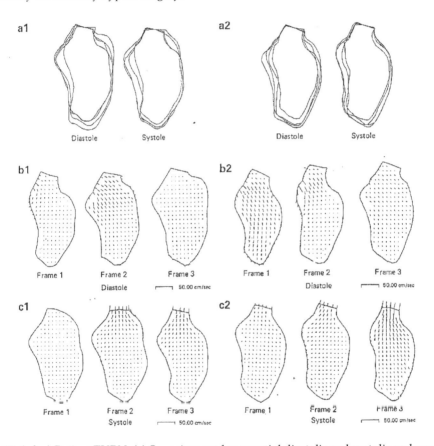

Fig. 11. (a,b,c) Patient TURN: (a) Superimposed sequential diastolic snd systolic endocardial frames (whose aortic valves centres and the long axis are matched) before (1) and after (2) administration of nitroglycerin. (b) Instantaneous intra-LV distributions of velocity during diastole, before (1) and after (2) administration of nitroglycerin. (c) Instantaneous LV distributions of velocity during ejection phase, before (1) and after (2) administration of nitroglycerin. (d) Instantaneous intra-LV distributions of pressure-differentials during diastole, before (1) and after (2) administration of nitroglycerin. (e) Instantaneous intra-LV distributions (pressure-differential) during ejection phase, before (1) and after (2) administration of nitroglycerin.

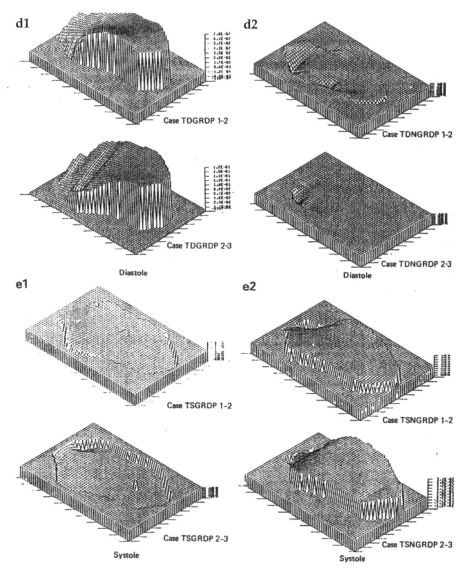

Fig. 11. (d,e) Patient TURN: (a) Superimposed sequential diastolic snd systolic endocardial frames(whose aortic valves centresand the long axis are matched) before (1) and after (2) administration of nitroglycerin. (b)Instantaneous intra-LV distributions of velocity during diastole, before (1) and after (2) administration of nitroglycerin. (c) Instantaneous LV distributions of velocity during ejection phase, before (1)and after (2) administration of nitroglycerin. (d) Instantaneous intra-LV distributions of pressure-differentials during diastole, before (1) and after (2) administration of nitroglycerin. (e) Instantaneous intra-LV distributions (pressure-differential) during ejection phase, before (1) and after (2) administration of nitroglycerin.

Physiological Nondimensional Indices in Medical Assessment: For Quantifying Physiological Systems and
Analysing Medical Tests' Data

289

11. Biomedical engineering concept of Heart Failure

Heart Failure, in biomedical engineering (BME) terms, can imply failure of the heart to:

i.   develop adequate contractility, due to a sizable amount of non-contractile infarcted myocardium (systolic heart failure)

ii. generate appropriate intra-LV pressure gradient during the ejection phase, to produce adequate LV outflow rate, stroke volume and cardiac output;

iii. produce adequate pressure increase during isovolumic contraction and ejection phases, to overcome the aortic systolic pressure;

iv. effect adequate stroke volume, due to poor contractility (systolic heart failure) or poor filling due to diseased and stiff myocardium (diastolic heart failure).

So the factors causing systolic heart failure may be summarized to be:

1.   excessive percentage amount of infarcted myocardium PMI (as determined by the procedure in section 10),

2.   resulting incapacity of the LV to produce appropriate intra-LV pressure gradient, for adequate LV outflow velocity and flow rate dV/dt (as determined by the methodology in section 10), as manifested by the contractility index CCI (formulated in section 2).

Now the resultant dV/dt factor in item 2 is incorporated in the formula for CCI (equation 3). No doubt, the PMI affects CCI, but there is no direct formulation connecting these two indices. So we can state that PMI and CCI are the two parameters that can be attributed to the occurrence of heart failure.

We can then define the Systolic Heart Failure index, as

$$HFIN = PMI(\text{in }\%)\ \text{x}HR\left(\text{in s}^{-1}\right) / CCI\left(\text{in s}^{-1}\right) \tag{46}$$

Now, in order to assess the terminal value of $HFIN$, we need to determine the terminal values of PMI, CCI, and HR, by studying normal subjects (as these indices are noninvasively determinable) as well as patients in different stages of heart failure.

Now, based on our studies (in Refs 4 & 5), let us (for the time being) adopt (i) the minimum acceptable value of CCI to be 3 s^{-1}, (ii) the maximum acceptable value of PMI to be 15 %, and (iii) the maximum resting HR to be 120/min or 2 s^{-1}. Substituting these values into equation (46) gives the terminal value of $HFIN$ to be 10. In other words, if the value of $HFIN$ exceeds the value 10, we can designate the patient to be in heart failure.

12. Concluding remarks

In this chapter, we have developed and presented noninvasive medical tests methodologies and associated NDPI(s), to make the case for reliable medical assessment of organ performance, physiological system function and dysfunction, anatomical structural property and pathology. The following tests and associated NDPI(s) have been presented:

3.   Determination of cardiac contractility, from measurement of LV chamber volume and myocardial volume, in terms of the cardiac contractility index CCI of $(d\sigma^*/dt)_{max}$, given by equation (3).

4.   Treadmill test to assess cardiac fitness and heart function, by means of the cardiac fitness index CFI, given by equation (8).

5. Lung Ventilation test, by monitoring lung volume by spirometry, for assessing lung ventilation and diagnosing lung disorders by means of lung ventilation index *LVPI-2* (equation 12).
6. Oral Glucose Tolerance test, to more reliably diagnose diabetes, by determining the diabetes index *DBI*, given by equation (16).
7. Noninvasive determination of arterial stiffness to detect arteriosclerosis, by ultrasound measurement of arterial dimensions and pulse wave velocity and auscultatory diastolic pressure, by means of the arteriosclerosis index *ART-NPI*, given by equation (20).
8. Noninvasive determination of (i) Aortic Pressure profile and (ii) Aortic normal vs disease state property in terms of the *Aortic number* given by equation (31), by monitoring the left ventricular outflow into the aorta and auscultatory diastolic and systolic pressures.
9. Noninvasive determination of Mitral valve pathology, by (i) monitoring its vibrational frequency from the first heart sound spectrum, and its size parameter from 2-d echocardiogram, and (ii) employing this data to structure its E^* *vs.* σ^* constitutive property (equations 36 and 37), and determine the alteration in the value of the constitutive index *m* given by equation (40).
10. Characterization of Osteoporosis, by determining the ulna bone vibratory resonance frequency, in terms of its flexural stiffness *EI*, given by equation (45).
11. Quantitative determination of (i) the amount of infarcted myocardial segment of the heart from echo-texture analysis of 2-d echo cardiograms (figure 10), and (ii) associated outcome of LV dysfunction in terms of the intra-LV blood-flow velocity and pressure distributions (figure 11).

Together these tests and their associated NDPI (s) can provide more reliable medical assessment. What now needs to be done is (i) application of these tests to large patient populations, and (ii) determination of the ranges of NDPI (s) for normal and abnormal states of organs, physiological systems and anatomical structures.

All of these tests can be employed in tertiary patient case, through the department of biomedical engineering (BME) in a tertiary-cave medical center. This makes a strong case for the institution of BME departments in tertiary case medical centers, which will revolution therapy health care.

13. References

[1] "Physiological Systems' Numbers in Medical Diagnosis and Hosipital Cost-effective Operation", by Dhanjoo N. Ghista, in *Journal of Mechanics in Medicine & Biology* 2005, vol 4, No.4.
[2] *Applied Biomedical Engineering Mechanics*, by Dhanjoo N. Ghista, CRC press (Taylor & Francis Group) Baton Rouge Florida 334872-2742, ISNBN 978-0-8247-5831-8,2008
[3] Measures and indices of intrinsic characterization of cardiac dysfunction during filling & systolic ejection," by Liang Zhong, Dhanjoo N. Ghista,Eddie Y-K. Ng, Lim SooTeik and Chua Siang Jin, in *Journal of Mechanics in Medicine & Biology* 2005, 5(2):37-322.

[4] "Validation of a novel noninvasive characterization of cardiac index of left ventricle contractility in patients", by Zhong L, Tan RS, Ghista DN, Ng E. Y-K, Chua LP, Kassab GS, *Am J Physiol Heart CircPhysiol*2007, 292:H2764-2772.

[5] "Effects of Surgical Ventricular Restoration on LV Contractility assessed by a novel Contractility index in patients with Ischemic cardiomyopathy", by Zhong L, Sola S, Tan RS, Le TT, Ghista DN, Kurra V, Navia JL, Kassab G.; in *Am J Cardiology*, 2009;103(5):674-679.

[6] Cardiac Fitness mathematical Model of Heart-rate response to V02 during and after Stress-Testing", Lim GeokHian, Dhanjoo N. Ghista, Koo TseYoong, John Tan Cher Chat, Philip EngTiew& Loo Chian Min; *International Journal of Computer Application in Technology(Biomedical Engineering & Computing Special Issue)*, Vol 21, No 1/2, 2004.

[7] "Glucose Tolerance Test Modeling & Patient-Simulation for Diagnosis", by Sarma Dittakavi & Dhanjoo N. Ghista, *Journal of Mechanics in Medicine & Biology*, Vol. 1, No.2, Oct.2001.

[8] "Determination of the In-vivo Elasticity of Blood Vessel and Detection of Arterial Disease", by D. N. Ghista, *Automedica*, Vol. 1, No. 3, 1974.

[9] "Determination of Aortic Pressure-time Profile , Along with Aortic Stiffness and Peripheral Resistance", by Liang Zhong, Dhanjoo N. Ghista, Eddie Y-K. Ng, Lim SooTeik and Chua Siang Jin, in *Journal of Mechanics in Medicine & Biology* 2004, 4(4):499-509.

[10] "Mechanics of the Mitral Valve: Stresses in the Membrane, Indirect Determination of the Instantaneous Modulus of the Membrane", by D.N. Ghista; *Journal of Biomechanis*, Vol. 5, No. 3, 1972.

[11]"Mitral Valve Mechanics – Stress/Strain Characteristics of Excise Leaflets, Analysis of its Functional Mechanics and its Medical Applications", by D. N. Ghista and A. P. Rao: *Medical and Biological Engineering*, Vol. 11, No. 6, 1973.

[12] In Vivo Measurement of the Dynamic Response of Bone, by J.M. Jurist, H.D. Hoeks, D.A. Blackketter, R. K. Snyder and E.R. Gardner, in *Orthopaedic Mechanics: Procedures and Devices (Volume III)*,ed by Dhanjoo N. Ghista and Robert Roaf, Academic Press (London), 1981.

[13] Noninvasive determination of Ulna stiffness from Mechanical Response---In Vivo comparison of stiffness and Bone mineral content in humans, by C.R. Steele, L.J. Zhou, D.Guido, R Marcus, W. T., Heinrichs and C. Cheema, in *Journal of Biomechanical Engineering*, Vol. 10 (87), 1988.

[14] Preliminary Clinical results using SOBSA for noninvasive determination of ulna bending stiffness, by C.R. Steele and A.F. Gordon, in *Advances in Bioengineering (ASME)*, edited by R. C Eberhand and A.H. Burstein, 1978, pp 85-87.

[15] Measures and Indices for Intrinsic Characterization of Cardiac Dysfunction during Filling and Systolic Ejection, by Liang Zhong, Dhanjoo N. Ghista, Eddie Y. Ng, Lim SooTeik, and Chua Siang Jin, in *Journal of Mechanics in Medicine and Biology*,Vol 5, No. 2, 2005.

[16] Detection of myocardial scars in neonatal infants form computerised echocardiographic texture analysis, by M.V Kamath, R.C Way, D.N. Ghista, T.M. Srinivasan, C. Wu, S Smeenk, C. Manning, J. Cannon, *Engineering in Medicine*, Vol. 15, Vo.3, 1986
[17] Intrinsic Indices of the Left Ventricle as a Blood Pump in Normal and Infarcted Left Ventricle, by K. Subbaraj, D.N. Ghista, and E. L. Fallen, in *J of Biomedical Engineering*, Vol 9, July issue, 1987

Permissions

The contributors of this book come from diverse backgrounds, making this book a truly international effort. This book will bring forth new frontiers with its revolutionizing research information and detailed analysis of the nascent developments around the world.

We would like to thank Prof. Dhanjoo N. Ghista, for lending his expertise to make the book truly unique. He has played a crucial role in the development of this book. Without his invaluable contribution this book wouldn't have been possible. He has made vital efforts to compile up to date information on the varied aspects of this subject to make this book a valuable addition to the collection of many professionals and students.

This book was conceptualized with the vision of imparting up-to-date information and advanced data in this field. To ensure the same, a matchless editorial board was set up. Every individual on the board went through rigorous rounds of assessment to prove their worth. After which they invested a large part of their time researching and compiling the most relevant data for our readers. Conferences and sessions were held from time to time between the editorial board and the contributing authors to present the data in the most comprehensible form. The editorial team has worked tirelessly to provide valuable and valid information to help people across the globe.

Every chapter published in this book has been scrutinized by our experts. Their significance has been extensively debated. The topics covered herein carry significant findings which will fuel the growth of the discipline. They may even be implemented as practical applications or may be referred to as a beginning point for another development. Chapters in this book were first published by InTech; hereby published with permission under the Creative Commons Attribution License or equivalent.

The editorial board has been involved in producing this book since its inception. They have spent rigorous hours researching and exploring the diverse topics which have resulted in the successful publishing of this book. They have passed on their knowledge of decades through this book. To expedite this challenging task, the publisher supported the team at every step. A small team of assistant editors was also appointed to further simplify the editing procedure and attain best results for the readers.

Our editorial team has been hand-picked from every corner of the world. Their multi-ethnicity adds dynamic inputs to the discussions which result in innovative outcomes. These outcomes are then further discussed with the researchers and contributors who give their valuable feedback and opinion regarding the same. The feedback is then collaborated with the researches and they are edited in a comprehensive manner to aid the understanding of the subject.

Apart from the editorial board, the designing team has also invested a significant amount of their time in understanding the subject and creating the most relevant covers. They scrutinized every image to scout for the most suitable representation of the subject and create an appropriate cover for the book.

The publishing team has been involved in this book since its early stages. They were actively engaged in every process, be it collecting the data, connecting with the contributors or procuring relevant information. The team has been an ardent support to the editorial, designing and production team. Their endless efforts to recruit the best for this project, has resulted in the accomplishment of this book. They are a veteran in the field of academics and their pool of knowledge is as vast as their experience in printing. Their expertise and guidance has proved useful at every step. Their uncompromising quality standards have made this book an exceptional effort. Their encouragement from time to time has been an inspiration for everyone.

The publisher and the editorial board hope that this book will prove to be a valuable piece of knowledge for researchers, students, practitioners and scholars across the globe.

List of Contributors

Andrea Mahn
Universidad de Santiago de Chile, Chile

Shuang-Qing Zhang
National Center for Safety Evaluation of Drugs, National Institutes for Food and Drug Control, China

Ying-Fang Yang
Biomedical Technology and Device Research Laboratories, Industrial Technology Research Institute, Republic of China

Jaroslav Turánek and Josef Mašek
Veterinary Research Institute, Czech Republic

Milan Raška
Palacky University, Czech Republic

Miroslav Ledvina
Institute of Organic Chemistry and Biochemistry, Czech Republic

Cai-Xia Yang and Jason W. Ross
Department of Animal Science, Iowa State University, USA

Ambrose Jon Williams and Vimal Selvaraj
Cornell University, USA

L. D. Jensen
Department of Microbiology, Tumor and Cell Biology, the Karolinska Institute, Stockholm, Sweden
Institution of Medicine and Health, Linköping University, Linköping, Sweden

Zisis Kozlakidis and John Cason
Department of Infectious Diseases, King's College London, 2nd Floor Borough Wing, Guy's Hospital, UK
National Institute of Health Research's (NIHR) comprehensive Biomedical Research Centre (cBRC) at Guy's and St Thomas' NHS Foundation Trust, UK

Robert J. S. Cason
School of Law, Birkbeck College London, UK

Christine Mant
Department of Infectious Diseases, King's College London, 2nd Floor Borough Wing, Guy's Hospital, UK

David Chee-Eng Ng
Department of Nuclear Medicine and PET, Singapore General Hospital, Singapore

Dhanjoo N. Ghista
Department of Graduate and Continuing Education, Framingham State University, Framingham, Massachusetts, USA

Liang Zhong and Ru San Tan
Department of Cardiology, National Heart Centre, Singapore

Leok Poh Chua
School of Mechanical and Aerospace Engineering, Nanyang Technological University, Singapore

Ghassan S. Kassab
Departments of Biomedical Engineering, Surgery, Cellular and Integrative Physiology, Indiana University-Purdue University Indianapolis, Indianapolis, Indiana, USA

Yi Su
Institute of High Performance Computing, Agency for Science, Technology and Research, Singapore

Kah Meng Koh
VicWell Biomedical, Singapore

Rohit Pasam
Quodient, Inc., USA

Printed in the USA
CPSIA information can be obtained
at www.ICGtesting.com
JSHW011503221024
72173JS00005B/1182